Organic Electro-Optics and Photonics

This definitive guide to modern organic electro-optic and photonic technologies provides critical insight into recent advances in organic electro-optic materials, from the underlying quantum and statistical concepts through to the practical application of materials in modern devices and systems.

- Introduces theoretical and experimental methods for improving organic electro-optic and photonic technologies
- Reviews the central concepts of nonlinear optics, focusing on multi-scale theoretical methods
- Provides clear insight into the structure and function relationships critical to optimizing the performance of devices based on organic electro-optic materials

Serving as a primer for the systematic nano-engineering of soft matter materials, this is an invaluable resource for those involved in the development of modern telecommunication, computing, and sensing technologies depending on electro-optic technology. It is also an indispensable work of reference for academic researchers and graduate students in the fields of chemistry, physics, electrical engineering, materials science and engineering, and chemical engineering.

Larry R. Dalton is the Founding Director of the National Science Foundation Science & Technology Center on Materials and Devices for Information Technology Research, Director of the DARPA MORPH program, and Director of two Department of Defense MURI Centers. He is a Fellow of the American Chemical Society, the Materials Research Society, the Optical Society of America, the SPIE, and the American Association for the Advancement of Science.

Peter Günter is Emeritus Professor at the Swiss Federal Institute of Technology (ETH) and a member of the board of Rainbow Photonics Ltd in Zürich. He is a Fellow of the Optical Society of America.

Mojca Jazbinsek is a member of the ETH spin-out, Rainbow Photonics AG, where she is currently working on applied research projects on organic nonlinear optical materials for high-speed electro-optics and THz-wave generation.

O-Pil Kwon is Associate Professor in the Departments of Molecular Science and Technology and of Applied Chemistry and Biological Engineering at the Ajou University in Korea.

Philip A. Sullivan is Assistant Research Professor in the Department of Chemistry and Biochemistry of Montana State University. He has worked in the area of organic materials for photonics applications for over ten years.

Organic Electro-Optics and Photonics

Molecules, Polymers, and Crystals

LARRY R. DALTON
University of Washington

PETER GÜNTER
ETH, Rainbow Photonics AG

MOJCA JAZBINSEK
ETH, Rainbow Photonics AG

O-PIL KWON
Ajou University

PHILIP A. SULLIVAN
Montana State University

MRS **MATERIALS RESEARCH SOCIETY**®
Advancing materials. Improving the quality of life.

CAMBRIDGE UNIVERSITY PRESS

CAMBRIDGE
UNIVERSITY PRESS

University Printing House, Cambridge CB2 8BS, United Kingdom

Cambridge University Press is part of the University of Cambridge.

It furthers the University's mission by disseminating knowledge in the pursuit of education, learning, and research at the highest international levels of excellence.

www.cambridge.org
Information on this title: www.cambridge.org/9780521449656

© Larry R. Dalton, Peter Günter, Mojca Jazbinsek, O-Pil Kwon, and Philip A. Sullivan 2015

This publication is in copyright. Subject to statutory exception
and to the provisions of relevant collective licensing agreements,
no reproduction of any part may take place without the written
permission of Cambridge University Press.

First published 2015

Printed in the United Kingdom by TJ International Ltd. Padstow Cornwall

A catalog record for this publication is available from the British Library

ISBN 978-0-521-44965-6 Hardback

Cambridge University Press has no responsibility for the persistence or accuracy of URLs for external or third-party internet websites referred to in this publication, and does not guarantee that any content on such websites is, or will remain, accurate or appropriate.

On September 30, 2008, Dr. Nicole Boand, wife of Larry Dalton, was struck by an out-of-control automobile, while sitting in a restaurant in Gig Harbor, Washington. She suffered, among other injuries, massive spinal column and spinal cord injuries, which required multiple operations and have led to a lifelong struggle with mobility and pain. Fortunately, Nicole's active lifestyle before the accident, which involved horseback riding, helped to prevent total muscle atrophy and with great courage she has struggled on a daily basis to maintain some level of use of her arms and legs. Despite the personal tribulations of her life, she has consistently encouraged completion of this book. It is in honor of her courage that the authors dedicate this book to her.

Contents

1	**Introduction**	1
	1.1 Motivation	1
	1.2 Overarching objective of this book	2
	1.3 Overview of topics covered	3
	1.4 A brief history	4
	1.5 Units and conversion factors	7
2	**Nonlinear optical effects**	11
	2.1 Nonlinear optical response and susceptibilities	11
	2.2 Second-order nonlinear optical effects and applications	15
	2.3 Third-order nonlinear optical effects and applications	23
3	**Electro-optic effects**	28
	3.1 Fundamentals of the electro-optic effect	28
	3.2 Frequency dependence	31
	3.3 Wavelength dependence	34
	3.4 Electro-optic modulation	35
	3.5 High-frequency modulation	37
4	**Molecular nonlinear optics**	40
	4.1 Microscopic and macroscopic nonlinearities of organic molecules	40
	4.2 Organic molecules for second-order nonlinear optics	40
	4.3 Numerical calculations	44
	4.4 Characterization methods	62
	4.5 Synthetic methods	64
5	**Acentric self-assembled films**	70
	5.1 Polar Langmuir–Blodgett films	70
	5.2 Acentric solution-deposited films	75
	5.3 Acentric vapor-deposited films	78
6	**Crystalline materials**	88
	6.1 Non-centrosymmetric organic crystalline packing: approaches	88
	6.2 Examples of organic electro-optic crystals	91

	6.3	Ionic crystals: stilbazolium salts	91
	6.4	Supramolecular hydrogen bonded crystals	101
	6.5	Molecular crystals: configurationally locked polyene crystals	104
	6.6	Crystal growth techniques	108
7	**Electrically poled organic materials and thermo-optic materials**		118
	7.1	Chromophore/polymer composites	119
	7.2	Covalently incorporated chromophore materials	126
	7.3	Matrix-assisted poling (MAP) materials	128
	7.4	Binary chromophore materials	133
	7.5	Complexity	139
	7.6	Thermal stability issues	141
	7.7	Optical loss issues	146
	7.8	Photochemical stability	147
	7.9	Experimental methods for evaluating r_{33} in poled EO materials	149
	7.10	Optical measurement of poling-induced order: VAPRAS and VASE	150
	7.11	Processing options	154
	7.12	Fabrication of all-organic devices	155
	7.13	Fabrication of silicon photonic, plasmonic, and photonic crystal hybrid devices	160
	7.14	Synthetic strategies for covalently incorporated chromophore materials	163
	7.15	Summary of macromolecular electro-optic materials	167
	7.16	Thermo-optic materials	169
8	**Overview of applications**		175
	8.1	Device parameters and materials requirements	175
	8.2	Applications	182
9	**Organic electro-optic waveguides, switches, and modulators**		188
	9.1	Light propagation in optical waveguides	188
	9.2	Integrated phase and amplitude electro-optic modulators	194
	9.3	Optical coupling between waveguides	195
	9.4	Microring resonators	198
	9.5	Light propagation in periodic media: photonic crystals	203
	9.6	Single-crystalline organic waveguides and modulators	207
	9.7	Polymer waveguides and modulators	216
	9.8	Silicon–organic hybrid waveguides	221
10	**Nonlinear optical infrared and terahertz frequency conversion**		228
	10.1	Nonlinear optical frequency conversion	228
	10.2	Terahertz-wave generation with organic nonlinear optical materials	232

11	**Photorefractive effect and materials**		250
	11.1 Theoretical models of the photorefractive effect		251
	11.2 Steady-state space-charge field		254
	11.3 Space-charge field dynamics		256
	11.4 Model for photo-induced refractive index changes in crystals and polymers		257
	11.5 Measurement of photo-induced refractive index changes		264
	11.6 Applications		266
	11.7 Materials requirements and figures-of-merit		266
	11.8 Photorefractive materials and their properties		268
12	**Conclusions and future prospects**		282
	12.1 General conclusions		282
	12.2 Future prospects: competing technologies for electrical-to-optical signal transduction		283
	12.3 Future prospects: fundamental issues facing the development and utilization of organic electro-optic materials		284
	12.4 Future prospects: optical sum and difference-frequency generation, optical rectification, and THz generation		286
	12.5 Future prospects: final comments		286
	Index		287

1 Introduction

1.1 Motivation

It has become increasingly recognized [1] that "photonic integration" is an important next step in the evolution of computing, telecommunications, sensing, transportation, medical, defense, and entertainment technologies. Such integration permits the best features of electronics and photonics to be exploited for information technology applications. There are important technological drivers of photonic/electronic integration, including realization of improved bandwidth and thermal management – transporting and manipulating information using photons avoids the high-frequency resistive losses and heating associated with movement of electrons in metal. Photonic/electronic integration is also evolving to include "chipscale" integration wherein both electronic and photonic circuitries are integrated on to the same chip, analogous to complementary metal-oxide semiconductor (CMOS) electronic integration, which has revolutionized computing. Potential advantages of chipscale integration include far-reaching improvements in size, weight, power consumption, performance, reliability, and cost. While chipscale photonic/electronic integration is not likely to be monolithic, as in CMOS electronic integration, and the problems to be faced will certainly be challenging, there can be little doubt that it will ultimately occur and will have considerable societal and economic impact. Such integration has been greatly advanced by recent developments in the field of silicon photonics, plasmonics, and metamaterial device architectures [2 to 21]. For example, the high index of refraction of silicon has permitted a striking reduction in the size of photonic circuits, making the dimensions of these circuits more compatible with chipscale integration. Further reductions in circuit dimensions appear possible by exploiting plasmonics.

A key component of photonic/electronic integration is the interconversion of signals between the electronic and photonic domains. This is where electro-optics comes into play. An electro-optic (EO) material is one in which the electrical fields associated with photons and electrons can communicate through a highly hyperpolarizable (easily perturbed by electric fields) charge distribution. For an electro-optic material to be optimum for the transduction of electronic signal information into photonic signal information, the charge distribution should be easily perturbed by small electric field potentials (ideally by millivolt electric field potentials to minimize power consumption) and should have very fast (ideally femtosecond) response to time-varying electrical fields. It is these features that have attracted attention to organic EO materials.

The fundamental response time of conjugated π-electron systems is the phase relaxation time and is almost universally of the order of tens of femtoseconds. Moreover, conjugated π-electron molecular systems exhibit exceptional hyperpolarizability, suggesting that a few millivolts applied to devices may be used to encode electronic information onto optical transmissions and to manipulate that information (e.g., route or steer, wavelength division multiplex or color code).

Unlike inorganic crystalline EO materials, which are typically naturally occurring, many organic (π-electron) EO materials are created by design, and a virtually endless stream of new materials can be created. This seemingly endless ability to create new organic EO materials is both an incredible opportunity and challenge. The opportunity lies in ever-increasing electro-optic activity (and thus reduced power requirements) and in realizing special properties such as conformal and flexible materials, materials that can be effectively integrated with a wide variety of disparate materials (metals, metal oxides, semiconductors, inorganic glasses, etc.), and materials amenable to a wide variety of processing options, including low-cost manufacture of complex device structures by techniques such as soft and nano-imprint lithography. The downside of the seemingly endless structural modifications possible with organic EO materials is the difficulty of deciding which materials to focus upon for device applications.

Two decades of defining structure/function relationships critical for device applications have already occurred. This effort has led to marked improvement of some properties. For example, during the first decade of the twenty-first century, the electro-optic activity of organic materials was improved at nearly a Moore's Law rate, while auxiliary properties were improved to satisfy Telcordia standards. Indeed, it is now likely that organic electro-optic materials yield the fastest response times (bandwidth performance to 15 THz has been demonstrated in difference-frequency generation [22], in all-optical modulation [5], and in pulsed studies [23]) and the largest electro-optic effect (∼500 pm/V compared with ∼30 pm/V for the commercial standard lithium niobate, $LiNbO_3$). With silicon–organic hybrid (SOH) devices, digital signal processing bandwidths greater than 100 Gbit/s with power consumption of less than 1 femtojoule/bit have been demonstrated. Thus, the commercial potential of organic electro-optics for next-generation information technology is becoming increasingly recognized. Electro-optic devices based on organic materials sold by Gigoptix have satisfied Telcordia requirements. Still further improvement in performance is likely required for chipscale integration applications and for certain other device applications such as spatial light modulation. This book provides important insight into how necessary improvements are likely to be achieved and what will be the ultimate limits to performance of EO materials.

1.2 Overarching objective of this book

The purpose of this book is to provide clear insight into the structure/function relationships critical to optimizing the performance of devices based on organic electro-optic materials. This explicitly includes those relationships critical for optimizing electro-optic activity, optical loss, stability, etc. The lessons learned from the past 20 years of research are reviewed, and suggestions are made as to how these lessons can be used to further advance the performance of organic electro-optic materials. An important objective of this book is to provide critical understanding of state-of-the-art theoretical methods that have been employed to guide the selection of nonlinear optical chromophore structures for large molecular first hyperpolarizability, and to nano-engineer component structures to achieve the non-centrosymmetric organization of chromophores necessary for large macroscopic electro-optic activity. An introduction to techniques for characterizing all of the relevant properties of electro-optic materials is also important, as these techniques provide the data tables desired by engineers utilizing materials for device and systems applications. Characterization data are also critical components of the feedback loop necessary for optimizing material performance. A unique feature of organic electro-optic materials is their exceptional processability, and this book attempts to provide insight into this diversity of processing methodologies. Finally, organic electro-optic materials provide an impressive array of new application possibilities, and an understanding of current and emerging device concepts is thus an important objective of this book.

This book is aimed at providing advanced undergraduate students, graduate students, postgraduate students and faculty, together with industrial scientists and engineers, with a comprehensive understanding of the field of organic electro-optics. A particular objective is to provide a knowledge base that facilitates individuals seeking to enter the field of organic electro-optics either as researchers or applications engineers. To that end, the book commences with an overview of the central concepts of nonlinear optics.

The book can also be an important reference text for those desiring to understand nano-engineering of materials dependent on spatially anisotropic interactions. It should also prove helpful to those desiring to learn the theoretical methods, such as Monte Carlo and molecular dynamics methods, used to simulate the assembly of individual molecules into macroscopically ordered lattices. Electro-optic materials are attractive model systems for understanding forces that dominate macroscopic order, since non-centrosymmetric (acentric) symmetry is required for finite electro-optic activity for materials based on dipolar chromophores. Study of electro-optic activity provides critical insight into which theoretical methods are most effective in designing improved materials and in understanding order/disorder processes associated with various processing protocols. The insights gained from considering organic electro-optic materials can be invaluable aids to improving organic electronic, photovoltaic, photorefractive, and light-emitting materials. In general, this book can serve as a primer for the systematic nano-engineering of soft-matter electroactive materials.

1.3 Overview of topics covered

Chapter 1 focuses on the perspective and motivation of this text and provides a brief introduction to topics covered and to a brief history of organic electro-optics. Although discovery of the Pockel's effect (electro-optics) pre-dates the laser, the introduction of the laser clearly paved the way for nonlinear optics because of the large electric fields associated with intense laser beams. The advent of fiber optic telecommunications opened avenues for commercial application of electro-optic materials. Chapter 1 provides a brief general introduction to fundamental phenomena of nonlinear optics, which are discussed in Chapters 2 and 3. Chapter 2 provides an important introduction to the terminology of nonlinear optics, which is necessary for understanding the subsequent chapters. Figures-of-merit introduced in this chapter are particularly important for engineers attempting to consider electro-optic materials for device and systems applications and in guiding the choice among different material options. This chapter also provides an introduction to the measurement techniques used in the characterization of various macroscopic optical nonlinearities and related physical properties. Chapter 3 continues in the spirit of Chapter 2, but narrows the focus to electro-optic phenomena. Chapter 4 considers molecular hyperpolarizabilities, including theoretical methods of computing and experimental techniques for measuring such hyperpolarizabilities. This chapter illustrates the fundamental approaches to the design of organic electro-optic molecules (chromophores). Chapter 5 deals with self-assembled materials; Chapter 6 deals with crystalline materials; and Chapter 7 focuses on poled polymers (with a brief section on thermo-optic polymers). These three chapters cover the three fundamental approaches to generation of macroscopic electro-optic materials. Chapter 8 provides an overview of applications while Chapter 9 considers specifics of device structures. Chapter 9 provides a more detailed discussion of various stripline and resonant electro-optic device structures. Chapter 10 focuses on other second-order nonlinear optical effects including frequency doubling, optical rectification, and difference-frequency generation. Chapter 11 deals with photorefractivity, which is an important application of the electro-optic effect. Chapter 12 provides conclusions and future prognosis, particularly in light of emerging competing technologies.

1.4   A brief history

The Pockel's effect [24] pre-dates the discovery of the laser; however, it was the discovery of the laser that opened the field of nonlinear optics and promoted identification and definition of various second-order (electro-optic modulation, second-harmonic generation, optical rectification, difference-frequency generation, etc.) and third-order (phase conjugation and four-wave mixing, third-harmonic generation, all-optical modulation, etc.) nonlinear optical effects. For his seminal contributions to nonlinear optics, Nicolaas Bloembergen was awarded the Nobel Prize in Physics in 1981 and is viewed as the father of nonlinear optics [25 to 27].

It is difficult to provide a meaningful history of organic nonlinear optics, as many hundreds of exceptional scientists have contributed tens of thousands of important papers contributing to the evolution and definition of the field. Thus, any brief history must be incomplete and to a great extent arbitrary in coverage. With that caveat, we attempt to provide a brief perspective on events and circumstances that led to the current state-of-the-art. Perhaps the first motivation for consideration of organic π-electron systems for nonlinear optics was provided by French theorists [28 to 32] who suggested that the long conjugation lengths of π-electron systems could give rise to large optical nonlinearities. Indeed, it was suggested that first molecular hyperpolarizabilities might increase with a fourth power dependence on π-conjugation length and that second molecular hyperpolarizabilities might increase with a sixth power dependence on conjugation length. Of course, early calculations were overly simplistic, and the combination of electron Coulomb and electron–phonon interactions resulted in limitations on delocalization and thus on the increase in optical nonlinearity with increasing π-conjugation (chromophore) length. Moreover, the charge transfer absorption band of chromophores undergoes a bathochromic solvatochromic shift with increasing length, and thus optical loss becomes a serious problem even at telecommunication wavelengths for the longest π-electron molecules (chromophores). This revision of expectations regarding organic electro-optic materials based upon improved understanding of the complexity of these materials was to become the hallmark of organic electro-optic materials research over the next two decades.

Major advancement in fields of science usually requires substantial financial investment, and in the late 1980s such an investment was made by international companies and by the US government in the area of nonlinear optics. In the United States, a major stimulus was provided by the Defense Advanced Projects Research Agency (DARPA) in a program managed by Dr. John Neff that involved Hoechst-Celanese Corporation, Lockheed Martin Corporation, and others, as well as a number of universities. A major motivation for DARPA interest was the potential for high-bandwidth time-division-multiplexing (TDM) possible with the ultrafast response times of π-electrons to time-varying electrical fields. By the late 1980s and early 1990s, outstanding research teams at Eastman Kodak, DuPont, IBM, 3M, Hochest-Celanese, Lockheed Martin, Boeing, Allied Signal, Eni Chem, Akzo, NTT, Battelle, British Telecom, etc. were producing organic electro-optic materials by electric field poling, sequential synthesis/self-assembly, and crystal growth. Crystals such as methylnitroaniline (MNA) and poled polymers based on incorporating stilbene and azobenzene chromophores into commercially available polymer hosts were extensively studied. The computation of molecular hyperpolarizabilities was rapidly advanced by groups such as those led by Anthony Garito (Univ. of Pennsylvania), Paras Prasad (SUNY-Buffalo), Brian Pierce (Hughes Aircraft Company), Tobin Marks and Mark Ratner (Northwestern), and many groups in Europe and Asia [33 to 36]. Auxiliary properties such as thermal stability were being extensively studied by researchers in Europe and the US (see Chapter 7). Individuals such as Seth Marder and Joseph Perry (then at the Jet Propulsion Laboratory/Cal Tech) popularized organic nonlinear optical materials with the chemistry community by describing structure/function relationships in terms familiar to organic chemists [35]. Proceedings of the SPIE (International Society of Optics and Photonics) and the Materials Research Society provide good snapshots of the remarkable evolution of organic nonlinear optics from the mid-1980s to the mid-1990s.

And then the bottom dropped out in the mid-1990s. Most of the industrial efforts were shut down, and funding from US Federal agencies was limited to small grants. One notable exception was Lockheed Martin Corporation, which continued research on chromophore/polymer composite materials and contributed greatly to the understanding of optical loss in these materials [37,38]. The reasons for this sudden "market crash" of organic electro-optic materials and device research are likely three-fold: (1) The driving force for interest in organic electro-optics in the late 1980s likely was related to anticipated need for high (TDM) bandwidth in telecommunications, and it was realized that organic materials afforded the potential for high bandwidth as well as large optical nonlinearity and novel processing options. By the early 1990s it was realized that techniques such as wavelength-division-multiplexing (WDM) and code-division-multiplexing (CDM) could provide much of the needed bandwidth. (2) Corporations realized that organic electro-optics did not provide the market potential initially anticipated and particularly not at the level of core corporate activities focused on organic fibers (e.g., DuPont) and liquid crystalline materials (e.g., Hoehsct-Celanese). (3) By the mid-1990s, the electro-optic activity of organic materials still had not exceeded that of lithium niobate, and questions about the thermal and photochemical stability of organic materials had risen to the point of high concern. Moreover, the production of organic electro-optic materials was seen to have little in common with traditional polymer processing. In short, the engineering of transformative organic electro-optic materials was turning out to be more difficult than anticipated and the market potential not as great as anticipated.

It should be kept in mind that the evolution of organic electro-optics occurred at a time when polyacetylene (discovered in the 1980s) was being considered as having great promise for battery applications. Moreover, organic light-emitting device (OLED), photorefractive, and photovoltaic materials were attracting great interest. Enthusiasm for each of these types of organic electroactive materials waxed and waned during the late 1980s and early 1990s. It is only at the present time that commercial application has become a reality for OLEDs, and the greatest potential still lies in the future for each of these technologies.

Fortunately, a few research (mostly academic) groups in Europe, Asia, and North America remained active in research related to organic electro-optic materials in the late 1990s, and during this period several events occurred that were to produce a Moore's Law rate of improvement in the performance of organic electro-optic materials in the first decade of the twenty-first century. One of the events was the introduction of the tricyanovinyl furan (TCF) acceptor into chromophore synthesis [39 to 41]. This acceptor permitted an improvement in both molecular first hyperpolarizability and photochemical stability. The introduction of the TCF acceptor was also accompanied by introduction of isophorone group(s) into polyene bridges to improve the photochemical stability of this very effective bridge moiety. For poled polymer materials, an even greater advance occurred with the utilization of multi-scale modeling (coupled quantum and statistical mechanical methods) to optimize the efficiency of electric field poling. In the late 1980s, only 1–4% of molecular optical nonlinearity (first hyperpolarizability) was being effectively translated to macroscopic electro-optic activity because of deleterious effects of chromophore–chromophore dipolar interactions. An article in *Science* in 2000 marked the turning point in use of theory-guided improvement in electrically poled organic electro-optic materials [42]. With these advances in chromophore design and processing, the electro-optic activity of organic materials exceeded that of lithium niobate and a wide range of prototype devices were successfully demonstrated.

The development of other photonic materials has certainly impacted the commercial potential of organic electro-optic materials. The development of low-loss silica fiber transformed telecommunications and stimulated the development of modulated lasers and electro-optic devices as methods of encoding electronic information onto long-range transmissions. As already noted in this chapter, the evolution of silicon photonics and demonstration of hybrid organic electro-optic/silicon photonic devices is having a significant effect, with promise of an even greater future impact, particularly on the chipscale integration of electronics and photonics.

New issues in material processing and device performance have been identified in the preceding decade, including the role that buffer layers play in controlling conductivity in organic electro-optic materials. Unlike organic electronic, photovoltaic, and light-emitting device materials, conductivity is best avoided in electro-optic materials. Interface engineering has become an important activity for all cases of organic electroactive materials. The introduction of thin interfacial layers of materials such as titanium dioxide to control charge injection and extraction has become commonplace in the fabrication of certain electro-optic devices.

Substantial progress has been made in the theoretical modeling of nonlinear optical effects (see Chapters 4 and 7), including the increased use of Møller–Plesset (MP) and density functional theory (DFT) quantum computational approaches for estimation of molecular hyperpolarizabilities and the dependence of hyperpolarizabilities on dielectric permittivity and optical frequency. Monte Carlo and molecular dynamics statistical mechanical methods have provided considerable insight into the role of various electrostatic interactions in influencing chromophore organization needed for macroscopic second-order optical nonlinearity. Recently, coarse-graining of statistical mechanical computations has permitted quantitative simulation to be extended to large systems.

Some additional seminal texts and reviews are cited in references [43 to 71], and more references are available in the subsequent chapters.

1.5 Units and conversion factors

A certain amount of confusion exists with respect to definitions (conventions) and units, and we address these in various chapters in this book. To start with, it has been common practice to report values for molecular hyperpolarizabilities in terms of centimeter-gram-second or cgs units (that is, in terms of electrostatic units or esu) while electro-optic coefficients are most commonly reported in SI (meter-kilogram-second or MKS) units (picometers/volt). To maintain connectivity with the substantial literature, we shall continue this duality. The conversion factors employed to go between SI and cgs units depend on the convention used for defining polarization in SI (or MKS) units. In the first convention, which we denote MKS(1), $P^{(n)} = \varepsilon_0 \chi^{(n)} E^{(n)}$ where $P^{(n)}$ is the polarization, ε_0 is the vacuum dielectric constant (= 8.8542×10^{-12} coulomb^2/Nm^2), $\chi^{(n)}$ is the nth-order optical susceptibility, and $E^{(n)}$ is the electric field. In the second convention, which we denote MKS(2), $P^{(n)} = \chi^{(n)}E^{(n)}$. This is also the convention relevant to cgs units. Thus, $(4\pi\varepsilon_0)(10^{-4}c)^{1-n}\chi^{(n)}_{\text{cgs}} = \varepsilon_0\chi^{(n)}_{\text{MKS(1)}} = \chi^{(n)}_{\text{MKS(2)}}$ where c is the speed of light (= 2.9979×10^8 m/s). With appropriate attention to conventions used, $\chi^{(2)}$ in cgs units can be related to $\chi^{(2)}$ in MKS(1) units, i.e., 1 statvolt (stV)/centimeter (cm) = 4.1888×10^{-4} volt (V)/meter (m). There is also confusion in terms of the conventions employed in the definition of the nonlinear susceptibility and molecular hyperpolarizability expressions appearing in power- or Taylor-series expansions of polarization in terms of electric field. The reader is referred to subsequent chapters for in-depth discussions of various conventions and conversion factors, or to Ref. [72].

References

[1] NSF Workshop on Very Large Scale Photonic Integration, Arlington, VA, March 19–20, 2007; Defense Science Board Advisory Group on Electron Devices Workshop on Photonic Integration, Naval Postgraduate School, Monterey, CA, July 24–26, 2007.
[2] B. Jalali and S. Fathpour, *IEEE J. Lightwave Technol.*, **24**, 4600 (2006).
[3] Q. Xu, V. R. Almeida, R. R. Panepucci, and M. Lipson, *Opt. Lett.*, **29**, 1626 (2004).
[4] V. R. Almeida, C. A. Barrios, R. R. Panepucci, and M. Lipson, *Nature*, **431**, 1081 (2004).
[5] M. Hochberg, T. Baehr-Jones, G. Wang, et al., *Nature Mater.*, **5**, 703 (2006).
[6] M. Hochberg, T. Baehr-Jones, G. Wang, et al., *Opt. Express*, **13**, 5216 (2005).
[7] J. Takayesu, M. Hochberg, T. Baehr-Jones, et al., *IEEE J. Lightwave Technol.*, **27**, 440 (2008).
[8] T. Baehr-Jones, B. Penkov, J. Huang, et al., *Appl. Phys. Lett.*, **92**, 163303 (2008).
[9] B. A. Block, T. R. Younkin, R. Reshotko, et al., *Opt. Express*, **16**, 18326 (2008).
[10] R. Ding, T. Baehr-Jones, Y. Liu, et al., *Opt. Express*, **18**, 15618 (2010).
[11] M. Gould, T. Baehr-Jones, R. Ding, et al., *Opt. Express*, **19**, 3952 (2011).
[12] R. Ding, T. Baehr-Jones, W.-J. Kim, et al., *IEEE J. Lightwave Technol.*, **29**, 1112 (2011).
[13] H. Figi, D. H. Bale, A. Szep, L. R. Dalton, and A. Chen, *J. Opt. Soc. Am. B*, **28**, 1191 (2011).
[14] C.-Y. Lin, X. Wang, S. Chakravarty, et al., *Appl. Phys. Lett.*, **97**, 093304 (2010).
[15] W. Wang, C.-Y. Lin, S. Chakravarty et al., *Opt. Lett.*, **36**, 882 (2011).
[16] J. H. Wulbern, J. Hampe, A. Petrov, et al., *Appl. Phys. Lett.*, **94**, 241107 (2009).
[17] J. H. Wulbern, S. Prorok, J. Hampe, et al., *Opt. Lett.*, **35**, 2753 (2010).
[18] J. H. Wulbern, A. Petrov, and M. Eich, *Opt. Express*, **17**, 304 (2009).
[19] J. Leuthod, W. Freude, J.-M. Brosi, et al., *Proc. IEEE*, **97**, 1304 (2009).
[20] S. Shi and D. W. Prather, *Adv. Optoelectron.*, **2011**, 714895 (2011).
[21] S. Shi and D. W. Prather, *Appl. Phys. Lett.*, **96**, 201107 (2010).
[22] C. V. MacLaughlin, L. M. Hayden, B. Polishak, et al., *Appl. Phys. Lett.*, **92**, 163303 (2008).
[23] K. A. Drenser, R. J. Larsen, F. P. Strohkendl, and L. R. Dalton, *J. Phys. Chem.*, **103**, 2290 (1999).
[24] F. Pockels, *Lehrbuch der Kristalloptik*, Leipzig, Teubner (1906).
[25] N. Bloembergen, *Nonlinear Optics,* New York, W. A. Benjamin (1965).
[26] N. Bloembergen, *Nonlinear Optics* 4th Edn, Singapore, World Scientific (1996).
[27] N. Bloembergen, W. K. Burns, and C. L. Tang, *Int. J. Quant. Chem.*, **5**, 555 (2009).
[28] C. Flytzanis, *Contribution à la theorie des susceptibilities non lineaires dans less solides par FLYTZANIS Christo*, Theses Sc. Phys., Orsay (1970).

[29] G. P. Agrawal and C. Flytzanis, *Chem. Phys. Lett.*, **44**, 356 (1976).
[30] G. P. Agrawal, C. Cojon, and C. Flytzanis, *Phys. Rev. B*, **17**, 775 (1978).
[31] C. Flytzanis and J. L. Oudar, *Nonlinear Optics: Materials and Devices*, Berlin, Springer-Verlag (1986).
[32] D. S. Chemla and J. Zyss, Eds., *Nonlinear Optical Properties of Organic Molecules and Crystals*, Orlando, Academic Press (1987).
[33] A. F. Garito, K. Y. Wong, Y. M. Cai, H. T. Man, and O. Zamani-Khamiri, *Proc. SPIE*, **682**, 1 (1986).
[34] B. M. Pierce and R. M. Wing, *Proc. SPIE*, **682**, 27 (1986).
[35] S. R. Marder, C. B. Gorman, F. Meyers *et al.*, *Science*, **265**, 632 (1994).
[36] I. D. L. Albert, T. J. Marks, and M. A. Ratner, in *Characterization Techniques and Tabulations for Organic Nonlinear Optical Materials*, M. G. Kuzyk and C. W. Dirk, Eds., New York, Marcel Dekker (1998), pp. 37–109, and references therein.
[37] R. Barto, Jr., C. W. Frank, P. V. Bedworth, S. Ermer, and R. E. Taylor, *J. Phys. Chem. B*, **108**, 8702 (2004).
[38] R. Barto Jr., P. V. Bedworth, C. W. Frank, S. Ermer, and R. E. Taylor, *J. Chem. Phys.*, **122**, 234907 (2005).
[39] F. Wang, *Photo- and Thermo-Stabilization of Second-Order Nonlinear Optical Polymers*. Unpublished Ph.D. dissertation, University of Southern California (1998).
[40] C. Zhang, *Novel Phenylpolyene-Bridged Second-Order Nonlinear Optical Chromophores and New Thermally Stable Polyurethanes for Electro-Optic Applications*. Unpublished Ph.D. dissertation, University of Southern California (1999).
[41] L. R. Dalton, in *Polymers for Photonics Applications I*, K. S. Lee, Ed., Berlin, Springer-Verlag (2002).
[42] Y. Shi, C. Zhang, H. Zhang, *et al.*, *Science*, **288**, 119 (2000).
[43] D. J. Williams, Ed., *Nonlinear Optical Properties of Organic and Polymeric Materials*, Washington, DC, ACS Symposium Series (1983).
[44] Y. R. Shen, *The Principles of Nonlinear Optics*, New York, John Wiley & Sons (1984).
[45] P. N. Prasad and D. Ulrich, Eds., *Nonlinear Optical and Electroactive Polymers*, New York, Plenum Press (1986).
[46] D. S. Chemla and J. Zyss, Eds., *Quantum Electronics: Principles and Applications*, New York, Academic Press (1987).
[47] P. N. Prasad and D. J. Williams, *Introduction to Nonlinear Optical Effects in Molecules and Polymers*, New York, John Wiley & Sons (1991).
[48] S. R. Marder, J. E. Sohn, and G. D. Stucky, Eds., *Materials for Nonlinear Optics: Chemical Perspectives*, Washington, DC, ACS Symposium Series (1991).
[49] J. Messier, F. Kajzar, and P. N. Prasad, Eds., *Organic Molecules for Nonlinear Optics and Photonics*, NATO ASI Series E, Boston, Kluwer Academic (1991).
[50] R. W. Boyd, *Nonlinear Optics*, New York, Academic Press (1992).
[51] L. A. Hornak, Ed., *Polymers for Lightwave and Integrated Optics*, New York, Marcel Dekker (1992).
[52] L. Y. Chaing, A. F. Garito, and D. J. Sandman, Eds., *Electrical, Optical, and Magnetic Properties of Organic Solid State Materials. Mater. Res. Soc. Proc.*, **247** (1992).
[53] K. Sienicki, Ed., *Molecular Electronics and Molecular Electronic Devices*, Boca Raton, FL, CRC Press (1993).
[54] J. Zyss, in *Molecular Nonlinear Optics: Materials, Physics, and Devices*, Boston, Academic Press (1993).

[55] M. G. Kuzyk and J. D. Swalen, Eds., *Progress in Nonlinear Optics: Organics and Polymeric Materials, Nonlinear Opt.*, **6** (1993).
[56] A. F. Garito, A. K. Y. Jen, C. Y. C. Lee, and L. R. Dalton, Eds., *Electrical, Optical, and Magnetic Properties of Organic Solid State Materials, Mater. Res. Soc. Proc.*, **328** (1994).
[57] G. A. Lindsay and K. D. Singer, Eds., *Polymers for Second-Order Nonlinear Optics*, Washington, DC, ACS Symposium Series (1995).
[58] Ch. Bosshard, K. Sutter, Ph. Pretre, *et al.*, *Organic Nonlinear Optical Materials*, Basel, Gordon & Breach (1995).
[59] A. K. Y. Jen, C. Y. C. Lee, L. R. Dalton, *et al.*, Eds., *Electrical, Optical, and Magnetic Properties of Organic Solid State Materials, Mater. Res. Soc. Proc.*, **413** (1996).
[60] F. Kajzar and J. D. Swalen, Eds., *Organic Thin Films for Waveguiding Nonlinear Optics*, Amsterdam, Gordon & Breach (1996).
[61] F. Kajzar, V. M. Agranovich, and C. Y. C. Lee, *Photoactive Organic Materials: Science and Applications*, Dordrecht, Kluwer (1996).
[62] H. S. Nalwa and S. Miyata, Eds., *Nonlinear Optics of Organic Molecular and Polymeric Materials*, Boston, CRC Press (1996).
[63] S. Miyata and H. Sasabe, Eds., *Poled Polymers and their Applications in SHG and EO Devices*, Amsterdam, Gordon & Breach (1997).
[64] S. A. Jenekhe and K. J. Wynne, *Photonic and Optoelectronic Polymers*, Washington, DC, ACS Symposium Series (1997).
[65] J. R. Reynolds, A. K. Y. Jen, M. F. Rubner, L. Y. Chiang, and L. R. Dalton, Eds., *Electrical, Optical, and Magnetic Properties of Organic Solid State Materials, Mater. Res. Soc. Proc.*, **488** (1998).
[66] M. G. Kuzyk and C. W. Dirk, Eds., *Characterization Techniques and Tabulations for Organic Nonlinear Optical Materials*, New York, Marcel Dekker (1998).
[67] D. J. Wise, G. E. Wnek, D. J. Trantolo, T. M. Cooper, and J. D. Gresser, Eds., *Electrical and Optical Polymer Systems*, New York, Marcel Dekker (1998).
[68] K. S. Lee, Ed., *Polymers for Photonics Applications I*, Berlin, Springer-Verlag (2002).
[69] T. A. Skotheim and J. R. Reynolds, *Handbook of Conducting Polymers. Conjugated Polymers: Theory, Synthesis, Properties, and Characterization*, 3rd edn, Boca Raton, FL, CRC Press (2007).
[70] W. N. Herman, S. R. Flom, and S. H. Foulger, *Organic Thin Films for Photonics Applications*, Washington, DC, ACS Symposium Series (2010).
[71] A. Chen and E. Murphy, Eds., *Broadband Optical Modulators: Science, Technology, and Applications*, New York, Taylor & Francis (2011).
[72] R. F. Shi and A. F. Garito, in *Characterization Techniques and Tabulations for Organic Nonlinear Optical Materials*, M. G. Kuzyk and C. W. Dirk, Eds., New York, Marcel Dekker (1998), pp. 1–36.

2 Nonlinear optical effects

Electro-optic and nonlinear optical effects, where the optical properties are changed by radio frequency (RF) or optical electric fields, are of fundamental importance for optoelectronic and photonic devices. For most applications, field-dependent refractive index changes are of high value because the incident optical intensity is not strongly altered. Therefore, only effects resulting from the real part of the complex optical susceptibility are considered here. Nonlinear effects associated with the imaginary component, such as two-photon absorption and electro-absorption (often exploited in semiconductor materials), are not considered in detail.

This chapter contains the fundamental descriptions of various second- and third-order nonlinear optical effects. The main material parameters and figures-of-merit for nonlinear optical frequency conversion and other nonlinear optical interactions are introduced. The electro-optic effect, as a special case of second-order nonlinear optical effects, is discussed in more detail in the next chapter of this book and will continue to be the primary focus.

2.1 Nonlinear optical response and susceptibilities

In this section, the terminology used to describe the different nonlinear susceptibilities and their associated nonlinear optical effects is introduced. The response of a dielectric material to an applied electric field, **E**, may be described by an induced polarization density, **P**. In a nonlinear optical medium, the relationship between **P** and **E** is nonlinear. The optical response of a nonlinear material can therefore be described in terms of the linear polarization, $\mathbf{P}^L$, and the nonlinear polarization, $\mathbf{P}^{NL}$, induced by **E** as

$$\mathbf{P}(t) = \mathbf{P}^L(t) + \mathbf{P}^{NL}(t) = \mathbf{P}^{(1)}(t) + \mathbf{P}^{(2)}(t) + \mathbf{P}^{(3)}(t) + \cdots, \qquad (2.1)$$

where $\mathbf{P}^{(1)}(t) = \mathbf{P}^L(t)$ is linear in $\mathbf{E}(t)$, $\mathbf{P}^{(2)}(t)$ is quadratic, $\mathbf{P}^{(3)}(t)$ is cubic, and so on. Note that the total material polarization in polar materials contains a static polarization $\mathbf{P}(0)$ in addition to the induced polarization $\mathbf{P}^{(1)}(t) = \mathbf{P}^L(t)$. The relation between $\mathbf{P}^{(n)}(t)$ and $\mathbf{E}^{(n)}(t)$ in the time domain is most generally expressed as [1]

$$\mathbf{P}^{(n)}(t) = \varepsilon_0 \int_{-\infty}^{\infty} dt_1 dt_2 \cdots dt_n \chi^{(n)}(t-t_1, t-t_2, \ldots, t-t_n) \mathbf{E}(t_1)\mathbf{E}(t_2)\cdots\mathbf{E}(t_n) \quad (2.2)$$

$$\chi^{(n)}(t_1, t_2, \ldots, t_n) = 0 \text{ if any of } t_i < 0, \quad (2.3)$$

which considers that the material response may not be instantaneous. The coefficients $\chi^{(n)}(t_1, t_2, \ldots, t_n)$ are time-dependent real susceptibilities of the nth order (in general tensors of $n+1$ rank). ε_0 is the vacuum dielectric constant (permittivity of free space). Equation (2.2) reduces to $\mathbf{P}^{(n)}(t) = \varepsilon_0 \chi^{(n)} \mathbf{E}^{(n)}(t)$ when the time delay (or material dispersion) can be neglected. Note that, in some cases, the response may also be non-local in space; in such a case, the spatial dependence of $\chi(t,\mathbf{r})$ should be considered. In nonlinear optics, spatial dispersion may largely be neglected because the polarizable units in common materials are much smaller than the wavelength of light [1].

Owing to the complexity of Eq. (2.2), the relationship between the electric field and induced polarization is more commonly expressed in the frequency domain. In nonlinear optics, electric fields are commonly encountered that oscillate at several discrete frequencies as

$$\mathbf{E}(t) = \sum_{\omega > 0} \mathbf{E}^\omega \cos(\omega t) = \frac{1}{2} \sum_{\omega > 0} (\mathbf{E}^\omega e^{-i\omega t} + \text{c.c.}), \quad (2.4)$$

where c.c. denotes the complex conjugate. In such a case, the induced polarization also contains discrete frequencies as given by

$$\mathbf{P}(t) = \sum_{\omega > 0} \mathbf{P}^\omega \cos(\omega t) = \frac{1}{2} \sum_{\omega > 0} (\mathbf{P}^\omega e^{-i\omega t} + \text{c.c.}). \quad (2.5)$$

In the frequency domain, the relation between the induced polarization at frequency ω and $\mathbf{E}$ (containing frequencies $\omega_1, \omega_2, \ldots, \omega_n$) can be written as

$$\mathbf{P}^{(n)\omega} = \varepsilon_0 K^{(n)}(-\omega, \omega_1, \omega_2, \ldots, \omega_n)\chi^{(n)}(-\omega, \omega_1, \omega_2, \ldots, \omega_n)\mathbf{E}^{\omega_1}\mathbf{E}^{\omega_2}\cdots\mathbf{E}^{\omega_n}, \quad (2.6)$$

where $\chi^{(n)}(-\omega, \omega_1, \omega_2, \ldots, \omega_n)$ is the Fourier transform of $\chi^{(n)}(t_1, t_2, \ldots, t_n)$ and $-\omega + \omega_1 + \omega_2 + \cdots + \omega_n = 0$. In components, Eq. (2.6) becomes

$$P_i^{(n)\omega} = \varepsilon_0 K^{(n)}(-\omega, \omega_1, \omega_2, \ldots, \omega_n)\chi^{(n)}_{ijk\ldots m}(-\omega, \omega_1, \omega_2, \ldots, \omega_n)E_j^{\omega_1}E_k^{\omega_2}\cdots E_m^{\omega_n}, \quad (2.7)$$

where the usual Einstein summation convention holds. The numerical factor $K^{(n)}(-\omega, \omega_1, \omega_2, \ldots, \omega_n)$ in the above equations depends on the definition of the electric fields and polarizations, according to Eqs. (2.4) and (2.5), where the ½ factors appear, but only for $\omega_i \neq 0$. This factor is often the source of errors in reporting the nonlinear optical susceptibilities, particularly because of many different definitions that are in use. For example, in the definition used here, the induced polarization is expressed as a power series in the electric field, but sometimes the definition with a Taylor series is used, in which additional $1/n!$ factors appear. The numerical factor $K^{(n)}(-\omega, \omega_1, \omega_2, \ldots, \omega_n)$ depends on the nonlinear optical process and can be calculated as [1]

$$K^{(n)}(-\omega, \omega_1, \omega_2, \ldots, \omega_n) = 2^{l+q-n} p \qquad (2.8)$$

where n is the order of nonlinearity, q is the number of input frequencies, $\omega_1, \ldots, \omega_n$, that are zero, p is the number of distinct permutations of the input frequencies, $\omega_1, \ldots, \omega_n$, and $l = 0$ and $l = 1$ for $\omega = 0$ and $\omega \neq 0$ respectively.

2.1.1 Simplified expressions

For simplicity, the induced polarization is often expressed as a function of the field assuming a dispersionless medium using [2]

$$\mathbf{P}(t) = \varepsilon_0 \chi^{(1)} \mathbf{E}(t) + \varepsilon_0 \chi^{(2)} \mathbf{E}(t)\mathbf{E}(t) + \varepsilon_0 \chi^{(3)} \mathbf{E}(t)\mathbf{E}(t)\mathbf{E}(t) + \cdots \qquad (2.9)$$

In this case it is relatively easy to evaluate the appropriate factors $K^{(n)}(-\omega, \omega_1, \omega_2, \ldots, \omega_n)$ for each process. For example, considering a second-order process in which only one input frequency, ω, is present, and inserting the input field from Eq. (2.4) with only one summation term at ω into the second term of Eq. (2.9), we obtain

$$\mathbf{P}^{(2)}(t) = \varepsilon_0 \chi^{(2)} \left[\frac{1}{2}(\mathbf{E}^{\omega} e^{-i\omega t} + \text{c.c.})\right]^2 = \varepsilon_0 \chi^{(2)} \frac{1}{4}(\mathbf{E}^{\omega}\mathbf{E}^{*\omega} + \text{c.c.}) + \varepsilon_0 \chi^{(2)} \frac{1}{4}(\mathbf{E}^{\omega}\mathbf{E}^{\omega} e^{-2i\omega t} + \text{c.c.}).$$
$$(2.10)$$

The resulting quadratic polarization contains a static term and a term oscillating at twice the input frequency, which can be expressed as

$$\mathbf{P}^{(2)}(t) = \mathbf{P}^0 + \frac{1}{2}(\mathbf{P}^{2\omega} e^{-2i\omega t} + \text{c.c.}). \qquad (2.11)$$

When the two equations above are compared, it is apparent that

$$\mathbf{P}^0 = \frac{1}{2}\varepsilon_0 \chi^{(2)}(0, \omega, -\omega)\mathbf{E}^{\omega}\mathbf{E}^{*\omega} \qquad (2.12)$$

$$\mathbf{P}^{2\omega} = \frac{1}{2}\varepsilon_0 \chi^{(2)}(-2\omega, \omega, \omega)\mathbf{E}^{\omega}\mathbf{E}^{\omega}. \qquad (2.13)$$

The first process, dependent on $\chi^{(2)}(0, \omega, -\omega)$, is known as optical rectification (OR) and the second, dependent on $\chi^{(2)}(-2\omega, \omega, \omega)$, is termed second-harmonic generation (SHG). Note that we can reconsider the dispersion of $\chi^{(2)}$ in the above relations, since the correct K factors were obtained (compare with Eq. (2.8) with $l = 0$, $q = 0$, $n = 2$, $p = 2$ for OR and $l = 1$, $q = 0$, $n = 2$, $p = 1$ for SHG); therefore the above equations also satisfy Eq. (2.6).

2.1.2 Symmetry properties and dispersion of nonlinear optical susceptibilities

Most general symmetry properties of the nonlinear optical susceptibilities follow their definition. The coefficient $\chi^{(2)}_{ijk}$ must be invariant to the exchange of the indices

j and k, because it is a multiplier of the symmetric product $E_j E_k$ in Eq. (2.7). Similarly, $\chi^{(3)}_{ijkl}$ must be invariant to any permutations of j, k and l.

Since the materials are dispersive, the susceptibility tensor elements may change considerably with the frequencies of the involved electric fields. In a frequency regime far from the material absorption range, we can also exchange the first two indices i and j of the susceptibilities $\chi^{(n)}_{ij...m}$. In this case, the tensors $\chi^{(1)}_{ij}$, $\chi^{(2)}_{ijk}$, $\chi^{(3)}_{ijkl}$ are invariant to any permutations of their indices. This is a so-called Kleinman symmetry.

In a dispersive frequency regime, the frequency dependence of the susceptibility tensors has to be included. For example, for $n = 2$, Eq. (2.7) is of the form

$$\varepsilon_0 K \chi^{(2)}_{ijk}(-\omega_3, \omega_1, \omega_2) E_j^{\omega_1} E_k^{\omega_2} \tag{2.14}$$

if the polarization at frequency $\omega_3 = \omega_1 + \omega_2$ is produced. Because the product $E_j(\omega_1) E_k(\omega_2)$ is commutative, the indices j and k in $\chi^{(2)}_{ijk}$ can be exchanged, but only if the frequencies are exchanged as well

$$\chi^{(2)}_{ijk}(-\omega_3, \omega_1, \omega_2) = \chi^{(2)}_{ikj}(-\omega_3, \omega_2, \omega_1). \tag{2.15}$$

Similarly, the third-order susceptibility tensor $\chi^{(3)}_{ijkl}(-\omega_4, \omega_1, \omega_2, \omega_3)$ is invariant under all three permutations of pairs (j,ω_1), (k,ω_2), (l,ω_3). If all the optical frequencies occurring in the susceptibility tensor dependence are far from the transition frequencies of the nonlinear medium (no absorption), this permutation includes the additional pair (i,ω_4).

As already mentioned, optical wavelengths are usually large compared with the dimensions of the polarizable units; therefore the local field may be considered uniform in most cases and the spatial dispersion (the **k** dependence) of the susceptibility neglected [1].

2.1.3 The nonlinear wave equation

To describe the propagation of light in a nonlinear medium, one has to solve the following wave equation that can be derived from Maxwell's equations for a non-magnetic dielectric medium [3,4]

$$\nabla^2 \mathbf{E}(\mathbf{r},t) - \frac{1}{c^2} \frac{\partial^2 \mathbf{E}(\mathbf{r},t)}{\partial t^2} = \mu_0 \frac{\partial^2 \mathbf{P}(\mathbf{r},t)}{\partial t^2}, \tag{2.16}$$

where $c = 1/\sqrt{\varepsilon_0 \mu_0}$ is the speed of light in a vacuum, μ_0 is the magnetic constant (permeability of a vacuum), $\mathbf{E}(\mathbf{r},t)$ is the electric field of the optical wave and $\mathbf{P}(\mathbf{r},t) = \varepsilon_0 \chi^{(1)} \mathbf{E}(\mathbf{r},t) + \mathbf{P}^{NL}(\mathbf{r},t)$ the induced material polarization.

In general, the optical field that fulfills the wave equation in a medium will contain several frequency components that may also propagate in different directions. We therefore generalize Eq. (2.4) to

$$\mathbf{E}(\mathbf{r},t) = \frac{1}{2} \sum_{\mathbf{k},\omega} \left(E(\mathbf{k},\omega) e^{i(\mathbf{k}\mathbf{r}-\omega t)} + \text{c.c.} \right), \tag{2.17}$$

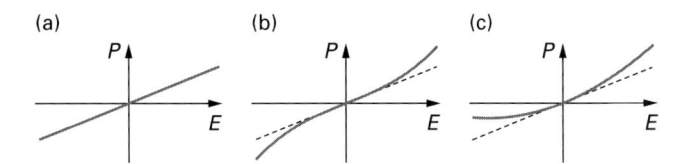

Figure 2.1 Induced polarization |**P**| in dielectrics for (a) a linear medium, (b) a nonlinear centrosymmetric medium, and (c) nonlinear non-centrosymmetric medium.

where **k** is the wave vector of the optical field component at frequency ω. For the case of a linear medium with $\mathbf{P}^{NL} = 0$, inserting Eq. (2.17) into the wave equation, Eq. (2.16), gives a well-known expression for the magnitude of the wave vector, i.e., the wavenumber $k(\omega) = |\mathbf{k}(\omega)| = \omega/c \cdot n(\omega)$, where the refractive index $n(\omega)$ is defined with $n^2(\omega) = \varepsilon(\omega) = 1 + \chi^{(1)}(\omega)$.

2.1.4 Second-order (quadratic) and third-order (cubic) nonlinear optical processes

The susceptibility tensors $\chi^{(n)}$ contain all information about the macroscopic optical properties of the respective material [5, 6, 7]. For symmetry reasons, the odd-order susceptibilities are present in any material, whereas the even-order ones only occur in non-centrosymmetric materials. The linear and nonlinear polarization response P, which can be simply (i.e. without considering the effect-dependent K factors, anisotropy and dispersion) expressed as

$$P = \varepsilon_0 \chi^{(1)} E + \varepsilon_0 \chi^{(2)} E^2 + \varepsilon_0 \chi^{(3)} E^3 + \cdots, \quad (2.18)$$

is schematically illustrated in Fig. 2.1.

The most important nonlinear optical effects of the second and third order are described by the nonlinear susceptibility tensors $\chi^{(2)}$ and $\chi^{(3)}$, respectively. Based on the characteristics of the involved electric fields, several second-order and third-order nonlinear optical effects can be distinguished. A schematic representation of the most important ones is shown in Figs 2.2 and 2.3.

In general, nonlinear optical properties allow for various frequency conversion effects (sum-frequency generation, difference-frequency generation, optical parametric generation, etc.), and generation of static or quasi-static electric fields (optical rectification, THz-wave generation), as well as for the field-induced change of the linear optical properties (electro-optic effect, optical Kerr effect), i.e., the field-induced change of the linear material susceptibility $\chi^{(1)}$ at optical frequencies. The electro-optic effect is a nonlinear optical effect, for which one of the applied fields is of low frequency (DC to THz) compared with the optical field.

2.2 Second-order nonlinear optical effects and applications

Based on the characteristics of the electric fields involved, several second-order nonlinear optical effects can be distinguished (see Fig. 2.2). The most important have their

Nonlinear optical effects

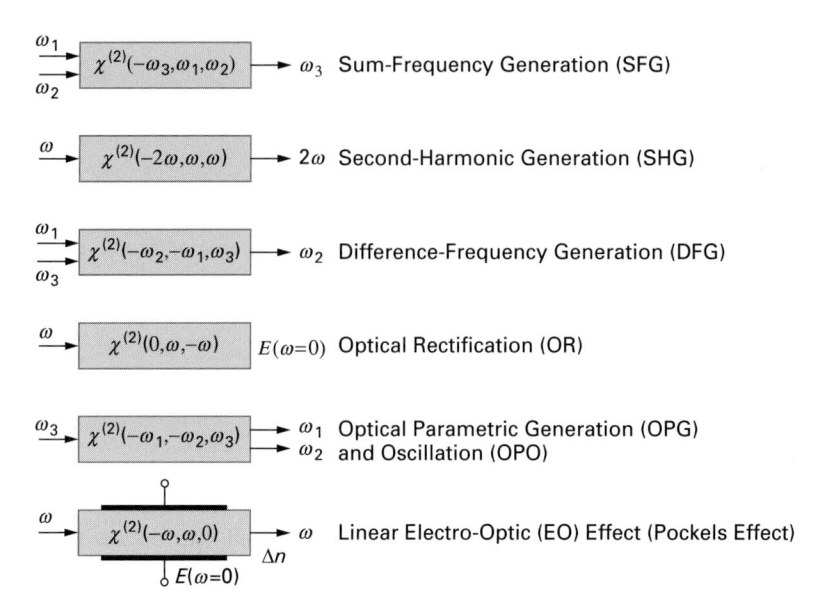

Figure 2.2 Schematic representation of important nonlinear optical effects of the second order, including the linear electro-optic effect.

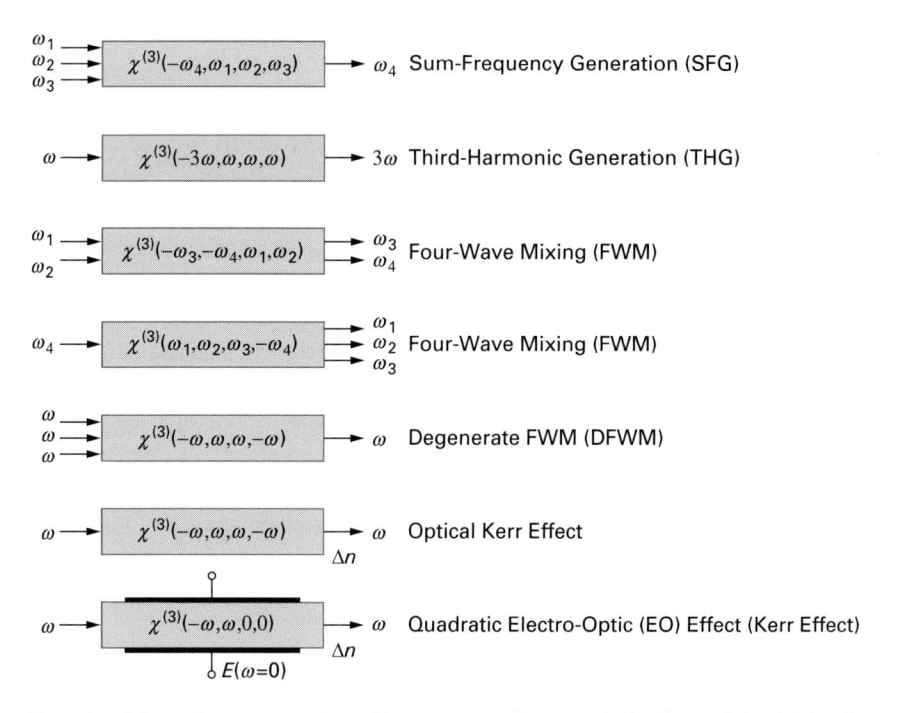

Figure 2.3 Schematic representation of important nonlinear optical effects of the third order, including the quadratic electro-optic effect.

own notation and will be introduced in the following sections. The linear electro-optic effect is considered separately in the next chapter of this book.

2.2.1 Sum-frequency generation and optical frequency doubling

Sum-frequency generation is the mixing of two incident light waves of frequencies ω_1 and ω_2 creating a wave of $\omega_3 = \omega_1 + \omega_2$. This situation is represented by the nonlinear polarization

$$P_i^{\omega_3} = \varepsilon_0 \chi_{ijk}^{(2)}(-\omega_3, \omega_1, \omega_2) E_j^{\omega_1} E_k^{\omega_2}. \tag{2.19}$$

Optical frequency doubling or second-harmonic generation (SHG) is a special case of sum-frequency generation. Only one light wave of frequency ω is incident and is "mixing with itself," thus generating a wave of twice the frequency. The nonlinear polarization for SHG can be expressed by the use of the nonlinear optical coefficient d_{ijk}, which is often used for the nonlinear optical characterization of macroscopic samples

$$P_i^{2\omega} = \frac{1}{2}\varepsilon_0 \chi_{ijk}^{(2)}(-2\omega, \omega, \omega) E_j^{\omega} E_k^{\omega} = \varepsilon_0 d_{ijk}(-2\omega, \omega, \omega) E_j^{\omega} E_k^{\omega}. \tag{2.20}$$

Sum-frequency and second-harmonic generation are standard techniques used to create a new coherent output from existing laser systems and especially to access the short-wavelength range towards the ultraviolet region.

The efficiency for harmonic generation can be calculated using the nonlinear wave equation (2.16) with the nonlinear polarizations given above. As an example, we consider here optical frequency doubling of a plane monochromatic pump wave at frequency ω with a nonlinear polarization (2.20) generated inside the material. For simplicity we consider that the interacting waves at ω and 2ω are polarized in the same direction, and that they all propagate along the z direction. Therefore we can consider scalar quantities and the electric field composed of two harmonic components

$$E(z,t) = \frac{1}{2}E_1(z)e^{i(k_1 z - \omega t)} + \frac{1}{2}E_2(z)e^{i(k_2 z - 2\omega t)} + \text{c.c.}, \tag{2.21}$$

where $E_1(z)$ and $k_1 = n^{\omega}(\omega/c)$ are the amplitude and the wave vector of the pump wave frequency ω, and $E_2(z)$ and $k_2 = n^{2\omega}(2\omega/c)$ are the amplitude and the wave vector of the generated second-harmonic wave at frequency 2ω. Here $n^{\omega} \equiv n(\omega)$ and $n^{2\omega} \equiv n(2\omega)$ denote the refractive indices of the fundamental and of the second-harmonic wave, which are usually not equal even for the same polarization of both waves, owing to dispersion. We will also consider the following common assumptions:

- *Undepleted pump approximation*, valid if only a small part of the pump beam energy is transferred to the second-harmonic beam, i.e., $dE_1/dz = 0$.
- *Slowly varying envelope approximation* for the generated wave, saying that $|d^2 E_2/dz^2| \ll |k_2 dE_2/dz|$, i.e., the envelope $E_2(z)$ varies slowly in the z direction in comparison with the wavelength $\lambda_2 = 2\pi/k_2$.

Using the above assumptions and inserting the nonlinear polarization from (2.20), $P^{NL}(z,t) = (1/2)\,\varepsilon_0 d_{SHG} E_1^2 e^{i(2k_1 z - 2\omega t)} + \text{c.c.}$, and the optical field (2.21) into the wave equation (2.16) results in

$$ik_2 \frac{dE_2(z)}{dz} e^{i(k_2 z - 2\omega t)} = -\frac{1}{2} \frac{d_{SHG} E_1^2 (2\omega)^2}{c^2} e^{i(2k_1 z - 2\omega t)}. \tag{2.22}$$

Introducing a phase mismatch $\Delta k \equiv k_2 - 2k_1$ we can rewrite the above equation as

$$dE_2 = i\omega \frac{1}{cn^{2\omega}} d_{SHG} E_1^2 e^{-i\Delta k z} dz \tag{2.23}$$

which can be easily integrated over the crystal (material) length L from $z = 0$ to $z = L$, giving

$$E_2(L) = i\omega \frac{1}{cn^{2\omega}} d_{SHG} E_1^2 L e^{-i\Delta k L/2} \text{sinc}\left(\frac{\Delta k L}{2}\right) \tag{2.24}$$

We can now calculate the intensity $I^{2\omega}(L) = (1/2)\varepsilon_0 cn^{2\omega} |E_2(L)|^2$ of the generated second-harmonic wave for an interaction region of length L. The efficiency η of second-harmonic generation is therefore

$$\eta = \frac{I^{2\omega}(L)}{I^\omega} = \frac{2\omega^2}{\varepsilon_0 c^3} \left(\frac{d_{SHG}^2}{n^{2\omega}(n^\omega)^2}\right) I^\omega L^2 \text{sinc}^2\left(\frac{\Delta k L}{2}\right). \tag{2.25}$$

We note the following:

- The efficiency of second-harmonic generation is proportional to the pump intensity I^ω, at least as long as the undepleted pump approximation is valid. We can therefore increase the efficiency by increasing the pump beam intensity, which can be done e.g. by using high-Q resonators and/or using waveguiding structures.
- Considering the material properties, the efficiency scales with d^2/n^3, which is a useful figure-of-merit for a second-harmonic generation. Figures-of-merit of different organic and inorganic materials are discussed in Chapter 10.
- For $\Delta k = 0$ the efficiency increases quadratically with the interaction length L.
- A non-zero phase-mismatch parameter $\Delta k \neq 0$ will considerably decrease the efficiency. This effect is illustrated in Fig. 2.4, where also the coherence length l_c is defined as

$$l_c = \frac{\pi}{\Delta k} = \frac{\lambda}{4(n^{2\omega} - n^\omega)} \tag{2.26}$$

and measures a useful material length for frequency doubling. If we do not carefully choose the conditions, l_c will be of the order of 1–100 µm, which will lead to a very small macroscopic efficiency η. The possibilities to achieve the so-called phase-matching condition with $\Delta k = 0$ are discussed in Section 2.2.3.1 of this chapter.

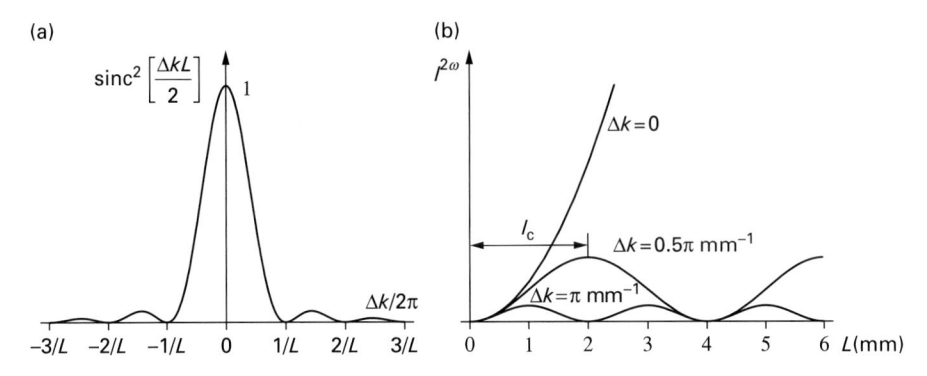

Figure 2.4 (a) Factor that reduces the efficiency of the second-harmonic generation, caused by the phase mismatch Δk. (b) Second-harmonic intensity as a function of the crystal thickness for different values of the phase mismatch Δk.

2.2.2 Difference-frequency generation, optical parametric generation/oscillation, and optical rectification

Difference-frequency generation (DFG) is characterized as the interaction of two input beams of frequencies ω_3 and ω_1 resulting in an optical field with the frequency $\omega_2 = \omega_3 - \omega_1$, i.e. the difference of the two. The nonlinear polarization for the difference-frequency generation can be written as

$$P_i^{\omega_2} = \varepsilon_0 \chi_{ijk}^{(2)}(-\omega_2, \omega_3, -\omega_1) E_j^{\omega_3} E_k^{*\omega_1} \tag{2.27}$$

The beam with the frequency ω_3 is typically the strongest and therefore referred to as the pump beam. Optical parametric generation (OPG) is a special case of difference-frequency generation, where only the pump beam is incident on the nonlinear material, generating two beams at the frequencies ω_1 and ω_2. These frequencies are selected based on the phase-matching condition. In order to enhance the efficiency of either process, the nonlinear medium can be placed inside a cavity with highly reflecting mirrors for the frequencies ω_1 and/or ω_2; in this case the process is referred to as optical parametric oscillation (OPO).

In contrast to sum-frequency generation, difference-frequency generation and optical parametric oscillation are well-suited to coherent light production at longer wavelengths, i.e., near and mid infrared region. Another application is optical parametric amplification where the strong pump beam at frequency ω_3 transfers energy to amplify an optical signal at frequency ω_1. In this case the output at frequency ω_2 remains unused and is thus often referred to as the idler beam. Optical parametric generation and oscillation are of particular importance because they allow a single frequency laser to be used to create optical output that is tunable over a large number of wavelengths. Tuning is accomplished by adjusting the phase-matching condition (2.30) using, for example, angle or temperature changes. This is discussed further in following sections.

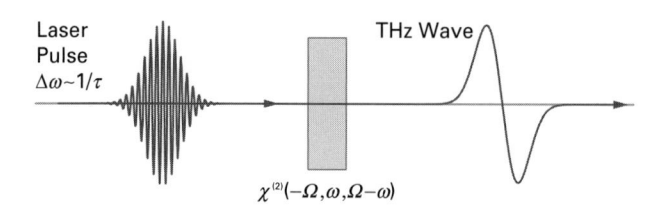

Figure 2.5 Schematic illustration of THz-wave generation via optical rectification in a second-order nonlinear optical crystal. Initial ultrashort optical pulse ($\tau \sim 100$ fs or less) contains several frequency components that mix to generate a nonlinear polarization oscillating in the THz frequency range.

Optical rectification (OR) is a special case of difference-frequency generation. Similarly, as in the process of SHG, only one optical field of frequency ω is incident. In OR, this degenerate wave mixing generates a static polarization in the material according to

$$P_i^0 = \frac{1}{2}\varepsilon_0 \chi_{ijk}^{(2)}(0, \omega, -\omega) E_j^\omega E_k^{*\omega}. \tag{2.28}$$

Optical rectification can be used, for example, to generate waves at THz frequencies using an ultrashort laser pulse with a large bandwidth as a pump source. The frequency components of such pulses are differenced with each other to produce nonlinear polarization from 0 to several THz (Fig. 2.5).

2.2.3 Conservation of energy and momentum (phase matching)

All of the nonlinear processes discussed previously conserve energy, which was already implicitly assumed:

$$\omega_3 = \omega_1 + \omega_2 \text{ (energy conservation)}. \tag{2.29}$$

Another common feature is that efficient nonlinear interaction can only occur if momentum is conserved. This requirement, also referred to as phase matching, is manifested in the following condition:

$$\mathbf{k}_{\omega_3} = \mathbf{k}_{\omega_1} + \mathbf{k}_{\omega_2} \text{ (momentum conservation or phase matching)}. \tag{2.30}$$

for collinear parametric interactions, where all wave vectors $\mathbf{k}_{\omega i}$ are parallel to one another, the phase-matching condition simplifies to

$$n^{\omega_3}\omega_3 = n^{\omega_1}\omega_1 + n^{\omega_2}\omega_2, \tag{2.31}$$

where $n^{\omega i}$ are the refractive indices of the waves at frequency ω_i. The nonlinear polarization for general directions in the crystal can be written as

$$|\mathbf{P}^{\omega_3}| = 2\varepsilon_0 d_{\text{eff}} |\mathbf{E}^{\omega_1}||\mathbf{E}^{\omega_2}| \text{ (sum} - \text{frequency generation)} \tag{2.32}$$

$$d_{\text{eff}} = \sum_{ijk} d_{ijk}(\omega_3, \omega_1, \omega_2)\cos(\alpha_i^{\omega_3})\cos(\alpha_j^{\omega_1})\cos(\alpha_k^{\omega_2}). \tag{2.33}$$

2.2.3.1 Phase-matching techniques

Here, we illustrate some possibilities of phase matching for the example of second-harmonic generation, which are also applicable to other nonlinear optical processes. For SHG, the condition (2.31) simplifies to

$$n^{2\omega} = n^{\omega} \qquad (2.34)$$

which will lead to $\Delta k = 0$ and a quadratic increase of the second-harmonic intensity as a function of the interaction length (see Fig. 2.4). Since most materials exhibit a positive dispersion of the refractive index and therefore $n^{2\omega} > n^{\omega}$, phase matching is most often achieved using birefringence, meaning that the polarizations of the interacting waves will not all be parallel. We distinguish type I and type II phase matching: in type I phase matching, the nonlinear polarization is generated using two photons of the same linear polarization, while for type II phase matching, it is generated using two photons of orthogonal linear polarization. Figure 2.6 illustrates the two types of phase matching for the case of an optically uniaxial crystal ($n_1 = n_2$) with $n_e^{\omega} > n_o^{2\omega} > n_o^{\omega}$, where $n_e = n_3$ is the so-called extraordinary refractive index and $n_o = n_1 = n_2$ is the ordinary index.

The phase-matching angle θ_{PM}^{I} for the phase matching of type I is characterized by $n_o^{2\omega} = n_e^{\omega}(\theta_{PM}^{I})$, where $n_e^{\omega}(\theta_{PM}^{I})$ can be calculated considering the optical indicatrix (see e.g. [2,4]) of the respective material as

$$\frac{1}{\left(n_e^{\omega}\left(\theta_{PM}^{I}\right)\right)^2} = \frac{\cos^2(\theta_{PM}^{I})}{(n_o^{\omega})^2} + \frac{\sin^2(\theta_{PM}^{I})}{(n_e^{\omega})^2} \qquad (2.35)$$

For type II phase matching, the two photons at ω are polarized orthogonally and therefore they travel with a different phase velocity. Considering also the condition (2.31) for the case of SHG we obtain

$$n_o^{2\omega} = \frac{1}{2}\left(n_o^{\omega} + n_e^{\omega}\left(\theta_{PM}^{II}\right)\right), \qquad (2.36)$$

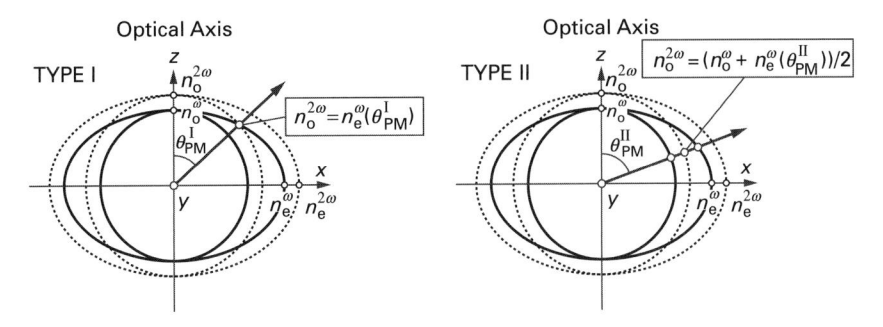

Figure 2.6 Type I (left) and type II (right) phase matching for a uniaxial crystal with $n_e > n_o$. The curves (circles, ellipses) are the so-called normal surfaces of the two orthogonal eigenwaves at ω (solid curves) and at 2ω (dotted curves), which give the refractive index as a function of the propagation direction (the direction of the wave vector **k**). The normal surfaces can be constructed from the usual optical indicatrix of the material.

where $n_e^\omega(\theta_{PM}^{II})$ can be expressed analogously as $n_e^\omega(\theta_{PM}^{I})$ in (2.35). We can tune the phase-matching condition for different fundamental frequencies ω by changing the propagation angle (angle tuning) or the temperature of the crystal through $n(T)$ (temperature tuning).

In the example above, the energy propagation directions of the interacting waves may not match, which will also reduce the efficiency. We must distinguish between critical and non-critical phase matching. Non-critical phase-matching is possible only if the phase-matching can be achieved for propagation along the dielectric axis of the crystal, i.e. $\theta_{PM} = 90°$. In this case, the so-called walk-off angle between the energy propagation direction of the fundamental and the second-harmonic beam will be zero, which will considerably increase the acceptance angle for the incoming fundamental beam.

Quasi-phase matching (QPM) is another possibility besides birefringence that may be used to increase the efficiency of frequency conversion processes. In the case of QPM, there is no restriction requiring propagation along a specific direction in the material in order to achieve $\Delta k = 0$. The diagonal nonlinear optical coefficients may also be employed because the polarizations of the interacting waves may all be parallel. However, QPM requires a certain periodic modulation of the nonlinear optical susceptibility, which is technologically much more challenging. The basic principle of QPM, exemplified here for the case of SHG, is as follows: we need a material in which the effective nonlinear optical coefficient d_{eff} changes sign after each coherence length l_c as defined in Eq. (2.26).

This can be done by so-called periodical poling of the material, which changes the polarization direction of the material in repeating domains with a period of $2l_c$. In a homogeneously poled material, after passing through one coherence length, destructive interference of the newly generated SHG light with SHG light generated earlier, starts to reduce $I^{2\omega}$ until after $2l_c$ it drops to zero. As a result, we have a periodic modulation of $I^{2\omega}$ with the thickness as shown in Fig. 2.7(a). For a perfectly poled material with a period of $2l_c$ (see Fig. 2.7(b)) it can be shown [8] that the average $I^{2\omega}$ signal increases

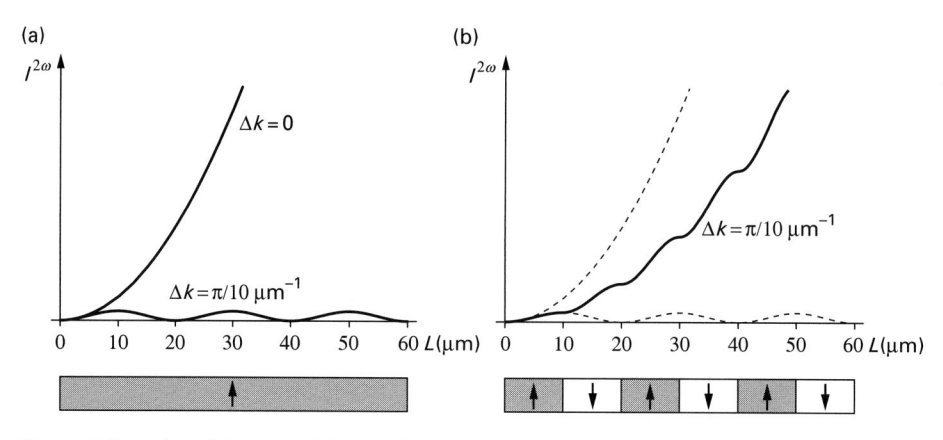

Figure 2.7 Intensity of the second-harmonic wave as a function of the propagation length L in (a) homogeneously poled material with $\Delta k = 0$ and $\Delta k = \pi/10~\mu m^{-1}$ and (b) periodically poled material with $\Delta k = \pi/10~\mu m^{-1}$ and domain lengths of 10 μm.

with $((2/\pi)d_{\text{eff}}L)^2$. This can be compared with the increase in the homogeneously poled material with $\Delta k = 0$, which is proportional to $(d_{\text{eff}}L)^2$ as given in Eq. (2.25).

Mode dispersion in optical waveguides is another possibility for achieving phase matching in nonlinear optical materials. Optical waveguides are basic units of integrated optical elements allowing for guided-wave propagation in micrometer-size cross-sections without experiencing diffraction. They consist of a core region with dimensions comparable to light wavelength, which has a higher refractive index than the surrounding cladding region, resulting in light confinement in the region of high refractive index (for more details, see Chapter 9 of this book). By choosing the refractive indices of the core and of the cladding region, we can tune the effective refractive index of the light propagating in a waveguide. Depending also on the dimensions of the core regions, several optical modes with different effective indices may propagate in a waveguide. Phase matching ($\Delta k = 0$) can be achieved by matching the effective index of one mode at frequency ω with another mode at frequency 2ω [6].

Waveguides are very important elements for increasing the efficiency of the frequency conversion mechanisms that scale with intensity I^ω as shown in Eq. (2.25). Because of the strong light confinement into areas of the order of $A = 1$ μm^2, we can achieve very high intensities $I^\omega = P^\omega/A$ even for low pump powers P^ω.

2.3 Third-order nonlinear optical effects and applications

In centrosymmetric media the second-order nonlinear term is absent. The dominant nonlinearity is of the third order

$$P_i^{NL} = \varepsilon_0 \chi_{ijkl}^{(3)} E_j E_k E_l \tag{2.37}$$

and gives rise to the third-harmonic generation and related mixing phenomena that are schematically illustrated in Fig. 2.2.

The third-order response of any medium to a monochromatic optical field of frequency ω consists of a nonlinear polarization containing ω and 3ω frequency components

$$P_i^{3\omega} = \frac{1}{4} \varepsilon_0 \chi_{ijkl}^{(3)}(-3\omega, \omega, \omega, \omega) E_j^\omega E_k^\omega E_l^\omega \tag{2.38}$$

$$P_i^\omega = \frac{3}{4} \varepsilon_0 \chi_{ijkl}^{(3)}(-\omega, -\omega, \omega, \omega) E_j^{*\omega} E_k^\omega E_l^\omega \tag{2.39}$$

The first process is known as third-harmonic generation (THG). Similar to the process of second-harmonic generation, the efficiency of THG is negligible unless the phase-matching conditions are satisfied.

The second effect, the generation of the polarization component at frequency ω, is equivalent to changing the effective refractive index according to

$$\Delta n(\omega) \cong \frac{3\chi^{(3)}(-\omega, -\omega, \omega, \omega)}{8n_0(\omega)} |E^\omega|^2 \tag{2.40}$$

where $n_0(\omega)$ is the unperturbed index at frequency ω (the tensor notation is omitted for simplicity). The relation (2.40) can be easily derived if we insert the nonlinear polarization (2.39) into the wave equation for the propagation of an optical field inside a medium [1,2]. The overall refractive index is therefore a linear function of the optical intensity $I \propto |E^\omega|^2$

$$n(I) = n_0 + n_2 I. \tag{2.41}$$

This effect is known as the optical Kerr effect. It is a self-induced effect, in which the phase velocity of the wave depends on the wave's own intensity. As a result, a wide variety of important processes occur, such as self-phase modulation and self-focusing effects.

2.3.1 Self-phase modulation

The result of the optical Kerr effect is that the optical wave traveling in a third-order nonlinear medium experiences an additional phase shift $\Delta\phi^{\mathrm{NL}}$, which is proportional to the optical intensity:

$$\phi = \frac{2\pi}{\lambda_0} n(I) L \quad \Delta\phi^{\mathrm{NL}} = \frac{2\pi}{\lambda_0} n_2 L I \tag{2.42}$$

where ϕ is the total phase shift, λ_0 is the light wavelength in vacuum, and L is the interaction length. Such self-phase modulation is useful in all-optical switching applications. An illustrative example for an optical switching device is shown in Fig. 2.8, where a nonlinear material is placed in a Fabry–Perot interferometer. At low incident intensity, the phase shift fails to satisfy the resonance condition for the interferometer. Therefore, the transmitted light intensity is blocked. As the intensity increases, the refractive index changes. When it reaches the value that satisfies the resonance condition, the transmitted beam is switched on. Device operation is controlled by the light beam itself.

2.3.2 Self-focusing and spatial Kerr solitons

In reality, optical beams are not in the form of plane waves, but have certain transverse intensity dependence. The laser output signal usually has a Gaussian transverse intensity

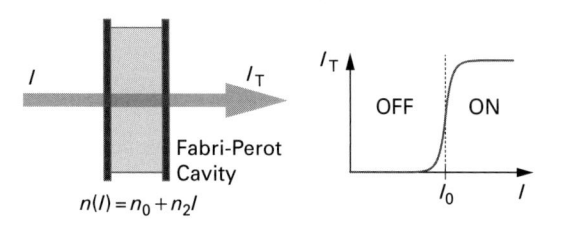

Figure 2.8 Optical switching behavior exhibited by a third-order nonlinear material in a Fabry–Perot interferometer.

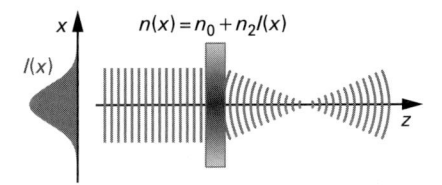

Figure 2.9 Self-focusing effect of a beam with transverse intensity dependence.

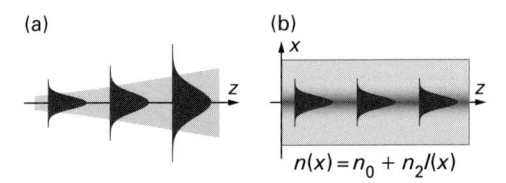

Figure 2.10 Comparison between (a) a Gaussian beam traveling in a linear medium, and (b) a spatial soliton traveling in a nonlinear medium.

distribution. If such a beam is transmitted through a film of material with nonlinear refractive index, the phase difference at the center of the beam, where the intensity is the highest, is larger than at the edge of the beam. The resulting wavefront curvature is similar to the ordinary lens effect. The nonlinear material therefore acts as a lens with intensity-dependent focal length (Fig. 2.9).

Consider now an intense optical beam traveling through a substantial thickness of a homogeneous medium instead of a thin sheet. In a linear medium, the beam of light will spread out owing to diffraction, as illustrated (Fig. 2.10(a)). In a nonlinear medium, the self-focusing effect counteracts diffraction. If the diffraction is exactly counteracted by the nonlinear focusing effect, the beam propagates without changing its shape. Such self-guided beams are called spatial solitons (Fig. 2.10(b)).

2.3.3 Optical phase conjugation

The response of a medium to a superposition of three monochromatic waves at frequencies ω_1, ω_3, and ω_4 is an example of four-wave mixing. Many different frequencies can be generated if frequency- and phase-matching conditions are satisfied. Widely used is the special case of degenerate four-wave mixing, where all four waves are of the same frequency $\omega_1 = \omega_2 = \omega_3 = \omega_4 = \omega$, but they propagate in different directions. If two waves (3 and 4, see Fig. 2.11) are uniform plane waves traveling in opposite directions

$$E_3(\mathbf{r}) = A_3 e^{i\mathbf{k}_3 \cdot \mathbf{r}}, \quad E_4(\mathbf{r}) = A_4 e^{i\mathbf{k}_4 \cdot \mathbf{r}}, \quad \mathbf{k}_4 = -\mathbf{k}_3 \tag{2.43}$$

then the induced polarization density is

$$P_2(\mathbf{r}) = \frac{3}{2} \varepsilon_0 \chi^{(3)} A_3 A_4 E_1^*(\mathbf{r}). \tag{2.44}$$

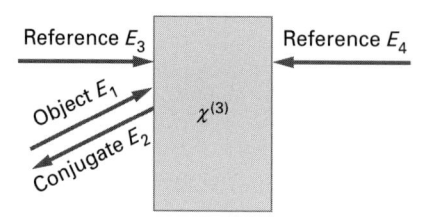

Figure 2.11 Degenerate four-wave mixing configuration, in which the generated wave 2 is the phase conjugate version of wave 1.

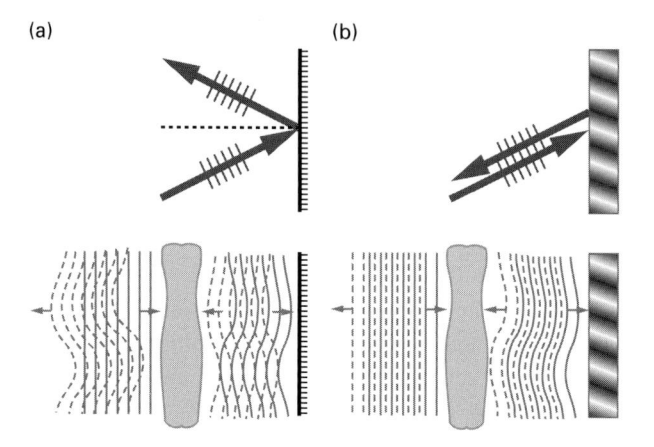

Figure 2.12 Comparison between (a) ordinary mirror and (b) phase conjugate mirror. A phase conjugate mirror reflects the incident wave back to itself, so that it propagates exactly in the opposite direction (above) and compensates distorted wavefronts (below).

Note that the numerical factor of 3/2 considers all possible permutations of the electric fields on the right-hand side that contribute to the induced polarization [1]. This polarization is the source emitting an optical wave

$$E_2(\mathbf{r}) \propto A_3 A_4 E_1^*(\mathbf{r}). \qquad (2.45)$$

In this case the generated wave is a conjugated version of wave 1. The device can be used as a phase conjugate mirror (see Fig. 2.12).

In this chapter, basic principles of second- and third-order nonlinear optical effects were introduced. This book is devoted in particular to second-order nonlinear optical effects in organic functional materials and their applications. Organic materials are also considered excellent candidates for third-order nonlinear optics; a recent review can be found in Ref. [9].

References

[1] P. N. Butcher and D. Cotter, *The Elements of Nonlinear Optics*, Cambridge, Cambridge University Press (1990).

[2] B. E. A. Saleh and M. C. Teich, *Fundamentals of Photonics*, New York, John Wiley & Sons, Inc. (1991).
[3] M. Born and E. Wolf, *Principles of Optics*, Oxford, Pergamon Press, Ltd. (1959).
[4] A. Yariv and P. Yeh, *Optical Waves in Crystals: Propagation and Control of Laser Radiation*, New Jersey, John Wiley & Sons (2003).
[5] A. Yariv, *Quantum Electronics*, 3rd Edn, New York, John Wiley & Sons (1989).
[6] Y. R. Shen, *The Principles of Nonlinear Optics*, New Jersey, John Wiley & Sons (2003).
[7] R. W. Boyd, *Nonlinear Optics*, 3rd Edn, San Diego, Academic Press (2008).
[8] M. M. Feyer, G. A. Magel, and D. H. Jundt, *IEEE J. Quantum Elect.* **28**, 2631 (1992).
[9] J. M. Hales and J. W. Perry, in *Introduction to Organic Electronic and Optoelectronic Materials and Devices*, S. Sun and L. Dalton, Eds., Boca Raton, Florida, CRC Press (2008), pp. 513–571.

3 Electro-optic effects

3.1 Fundamentals of the electro-optic effect

The electro-optic (EO) effect is often considered separately from other nonlinear optical effects, because one of the applied fields E^0 is not an optical but a static (or a quasi-static) electric field. The electro-optic effect can be expressed using the nonlinear $\chi^{(2)}$ and $\chi^{(3)}$ tensors as

$$P_i^\omega = 2\varepsilon_0 \chi_{ijk}^{(2)}(-\omega,\omega,0) E_j^\omega E_k^0 + 3\varepsilon_0 \chi_{ijkl}^{(3)}(-\omega,\omega,0,0) E_j^\omega E_k^0 E_l^0 \qquad (3.1)$$

The resulting nonlinear polarization oscillates at the same frequency ω as the linear polarization. Therefore, the refractive index n of the material will not only depend on the linear susceptibility $\chi^{(1)}$, but will contain a contribution from the nonlinear susceptibility induced by the external electric field. Note that the numerical factors 2 and 3 correspond to the K factors given in (2.7) and (2.8).

Most often the electro-optic effect is considered in terms of the change of the optical indicatrix, i.e. the inverse dielectric tensor at optical frequencies $\varepsilon_{ij}^{-1} = (n^2)_{ij}^{-1}$ as

$$\Delta\left(\frac{1}{n^2}\right)_{ij} = r_{ijk} E_k^0 + R_{ijkl} E_k^0 E_l^0 \qquad (3.2)$$

where the first term describes the linear electro-optic effect or the Pockels effect and the second term describes the quadratic electro-optic effect or the Kerr effect, according to whether the change of the indicatrix is a linear or a quadratic function of the applied field.

The above equation defines the linear electro-optic tensor r_{ijk} and the quadratic electro-optic tensor R_{ijkl}. For small changes and anisotropic materials, the refractive index change Δn_i along the main direction of the optical indicatrix can be approximated by

$$\Delta n_i \cong -\frac{1}{2} n_i^3 (r_{ijk} E_k + R_{ijkl} E_k E_l). \qquad (3.3)$$

For diagonal components r_{iii} and R_{iiii}, which describe the index change along the same direction as the applied field, the indices may be omitted, leading to the following change of the refractive index along the same direction

$$\Delta n = -\frac{1}{2} n^3 (rE + RE^2). \qquad (3.4)$$

3.1 Fundamentals of the electro-optic effect

We can define the material figure-of-merit (FOM) for the electro-optic effect as

$$\text{FOM} = n^3 r \quad \text{(Pockels effect)} \tag{3.5}$$

$$\text{FOM} = n^3 R \quad \text{(Kerr effect)} \tag{3.6}$$

i.e. higher FOM will result in a larger refractive index change with respect to an applied voltage. Note that for some electro-optic tensor components the optical indicatrix may also rotate and therefore Eq. (3.2), instead of Eq. (3.3), should be taken as a general case.

Inserting the nonlinear polarization (Eq. (3.1)) into the wave equation (2.16) and assuming a small refractive index change, the electro-optic tensor coefficients defined above are related to the second- and third-order susceptibility coefficients by

$$\chi^{(2)}_{ijk}(-\omega, \omega, 0) = -\frac{1}{2} n_i^2(\lambda) n_j^2(\lambda) r_{ijk}(\lambda), \tag{3.7}$$

$$\chi^{(3)}_{ijk}(-\omega, \omega, 0, 0) = -\frac{1}{2} n_i^2(\lambda) n_j^2(\lambda) R_{ijk}(\lambda) \tag{3.8}$$

where the (angular) optical-frequency (ω) dependence of the second-order susceptibility $\chi^{(n)}_{ijk}$ is reflected in the wavelength ($\lambda = 2\pi c/\omega$) dependence of the corresponding electro-optic coefficients r and R, as well as refractive indices n.

As discussed in Section 3.2, the applied electric fields can oscillate at any angular frequency Ω below optical molecular or lattice vibrations, where $\Omega \ll \omega$. Depending on these resonance frequencies, the bandwidth of, for example, electro-optic modulation can be limited to a few GHz (inorganic materials) or to more than a few hundred GHz (organic crystals and even higher for some polymers). Depending on the frequency of the applied field and these resonance frequencies, the dispersion of the susceptibility should be considered, e.g., for the second-order case, as $\chi^{(2)}_{ijk}(-\omega - \Omega, \omega, \Omega) \cong \chi^{(2)}_{ijk}(-\omega, \omega, \Omega)$, and the corresponding electro-optic coefficients will also depend on the frequency $v = \Omega/(2\pi)$ of the applied electric field. The frequency v and the wavelength λ dependence of the electro-optic coefficients $r_{ijk}(\lambda, v)$ and $R_{ijkl}(\lambda, v)$ are discussed in Sections 3.2 and 3.3.

3.1.1 The polarization-optic effect

The electro-optic effect is defined by Eq. (3.2) as the change of the optical indicatrix/refractive indices as a function of the externally applied electric field. Although this definition is phenomenologically the most useful one (the magnitude of the applied field is normally known), the so-called polarization-optic description is often more appropriate from the point of view of fundamental understanding of the material response. In this case, the change of the optical indicatrix is defined as a function of the induced polarization in the materials P^0 as

$$\Delta\left(\frac{1}{n^2}\right)_{ij} = f_{ijk} P^0_k + g_{ijkl} P^0_k P^0_l. \tag{3.9}$$

Table 3.1 Selection of electro-optic materials and their parameters relevant for the electro-optic response*: the electro-optic coefficient r, refractive index n, electro-optic figure-of-merit n^3r, low-frequency dielectric constant ε, and polarization-optic coefficient $f\varepsilon_0 = r/(\varepsilon - 1)$ scaled by ε_0.

	r (pm/V)	n	n^3r (pm/V)	ε	$f\varepsilon_0$	$(n^3r)^2/\varepsilon$ 10^3(pm/V)^2	λ (nm)	Ref.
Inorganic crystals:								
LiNbO$_3$	31.5	2.2	340	28	1.2	4.1	633	[1]
GaAs	1.2	3.5	51	13.2	0.1	0.2	1020	[2]
KNbO$_3$	63.3	2.2	650	44	1.5	9.6	633	[3]
KNb$_{0.55}$Ta$_{0.45}$O$_3$	(1400)	(2.2)	(15,000)	(2500)	(0.6)	90	633	(a)
Sn$_2$P$_2$S$_6$	170	2.8	4000	230	0.74	70	1313	[4]
Organic crystals:								
DAST^b	92	2.5	1470	5.2	22	420	720	[5]
	53	2.2	530	5.2	13	54	1313	[5]
Polymers:								
A-095.11^c	20	1.66	92	2.8	11	3	1313	[6, 7]
CLD-1^d	130	1.65	584	3.5	52	97	1313	[8, 9]

^{a} Estimations based on preliminary results at room temperature for a compound with the para- to ferroelectric phase transition at 65 °C
^{b} Organic crystal DAST (4-N,N-dimethylamino-4′-N′-methylstilbazolium tosylate)
^{c} Polyimide side-chain polymer based on Disperse Red
^{d} Phenyltetraene chromophore in a guest–host polymer system
* The reported quantities present unclamped values (see Section 3.2) at the optical wavelength λ.

where f_{ijk} and g_{ijkl} are the linear and quadratic polarization-optic tensors, respectively. The material polarization P^0 is induced by the applied static (or quasi-static) electric field E^0, as described by the basic material relation $P_i^0 = \varepsilon_0(\varepsilon_{ij}^0 - \delta_{ij})E_j^0$, where ε_{ij}^0 is the static dielectric tensor. This results in a simple relation between the electro-optic coefficients and the polarization-optic coefficients

$$r_{ijk} = \varepsilon_0 f_{ijm}(\varepsilon_{mk}^0 - \delta_{mk}) \tag{3.10}$$

$$R_{ijkl} = \varepsilon_0 g_{ijmn}(\varepsilon_{mk}^0 - \delta_{mk})(\varepsilon_{nl}^0 - \delta_{nl}). \tag{3.11}$$

From the above equations we can see that the effective electro-optic response may be large owing to a high static dielectric constant and/or a high polarization-optic response. This relation is particularly noticeable in inorganic ferroelectric crystals, whose static dielectric constants may reach very high values, typically of the order of 10^2, but sometimes as high as 10^4. Organic materials usually have low dielectric constants of 5 or under, but can achieve electro-optic coefficients that are much larger than those of the best inorganic materials, because of the essentially larger polarization-optic response (see Table 3.1). The electro-optic response of inorganics may be enhanced by the static dielectric response (leading also to enhanced local electric fields). This is particularly evident in the case of ferroelectric crystals close to a phase transition, where the electro-optic coefficient may reach values as high as $r = 10^5$ pm/V because of diverging dielectric constants. However, as described later on, low dielectric constants (i.e., organic materials) have several benefits in high-speed electro-optic modulation and electric field detection applications, especially where low drive voltages are required. Materials with low dielectric constants are also beneficial for other nonlinear optical applications such as frequency conversion and THz-wave generation. Table 3.1 illustrates that polarization-optic effects can be more than one order of magnitude larger in organic materials than in inorganic materials.

3.2 Frequency dependence

The nonlinear optical susceptibilities $\chi^{(n)}(\omega_1, \omega_2, \omega_3, \ldots, \omega_n)$ may strongly depend on the frequencies ($\omega_1, \omega_2, \omega_3, \ldots, \omega_n$) of the interacting fields, as already mentioned in Section 2.1.2. In this section, the two simplest examples are considered where $n = 1$ and $n = 2$, i.e., the linear susceptibility $\chi^{(1)}(\Omega)$ and the second-order nonlinear optical susceptibility $\chi^{(2)}(-\omega-\Omega, \omega, \Omega)$ at one fixed optical angular frequency ω, both as a function of the other angular frequency Ω. For lower Ω, $\chi^{(1)}(\Omega)$ relates to the usual quasi-static dielectric constant and $\chi^{(2)}(-\omega-\Omega, \omega, \Omega)$ describes the electro-optic effect. For optical Ω, $\chi^{(1)}(\Omega)$ relates to the refractive index and $\chi^{(2)}(-\omega-\Omega, \omega, \Omega)$ describes sum-frequency generation (or second-harmonic generation for $\Omega = \omega$). In other words, we consider the dielectric constant $\varepsilon(\nu)$ and the electro-optic coefficient $r(\lambda, \nu)$ at a certain optical wavelength $\lambda = 2\pi c / \omega$ as a function of the frequency $\nu = \Omega/(2\pi)$ of the applied quasi-static, high-frequency, or optical electric field.

The frequency dependence (dispersion) of the dielectric constant $\varepsilon(\nu)$ and the electro-optic coefficient $r(\lambda, \nu)$ is schematically illustrated in Fig. 3.1. The different regimes observed are a consequence of different physical mechanisms in the material leading to its polarization under the influence of an external electric field. These include the electronic polarizabilities of molecular units and contributions of ions (lattice vibrations). Eventual dielectric relaxation, which may occur at low frequencies, is not discussed here. The electronic and ionic material polarizabilities show oscillatory behavior, and therefore resonances can usually be observed in certain frequency regions; these resonances are used in spectroscopy for material characterization. The frequency dependence therefore reflects the different dynamics of the involved polarizable units, e.g., the dynamics of electrons are much faster than that of ions. At very high frequencies (above 10^{16} Hz) the material polarizability is reduced because the electrons cannot respond quickly enough to follow the field oscillations.

The free (unclamped or zero stress) dielectric constant ε^T and the free electro-optic coefficient r^T in the low-frequency region contain three different contributions: acoustic lattice vibrations (acoustic phonons) ε^a and r^a, optical lattice vibrations ε^o and r^o, as well as electronic contributions ε^e and r^e

$$\varepsilon^T = \varepsilon^a + \varepsilon^o + \varepsilon^e = \varepsilon^a + \varepsilon^S \tag{3.12}$$

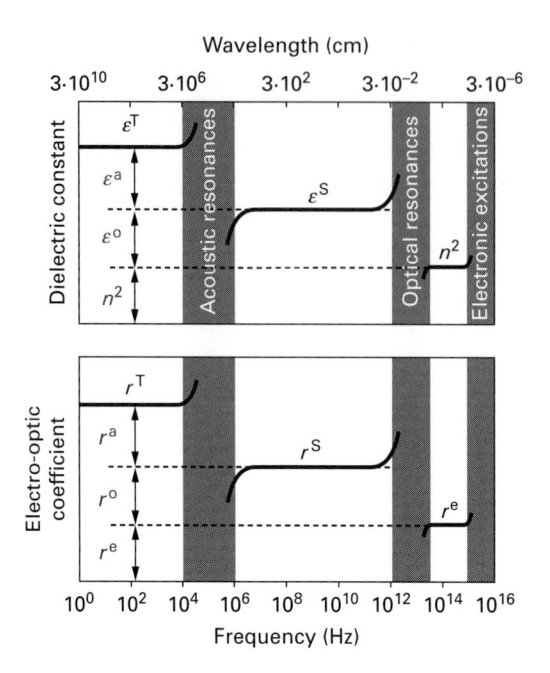

Figure 3.1 Schematic illustration of the material dielectric constant and the electro-optic coefficient as a function of the frequency of the applied electric field.

$$r^T = r^a + r^o + r^e = r^a + r^S \tag{3.13}$$

where $\varepsilon^S = \varepsilon^o + \varepsilon^e$ is the clamped (zero-strain) dielectric constant and $r^S = r^o + r^e$ the clamped electro-optic coefficient. The clamped coefficients describe the material response above the acoustic ionic resonances and below the optical resonances, roughly from about 100 MHz to the THz frequency range. At optical frequencies, the dielectric constant ε^e can be simply related to the refractive index, and the electro-optic coefficient r^e to the nonlinear optical coefficient for second-harmonic generation $d = 1/2\chi^{(2)}(-2\omega, \omega, \omega)$ as

$$\varepsilon^e = n^2 \tag{3.14}$$

$$r^e = -\frac{4}{n^4} d. \tag{3.15}$$

Organic and inorganic materials have considerably different dispersion relationships for the dielectric constant and electro-optic coefficient. While in the case of organic materials the electronic contribution is usually dominant, for inorganic materials the greater part of the response comes from acoustic and optical lattice vibrations. This is clearly evident from Table 3.2, in which measured dielectric, optical, electro-optic, and nonlinear optical coefficients of the organic crystal DAST are compared against inorganic crystals $LiNbO_3$ and $KNbO_3$.

The most elegant technique used to measure the different contributions from r^T, r^S, r^a has been described in Ref. [11]. As shown in Fig. 3.2, a short pulse is applied to the

3.2 Frequency dependence

Table 3.2 Comparison of inorganic crystals LiNbO$_3$ and KNbO$_3$, and the organic crystal DAST.*

	ε^T	n^T	r^T (pm/V)	r^S (pm/V)	r^a (pm/V)	r^o (pm/V)	r^e (pm/V)	d (pm/V)	Ref.
LiNbO$_3$	28	4.5	30	30	0	24.2	5.8	34	[1]
KNbO$_3$	44	4.5	63	34	29	31.3	3.7	21	[3]
DAST	5.2	4.6	47	48	−1	12	36	290	[10]

* Unclamped dielectric constant ε^T, refractive index n, unclamped r^T, clamped r^S, acoustic r^a, optical r^o and electronic r^e electro-optic coefficients, and nonlinear optical coefficient d. The values for LiNbO$_3$ and KNbO$_3$ are given at 1.06 μm, and for DAST at 1.535 μm.

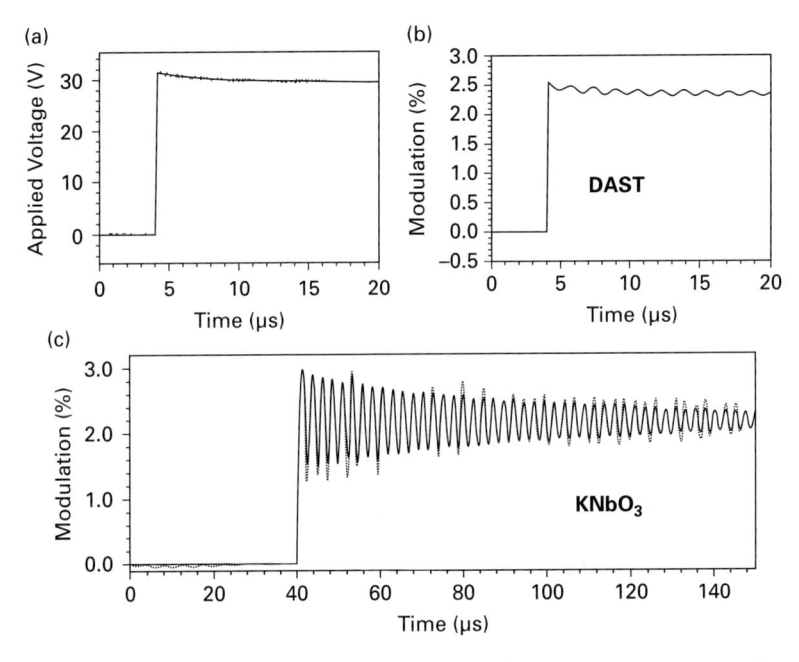

Figure 3.2 (a) The applied step voltage and (b) the electro-optic response, r_{111}, in DAST at 1535 nm [11]. A small and negative contribution resulting from acoustic excitation of the crystal can be seen. For comparison, the electro-optic response (containing a large acoustic phonon contribution) of KNbO$_3$ is also shown (c).

sample, and the time evolution of the refractive index change is measured by using an interferometer and a fast digital scope. The results of such measurements are shown in Fig. 3.2 for KNbO$_3$ (one of the best inorganic ferroelectric crystals with large acoustic lattice contributions), and for DAST. Large elastic contributions and piezo-electric ringing are observed in KNbO$_3$, while these effects are negligible in DAST. From these measurements, the acoustic contributions to the electro-optic effect r^a can be calculated from the measured nonlinear optical susceptibility for SHG $\chi^{(2)}_{111}(-2\omega, \omega, \omega) = 2d_{111}(\omega)$ using the oriented-gas and two-level models [12 to 15].

If we compare the measured optical and nonlinear optical parameters and the frequency dependence in DAST and LiNbO$_3$ / KNbO$_3$ (Table 3.2) we can see that in

DAST, the electronic contribution is dominant, whereas in the inorganics, lattice vibrations constitute the greater part of the response. The low dispersion characteristic found in organic materials is an essential advantage over inorganic materials for high-speed electro-optic applications.

In organic materials, the electric wave travels at about the same speed as the optical wave, because of the low dielectric constant in the low-frequency regime, i.e., $\varepsilon = n^2$, which is not the case for most inorganic electro-optic materials in which $\varepsilon \gg n^2$. This kind of velocity matching is important when building high-frequency electro-optic modulators. The low dielectric constant displayed by organic materials will also decrease the power requirements of such devices. Another advantage of organic materials over inorganics is the almost constant electro-optic coefficient over an extremely wide frequency range. This property is essential when building broadband devices such as modulators and field detectors. The almost purely electronic response is also advantageous for THz-wave generation by optical rectification of difference-frequency generation. A high THz generation efficiency can be obtained if the group velocity of optical pump pulses matches the phase velocity of the generated THz waves. This is easier to achieve in organic materials because of the relatively low dispersion compared with inorganic materials.

3.3 Wavelength dependence

The linear and the nonlinear susceptibility elements $\chi^{(n)}$ depend on the angular frequencies of all the involved (optical or quasi-static) fields. The dependence of $\chi^{(n)}$ on the angular frequency or wavelength λ of the involved optical field(s) is called dispersion. For the electro-optic effect this means that both the refractive index $n(\lambda)$ and the electro-optic coefficient $r(\lambda)$ may be wavelength-dependent. This dependence is particularly strong close to the electronic absorption bands of materials. Since, for many organic electro-optic materials, these bands are in the visible and even near-infrared wavelength range, the dispersion is an important parameter for applications, in particular where the velocities of various interacting fields in the material should be matched (phase matching, velocity matching). The wavelength dependence of the linear susceptibility $\chi^{(1)} = n^2 - 1$ is most often modeled by the phenomenological Sellmeier equation [16]

$$n^2(\lambda) = 1 + \sum_{i=1}^{N} \frac{S_i}{1 - (\lambda_{0,i}/\lambda)^2}, \qquad (3.16)$$

which can be also derived by assuming a simple harmonic oscillator model for the electrons [17]; S_i is the oscillator strength and $\lambda_{0,i}$ the resonant wavelength of the particular oscillator i. The number of oscillators N in the above equation depends on the precision of the available data. Often only one oscillator is considered, while for more accurate determination of the dispersion (e.g., for the evaluation of phase-matching conditions) usually two or three oscillators are necessary.

The wavelength dependence for the nonlinear susceptibility elements $\chi^{(2)}(\lambda)$ and the electro-optic coefficients can be described by different models, such as anharmonic-potential models or polarization potential models, both of which are mostly employed for inorganic electro-optic materials [17]. For organics, most often the two-level model [12] is considered to describe dispersion, as discussed in more detail in Chapter 4.

3.4 Electro-optic modulation

Various applications of the electro-optic effect will be discussed in more detail in several chapters of this book. Here, electro-optic modulation is briefly introduced in order to define the important material and device figures-of-merit used in subsequent chapters. Electro-optic modulators are optical devices that can modulate a beam of light, which can be used to transmit information. Because of the high frequencies of optical waves, very high modulation bandwidths are possible with light as the carrier of information. The optical beam may be modulated in phase, frequency, polarization, amplitude, and direction. In the following paragraphs we briefly describe most common modulation geometries including integrated-optics and high-frequency modulators.

Phase modulators involve variations of the phase of the carrier wave as shown in Fig. 3.3. As shown above in Eq. (3.3), the refractive index change of a biased electro-optic material will depend on the effective electro-optic coefficient r (dependent on the orientation of the material and the light polarization) and the magnitude of the applied field $V(t)$ as

$$\Delta n = -\frac{1}{2} n^3 r \frac{V(t)}{d}. \tag{3.17}$$

where d is the thickness of the electro-optic material. The phase of the wave after passing through a length of the material, l, will change with the voltage as

$$\Delta \phi = k(V)l - k(0)l = \frac{2\pi}{\lambda} \Delta nl = -\frac{\pi n^3 r}{\lambda} V(t) \frac{l}{d} \tag{3.18}$$

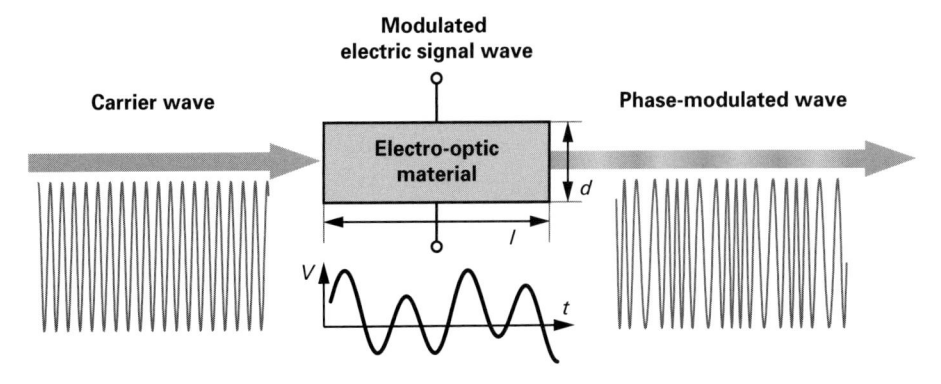

Figure 3.3 Phase modulator; the phase of the output beam is controlled by the applied electric field.

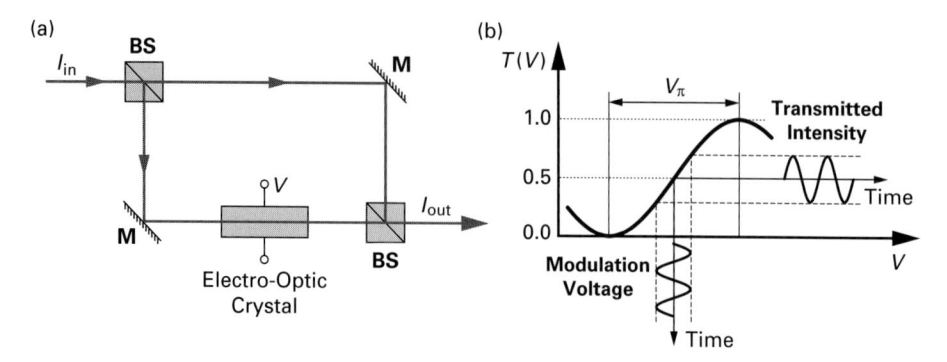

Figure 3.4 (a) Mach–Zehnder interferometer with a phase electro-optic modulator in one arm acts as an amplitude electro-optic modulator. BS: beam-splitter; M: mirror. (b) Transmittance of the interferometer as a function of the applied voltage.

In the case of simple sinusoidal modulation of the applied voltage $V(t) = V_m \sin \omega_m t$, the optical output wave can be written as

$$E(t) = A \cos(\omega t - \Delta\phi) = A \cos\left(\omega t + \frac{\pi n^3 r}{\lambda} \frac{l}{d} V_m \sin \omega_m t\right) = A \cos(\omega t + \phi_m \sin \omega_m t), \quad (3.19)$$

i.e. the output field is phase modulated with the modulation index

$$\phi_m = \frac{\pi n^3 r}{\lambda} \frac{l}{d} V_m, \quad (3.20)$$

Amplitude (intensity) modulation can be achieved, for example, by putting a phase modulator in one branch of an interferometer, most commonly a Mach–Zehnder interferometer as shown in Fig. 3.4(a). A beam-splitter is used to divide the laser light into two beams, one of which passes through a phase electro-optic modulator. The two beams are combined again with another beam-splitter.

If the incoming beam has an amplitude A and the beam-splitters are dividing a wave into two equal parts, the outgoing field is the sum of the two interfering beams:

$$E_{\text{out}}(t) = E_1(t) + E_2(t) = \frac{1}{2} A \cos(\omega t) + \frac{1}{2} A \cos(\omega t - \Delta\phi) = A \cos\left(\frac{\Delta\phi}{2}\right) \cos\left(\omega t - \left(\frac{\Delta\phi}{2}\right)\right) \quad (3.21)$$

The outgoing signal beam intensity is proportional to $E^2(t)$ averaged in time:

$$I_{\text{out}} = I_{\text{in}} \cos^2\left(\frac{\Delta\phi}{2}\right) = I_{\text{in}} \cos^2\left(\frac{\pi n^3 r}{\lambda} \frac{l}{d} \frac{V}{2}\right) = I_{\text{in}} \cos^2\left(\frac{\pi}{2} \frac{V}{V_\pi}\right), \quad (3.22)$$

where I_{in} is the input beam intensity and the half-wave voltage V_π has been defined as

$$V_\pi = \left(\frac{\lambda}{n^3 r}\right) \left(\frac{d}{l}\right). \quad (3.23)$$

The resulting transmittance of the interferometer $T = I_{out} / I_{in}$ as a function of $V(t)$ is shown in Fig. 3.4(b). We see that half-wave voltage V_π is the applied voltage needed to change the transmittance from 0 to 1, i.e., to operate the device as an optical switch. By operating the device in a limited region around $T(V) = 0.5$ (which can be adjusted simply by changing the length of one arm of the interferometer), we obtain a linear intensity modulator. In this case, the modulated transmitted light intensity is directly proportional to the modulation of the electric signal $V(t)$, as illustrated in Fig. 3.4(b) as

$$T(\text{small amplitude}) \approx \frac{1}{2}\left(1 + \frac{\pi}{V_\pi}V(t)\right). \tag{3.24}$$

The half-wave voltage V_π given by Eq. (3.23) is an important parameter for device applications. It is minimized for each device geometry and carrier wavelength by large $n^3 r$, which is the material figure-of-merit for electro-optics, as already defined in Eq. (3.6). Figures-of-merit of some selected organic and inorganic materials are listed in Table 3.1. On the other hand, half-wave voltage can be reduced by optimizing the geometry, i.e., long propagation distances l and short electrode distances d greatly reduce V_π. However, owing to optical diffraction, the reduction of d is limited: light focused down to small spot sizes (of the order of μm^2) will be diffracted so that after a few micrometers of propagation length the lateral dimension is doubled. The solution to this challenge may be found in integrated optical modulators, in which light is confined to waveguiding cores with dimensions comparable to the light wavelength. Organic waveguiding modulators are discussed in more detail in Chapter 9 of this book.

3.5 High-frequency modulation

One of the most important applications of organic electro-optic materials is high-frequency modulation of optical carriers using organic waveguides and hybrid organic–inorganic materials. A series of modulation arrangements and structures are discussed in Chapter 9.

Here we discuss two of the main basic requirements that have to be considered for such modulators: the modulator driving electrical power per unit bandwidth, and effects resulting from the different propagation velocity of electrical modulation and optical fields.

For simplicity we consider a bulk or waveguide configuration with both electrode spacing and waveguide width d and propagation distance l as in Fig. 3.3.

In the case of bulk modulators, Wemple and Didomenico [17] have evaluated the power requirement per unit bandwidth P/v as:

$$\frac{P}{v} = \frac{\pi\varepsilon_0\varepsilon\lambda^2 d^2}{4n_3^6 r_{33}^2 l} = \frac{\pi\varepsilon_0\lambda^2 d^2}{4\text{FOM}(P/v)l} \tag{3.25}$$

with the material figure-of-merit

$$\text{FOM}(P/v) = \frac{(n_3^3 r_{33})^2}{\varepsilon} \qquad (3.26)$$

This figure-of-merit is shown in Table 3.1, showing a roughly 20- to 100-fold improvement as compared with LiNbO$_3$ for CLD-1 polymers and DAST crystals.

Besides the electrical power requirements, one also has to consider the rapid changes of the high-frequency electric field during the propagation of the optical beam in the modulator. These changes will limit the bandwidth of the modulator.

In a high-frequency bulk amplitude modulator, the modulation amplitude will be reduced by the factor [18]

$$|r| = |\text{sinc}\,(\pi T v)| \qquad (3.27)$$

where $T = \frac{n_3 l}{c}$ is the propagation time of the optical wave through the modulator, and sinc $(x) = \sin(x)/x$.

Beam propagation effects are negligible ($|r| > 0.9$) for frequencies $v < v_L = 1/(4T)$.

For a 1-cm-long DAST modulator at wavelength $\lambda = 1.313$ μm and $n_3 = 2.2$, for example, one obtains a transit time $T = 73$ ps and a frequency

$$v_L = \frac{c}{4 l n_3} = 3.4 \text{ GHz}.$$

For modulators operating at higher frequencies, one uses traveling wave modulators where the electrical modulation signal travels in a transmission line with the optical modulator material in between the electrodes and with the optical beam traveling in the same direction as the electrical signal. In such an arrangement, the limiting frequency is given by the velocity mismatch between the optical velocity $c_o = \frac{c}{n_3}$ and the propagation velocity of the electrical wave $c_{el} = \frac{c}{\sqrt{\varepsilon}} = \frac{c}{n_{el}}$.

The maximum frequency of the traveling wave electro-optic modulator is given by [18]

$$v_L = \frac{c_o}{4 l |1 - c_o/c_{el}|} \qquad (3.28)$$

For DAST at $\lambda = 1.47$ μm one obtains with $c/c_{el} = 2.28/2.2$ a limiting frequency or bandwidth of $v_L = 93$ GHz.

A great advantage of using organic materials is that the dielectric constants at optical (n^2) and electrical frequencies ($\varepsilon = n^2$), and therefore their propagation velocities, are usually very close to each other since the dielectric response is almost exclusively due to the molecular electronic contributions; lattice or intermolecular contributions leading to dispersion effects are much smaller than in inorganic materials. This allows the realization of larger-bandwidth modulators.

Besides the simple modulator configuration discussed here to illustrate the materials requirements, a series of more complicated modulator structures have been realized for different applications. In Chapter 9, the basic principles of some of them are described in more detail.

References

[1] M. Jazbinsek and M. Zgonik, *Appl. Phys. B*, **74**, 407 (2002).
[2] I. Shoji, T. Kondo, A. Kitamoto, M. Shirane, and R. Ito, *J. Opt. Soc. Am. B*, **14**, 2268 (1997).
[3] M. Zgonik, R. Schlesser, I. Biaggio, *et al. J. Appl. Phys.*, **74**, 1287 (1993).
[4] D. Haertle, G. Caimi, A. Haldi, *et al. Opt. Commun.*, **215**, 333 (2003).
[5] F. Pan, G. Knopfle, C. Bosshard, *et al. Appl. Phys. Lett.*, **69**, 13 (1996).
[6] C. Bosshard, M. Bösch, I. Liakatas, M. Jäger, and P. Günter, in *Nonlinear Optical Effects and Materials*, P. Günter, Ed., Springer Series in Optical Science, Vol. 72, Berlin/Heidelberg/New York, Springer (2000), pp. 163–286.
[7] P. Pretre, P. Kaatz, A. Bohren, *et al. Macromolecules*, **27**, 5476 (1994).
[8] C. Zhang, L.R. Dalton, M.C. Oh, H. Zhang, and W.H. Steier, *Chem. Mater.*, **13**, 3043 (2001).
[9] L.R. Dalton, *J. Phys.: Condens. Matter*, **15**, R897 (2003).
[10] C. Bosshard, R. Spreiter, L. Degiorgi, and P. Gunter, *Phys. Rev. B*, **66**, 205107 (2002).
[11] R. Spreiter, C. Bosshard, F. Pan, and P. Gunter, *Opt. Lett.*, **22**, 564 (1997).
[12] J. L. Oudar and D. S. Chemla, *J. Chem. Phys.*, **66**, 2664 (1977).
[13] J. Zyss and J. L. Oudar, *Phys. Rev. A*, **26**, 2028 (1982).
[14] K. D. Singer, M. G. Kuzyk, and J. E. Sohn, *J. Opt. Soc. Am. B*, **4**, 968 (1987).
[15] C. Bosshard, K. Sutter, P. Prêtre, *et al. Organic Nonlinear Optical Materials*, Amsterdam, Gordon & Breach Science (1995).
[16] W. Sellmeier, *Ann. Phys. Chem.*, **219**, 272 (1871).
[17] S. H. Wemple and M. Didomenico, in *Applied Solid State Science*, Vol. 3, R. Wolfe, Ed., New York, Academic Press (1972), pp. 263–380.
[18] A. Yariv, *Quantum Electronics*, New York, John Wiley (1975).

4 Molecular nonlinear optics

4.1 Microscopic and macroscopic nonlinearities of organic molecules

The nonlinear optical effect on the microscopic (molecular) scale involves the response of individual polarizable units. For molecular and macromolecular materials, the single molecule is a natural choice for a microscopic polarizable unit. Equation (2.9) describes the nonlinear optical effect on a macroscopic scale, relating bulk material polarizability and interacting electric fields. In a similar way, we can consider these effects on the molecular scale. The dipole moment of the molecule **P** consists of its ground-state dipole moment μ_g and the induced contribution. The corresponding expansion

$$p_i = \mu_{g,i} + \varepsilon_0 \alpha_{ij} E_j + \varepsilon_0 \beta_{ijk} E_j E_k + \varepsilon_0 \gamma_{ijkl} E_j E_k E_l + \cdots \tag{4.1}$$

defines the molecular coefficients: linear polarizability α_{ij}, first-order hyperpolarizability β_{ijk}, and second-order hyperpolarizability γ_{ijkl}. The electric field **E** in the above equation represents the local electric field acting on the individual polarizable molecule. This field differs from the externally applied field because the dielectric properties of the molecules can be in marked contrast to that of their environment. Symmetry and dispersion considerations for the microscopic parameters are similar to those for macroscopic susceptibilities $\chi^{(n)}$ (see Section 3.1.2).

For organic materials, the macroscopic nonlinearities can be predicted from the microscopic ones because molecular polarizability represents the dominant component. In contrast, in the case of inorganic materials, the polarizability of atoms or atomic units is usually weak with respect to the polarizability originating from interactions between subunits (field-induced lattice distortions, see Section 3.2). Quantum-mechanics-based computer simulations can be used to evaluate the dipole moment ($\mu_{g,i}$) and polarizabilities and hyperpolarizabilities (α_{ij}, β_{ijk}, and γ_{ijkl}) of organic molecules and help provide guidance for their optimization. In the following sections, the basic design of molecular units having large nonlinear polarizabilities is described. Theoretical and numerical models as well as experimental methods used to evaluate these properties are also introduced and discussed.

4.2 Organic molecules for second-order nonlinear optics

The basic design of organic nonlinear optical molecules is based on π-conjugated electrons. Extended π-conjugation in organic molecules consists of alternating single

4.2 Organic molecules for second-order nonlinear optics

Figure 4.1 Typical dipolar organic molecules designed for large second-order nonlinear optical properties are shown. The π-bridge is functionalized with an electron donor (D) and an electron acceptor (A) in order to induce ground-state electron density asymmetry. The most common systems are those containing aromatic, heteroaromatic, or polyene bridges. R_1 and R_2 are most commonly carbon or nitrogen.

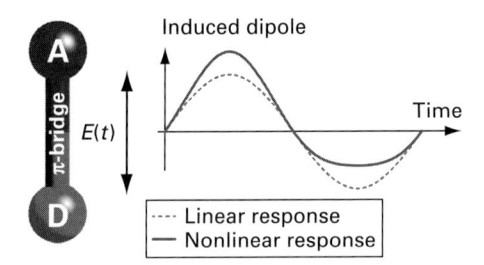

Figure 4.2 Simple picture of the physical mechanisms of the nonlinearity of donor–acceptor π-conjugated molecules. If such molecules are polarized by an electric field, an asymmetric electronic polarization response is induced owing to the asymmetry of the ground-state electron density distribution.

and double bonds. This alternating pattern of single and double bonds allows the delocalization of electronic charge distribution by way of π-orbital overlap. Such electronic delocalization leads to high electronic polarizability because the loosely bound π-electrons may be easily perturbed by application of an external electric field, leading to redistribution of the ground-state electron density. The extent of the linear redistribution is measured by the linear polarizability, α, and the extent of the asymmetric nonlinear redistribution is denoted by the first-order hyperpolarizability, β. The first-order hyperpolarizability displayed by organic molecules can be increased either by increasing the conjugation length or by attaching the appropriate electron donor and electron acceptor groups (Fig. 4.1). Typical electron donor substituents are exemplified by CH_3, NH_2, OH, $N(CH_3)_2$, OCH_2CH_3, and OCH_3. Typical early examples of electron acceptor groups are CN, NO_2, NO, and CHO; more recent acceptors leading to higher hyperpolarizability will be discussed later.

The addition of appropriate functionality at the ends of the π-system creates the asymmetric electron density distribution essential to realize a non-zero β. Another important aspect of molecular design is the planarity of the molecule, which will influence the effective size of the π-electron system and the mobility of electrons; twist angles between various components of the π-electron system may considerably reduce the charge transfer contribution and inhibit molecular nonlinearity.

For many molecules, microscopic second-order nonlinear optical behavior is adequately described by a simple two-level model [1], which considers only the two levels in the energy diagram: the ground state g, and the first excited charge-transfer state e. This approximation holds quite well as long as the operating wavelength is far away from electronic excitation (absorption) wavelengths i.e. $\lambda_{max}(\omega_{ge})$. In this limit, higher energy excited states lie much higher in the energy diagram than the photon energy of the driving field. Additionally, for most systems (dipolar molecules), with a strong nonlinearity along the charge transfer axis, it is reasonable to consider only the dominant hyperpolarizability tensor component β_{zzz}, oriented along this axis. Using the two-level model for $\beta_{zzz}(-\omega_3,\omega_1,\omega_2)$ we obtain [1]

$$\beta_{zzz}(-\omega_3,\omega_1,\omega_2) = \frac{1}{2\varepsilon_0\hbar^2} \frac{\omega_{eg}^2\left(3\omega_{eg}^2 + \omega_1\omega_2 - \omega_3^2\right)}{\left(\omega_{eg}^2 - \omega_1^2\right)\left(\omega_{eg}^2 - \omega_2^2\right)\left(\omega_{eg}^2 - \omega_3^2\right)} \Delta\mu\mu_{eg}^2 \quad (4.2)$$

where ω_{ge} denotes the resonant frequency of the transition, i.e. $\hbar\omega_{ge}$ denotes the energy difference between the ground state g and the excited state e, $\Delta\mu = \mu_e - \mu_g$ the difference between the excited and the ground-state dipole moments, $\mu_{eg} = \langle e|\mu_z|g\rangle$ the transition dipole moment between the excited and the ground state, and $\omega_1,\omega_2,\omega_3$ the light frequencies involved which are all below ω_{ge}.

We can also introduce a dispersion-free hyperpolarizability β_0 that is extrapolated to infinitely long wavelengths (low frequencies), away from electronic resonances

$$\beta_0 = \frac{3}{2\varepsilon_0\hbar^2} \frac{\Delta\mu\mu_{eg}^2}{\omega_{eg}^2} \quad (4.3)$$

This parameter is very useful when comparing the nonlinearity of molecules with different resonant frequencies ω_{eg}. For the case of second-harmonic generation we then obtain the following dispersion relation

$$\beta_{zzz}(-2\omega,\omega,\omega) = \frac{\omega_{eg}^4}{\left(\omega_{eg}^2 - 4\omega^2\right)\left(\omega_{eg}^2 - \omega^2\right)}\beta_0 \quad (4.4)$$

and for the linear electro-optic effect

$$\beta_{zzz}(-\omega,\omega,0) = \frac{\omega_{eg}^2\left(3\omega_{eg}^2 - \omega^2\right)}{3\left(\omega_{eg}^2 - \omega^2\right)^2}\beta_0 \quad (4.5)$$

The relationship between the macroscopic and the molecular coefficients is non-trivial because of interactions between neighboring molecules. However, in many cases the macroscopic second-order nonlinearities of organic materials are well explained by the nonlinearities of the constituent molecules (independent particle or oriented-gas model [2]). For example, the nonlinear optical and the electro-optic coefficients can be expressed, assuming only one dominant tensor element β_{zzz}, as

$$d_{ijk}(-2\omega,\omega,\omega) = \frac{1}{2}\chi_{ijk}^{(2)}(-2\omega,\omega,\omega)$$

$$= \frac{1}{2}N\frac{1}{\eta(g)}f_i^{2\omega}f_j^{\omega}f_k^{\omega}\sum_s\sum_{mnp}^{\eta(g)\ 3}\cos(\theta_{im}^s)\cos(\theta_{jn}^s)\cos(\theta_{kp}^s)\beta_{mnp}(-2\omega,\omega,\omega) \quad (4.6)$$

$$= \frac{1}{2}N\frac{1}{\eta(g)}f_i^{2\omega}f_j^{\omega}f_k^{\omega}\sum_s^{\eta(g)}\cos(\theta_{iz}^s)\cos(\theta_{jz}^s)\cos(\theta_{kz}^s)\beta_{zzz}(-2\omega,\omega,\omega)$$

and

$$r_{ijk}^e(-\omega,\omega,0) = -\frac{2}{n_i^2(\omega)n_j^2(\omega)}\chi_{ijk}^{(2)}(-\omega,\omega,0)$$

$$= -N\frac{1}{\eta(g)}\frac{2f_i^{2\omega}f_j^{\omega}f_k^{0}}{n_i^2(\omega)n_j^2(\omega)}\sum_s\sum_{mnp}^{\eta(g)\ 3}\cos(\theta_{im}^s)\cos(\theta_{jn}^s)\cos(\theta_{kp}^s)\beta_{mnp}(-\omega,\omega,0) \quad (4.7)$$

$$= -N\frac{1}{\eta(g)}\frac{2f_i^{2\omega}f_j^{\omega}f_k^{0}}{n_i^2(\omega)n_j^2(\omega)}\sum_s^{\eta(g)}\cos(\theta_{iz}^s)\cos(\theta_{jz}^s)\cos(\theta_{kz}^s)\beta_{zzz}(-2\omega,\omega,\omega)$$

where θ_{im}^s is the angle between the dielectric axis i and the molecular axis m, N is the number of molecules per unit volume (number density), $\eta(g)$ is the number of equivalent positions in the unit cell, s denotes a site in the unit cell, β_{mnp} is the molecular first hyperpolarizability, and f_i^{ω} are local field corrections that are usually simplified using the Lorentz approximation

$$f_i^{\omega} = \frac{n_i^2(\omega)+2}{3}, \quad (4.8)$$

where $n_i(\omega)$ is the refractive index at angular frequency ω. Note that Eq. (4.7) gives only the electronic contribution r^e to the electro-optic coefficient r^T (the dominant contribution in organic materials). For diagonal nonlinear optical tensor components, d_{kkk} and r_{kkk}^e respectively, the above expressions (4.6) and (4.7) can be simplified to

$$d_{kkk} = \frac{1}{2}NF^{\text{SHG}}\langle\cos^3\theta_{kz}\rangle\beta_{zzz}; \quad F^{\text{SHG}} = f_k^{2\omega}\left(f_k^{\omega}\right)^2 \quad (4.9)$$

$$r_{kkk}^e = NF^{\text{EO}}\langle\cos^3\theta_{kz}\rangle\beta_{zzz}; \quad F^{\text{EO}} = \frac{2f_k^0\left(f_k^{\omega}\right)^2}{n_k^4(\omega)} \quad (4.10)$$

where $\langle\cos^3\theta_{kz}\rangle$ is the so-called (acentric) order parameter, which is an important figure-of-merit measuring the polar alignment of the chromophores in the solid state: polarity of the crystalline arrangement in the case of crystals, and the poling efficiency in the case of poled polymers. (Poling refers to electric field poling, which introduces some degree of acentric (non-centrosymmetric) chromophore order (chromophore alignment in the direction of the poling field); poling efficiency refers to the extent of acentric order induced for a specific poling field strength.)

Figure 4.3 schematically shows the molecules and projection angles θ_{kz} between the polar axis k and the molecular axis z. To maximize the diagonal electro-optic or nonlinear optical coefficient along the polar axis, r_{kkk} and d_{kkk}, respectively,

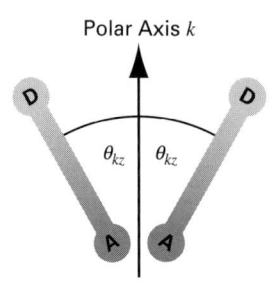

Figure 4.3 Nonlinear optical molecules in the oriented-gas model. For phase-matched nonlinear optical applications, the optimal angle θ_{kz} between the polar axis of the crystal and the charge transfer axis of the molecules is 54.7°, while for electro-optics the optimal $\theta_{kz} = 0°$.

the projection angles θ_{kz} should be close to zero, i.e. the charge transfer axes of the molecules should be close to parallel. For nonlinear optical applications such as frequency doubling, phase matching should also be considered. It was shown that for phase-matched applications, the optimal angle between the charge transfer axes of the molecules and the polar axis of the crystal is 54.7° for most crystallographic classes [2].

The relationship between the micro- and the macroscopic nonlinearities given above depicts a basic idea about the optimized packing of molecules. However, it still does not permit the precise determination of the macroscopic coefficients from the known microscopic nonlinearities, since intermolecular interactions (and the dielectric environment) may considerably influence the molecular nonlinearity β in the solid state. Although the relationship between macroscopic nonlinearity and molecular structure is not yet fully understood, the mechanisms leading to large microscopic effects are well known. Over the past two decades, a large number of various nonlinear optical molecules have been designed, prepared, and investigated. The resulting body of knowledge has allowed scientists to gain insight into the chemistry and physics of optical hyperpolarizabilities, and demonstrate molecules with extremely large β values. Theoretical calculations of the optimal non-resonant hyperpolarizabilies have suggested that they depend only on the number of electrons in the molecule N and the resonant frequency of the transition ω_{eg} and $\beta_0^{OPT} \propto N^{3/2}\omega_{eg}^{-7/2}$ [3].

4.3 Numerical calculations

Numerical computations of molecular hyperpolarizabilities are potentially useful for guiding the synthesis of new and improved chromophores, and for the analysis of the various structural factors that determine microscopic electro-optic activity. If molecular hyperpolarizability can be reliably estimated from theoretical calculations, then the poling-induced acentric order parameter can be estimated from measurement of the electro-optic coefficient, and insight can be gained into the roles played by various intermolecular electrostatic interactions in defining the organization of chromophores in the solid state. As will be discussed later (Chapter 7), optimization of the acentric order parameter is important for producing large macroscopic electro-optic activity, and currently measurement of electro-optic coefficients is the only way to characterize the acentric order parameter experimentally for comparison with statistical mechanical calculations of poling-induced order.

Calculation of molecular hyperpolarizability is challenging in many respects, including the requirements of correctly accounting for the influence of dielectric environment that surrounds chromophores of interest and accounting for the dependence of hyperpolarizability on optical frequency (resonance effects). Also, excited states and electron correlation contribute significantly to molecular hyperpolarizability.

Understanding the accuracy and utility of numerical calculations of molecular (chromophore) hyperpolarizability ultimately depends on the comparison of theoretical and experimental data. This is not a simple matter. Theoretical calculations are most easily effected for isolated molecules (*in vacuo*) and in the long-wavelength (zero-frequency) limit, while experimental measurements are carried out at finite wavelengths (typically 1–2 μm) and on chromophores surrounded by media of finite dielectric permittivity. Moreover, achieving high accuracy in experimental measurements is very difficult. For example, measurements of absolute first molecular hyperpolarizability β (for example by employing integrating sphere methods in hyper-Rayleigh scattering, HRS) are almost never done. More frequently, molecular hyperpolarizability of a given material is measured relative to another extensively studied material taken as a measurement (reference) standard. However, considerable disagreement exists as to the absolute value of molecular first hyperpolarizability (β_{HRS}) of the commonly employed reference standards, e.g., chloroform or CH_3Cl. Moreover, further confusion exists as the convention used to define β (discussed later in this chapter). Even when a consistent convention for definition of β is employed and data are reported as the ratio of hyperpolarizability of two materials, uncertainty can exist because of difficulty associated with a given experimental technique, e.g., two-photon fluorescence complications in HRS measurements and third-order nonlinear optical contributions to electrical field induced second-harmonic (EFISH) generation measurements. The difficulties encountered in the correlation of experimental and theoretical data make discussion of the efficacy of various theoretical methods difficult. Comparison of relative molecular hyperpolarizabilities for the purpose of guiding the decision whether or not to synthesize a new molecule is almost certainly the most meaningful use of theoretical methods. However, the accuracy of a given theoretical method is not the only important consideration in deciding which method to utilize. The time and cost of numerical calculations are also important considerations. In the following we provide a brief introduction to various theoretical methods used to calculate molecular first hyperpolarizabilities. Some paragraphs of this discussion are adapted from Ref. [4] with permission of the American Chemical Society.

4.3.1 Molecular quantum mechanics

The numerical computation of the properties of individual molecules can be approached by several different ways (e.g., orbital based methods such as Hartree–Fock (HF) and electron density based methods such as density functional theory (DFT)), but all begin with the Hamiltonian, H_0, for a molecule, which is augmented or perturbed by an external potential. External fields for study of organic nonlinear optical (ONLO) materials are usually no larger than 100 V/μm ≈ 2×10^{-4} au (where "au" is the atomic unit of electric field). For practical applications, the fields are much smaller and wavelengths are generally of the order of 1 μm. Thus, the perturbation Hamiltonian can be approximated as

$$H = H_0 + \sum_{i=1}^{N} q_i V(r_i) = H_0 - \sum_{i=1}^{N} q_i \mathbf{r}_i \cdot \mathbf{F} = H_0 - \boldsymbol{\mu} \cdot \mathbf{F} \qquad (4.11)$$

where q_i is the charge on the ith particle (electrons and nuclei) located at $\mathbf{r}_i$, $\boldsymbol{\mu}$ is the dipole moment, and $\mathbf{F}$ is the uniform electric field. This is the long wavelength limit (dipole approximation), i.e., the field strength is uniform over molecular dimensions. If the frequency dependence is required, the electric field must be written as the time-dependent field $\mathbf{F} = \mathbf{F}(t) \rightarrow \mathbf{F}\exp(i\omega t)$. If several different fields act at once, then

$$\mathbf{F} = \sum_j \mathbf{F}_j \exp(i\omega_j t). \qquad (4.12)$$

A molecular system will not react instantaneously to an oscillating field; therefore, the time evolution must be used in real-time time-dependent theory. Various methods are distinguished from one another by the approach to solving the wave equation.

$$H\psi = i\hbar \partial \psi / \partial t \qquad (4.13)$$

The time-independent theory is the easiest to consider.

4.3.2 Time-independent perturbation theory

The Hellmann–Feynman theorem states

$$\partial E / \partial \mathbf{F} = -\langle \psi | \boldsymbol{\mu} | \psi \rangle \qquad (4.14)$$

where E is the energy, i.e., $E = \langle \psi | H | \psi \rangle$. The energy can be differentiated any number of times to yield

$$\begin{aligned} E &= E_0 + \frac{\partial E}{\partial F_a} F_a + \frac{1}{2!} \frac{\partial^2 E}{\partial F_a \partial F_b} F_a F_b + \frac{1}{3!} \frac{\partial^3 E}{\partial F_a \partial F_b \partial F_c} F_a F_b F_c + \cdots \\ &= E_0 - \mu_a F_a - \frac{1}{2!} \alpha_{ab} F_a F_b - \frac{1}{3!} \beta_{abc} F_a F_b F_c - \frac{1}{4!} \gamma_{abcd} F_a F_b F_c F_d + \cdots \end{aligned} \qquad (4.15a)$$

From Eq. (4.15a), it follows that

$$-\partial E / \partial F_a = \mu_a(\mathbf{F}) = \mu_a^0 + \alpha_{ab} F_b + \frac{1}{2!} \beta_{abc} F_b F_c + \frac{1}{3!} \gamma_{abcd} F_b F_c F_d + \cdots \qquad (4.15b)$$

The second version of Eq. (4.15) defines electric field coefficients, polarizability, α_{ab}, first hyperpolarizability, β_{abc}, second hyperpolarizability, γ_{abcd}, etc., in the "Taylor series convention." If the factorials in Eq. (4.15b) are absorbed into the coefficients so that $\beta(\text{Taylor})/2! = \beta(\text{Perturbation})$, etc. in Eq. (4.15b), the "perturbation series convention" is used. The different choices for numerical coefficients have caused considerable confusion in the literature [5,6].

Using Rayleigh–Schrödinger perturbation theory, where $\langle \psi_r | \mu_a | \psi_s \rangle = \mu_{rs,a}$, gives

$$\alpha_{ab} = 2 \sum_e \frac{\langle \psi_0 | \mu_a | \psi_e \rangle \langle \psi_e | \mu_b | \psi_0 \rangle}{E_e - E_0} = 2 \sum_e \frac{\mu_{0e,a} \mu_{e0,b}}{E_e - E_0} \quad (4.16)$$

and

$$\beta_{abc} = \underset{abc}{\text{sym}} \sum_e \sum_{e'} \frac{\mu_{0e,a}(\mu_{ee',b} - \mu_{00,b}\delta_{ee'})\mu_{e'0,c}}{(E_e - E_0)(E_{e'} - E_0)} \quad (4.17)$$

in atomic units. The matrix elements $\langle \psi_r | \mu_a | \psi_s \rangle$ are the components of the dipole vector in state r for $r = s$, and of the transition moment vector for $r \neq s$. The *sym* operator is shorthand for terms arising from permutation of the tensor indices. The sums in Eq. (4.16) and Eq. (4.17) are over all excited states of the unperturbed system.

4.3.3 Sum over states

Perturbation theory provides the framework for computation of NLO properties by summation over excited states (SOS). This method has been used successfully by several groups [7 to 11] within the INDO (intermediate neglect of differential overlap) method to calculate α, β, and γ. The seminal papers of Marder and co-workers discussing the relationship between the polarization functions and bond length alternation (BLA) of merocyanine dyes were based on SOS methods [7,11].

4.3.4 Finite-field methods

The perturbation of the energy or charge density of a molecule that is induced by an external field is formalized in Eq. (4.15). Both nuclei and electrons move in response to an external field, but it is common practice to hold nuclei fixed when calculating the changes in dipole moment that are induced by the application of a field [12]. The separation of electronic and nuclear responses to external fields is an adiabatic approximation. Molecular geometry is sensitive to the solvent reaction field. For example, many calculations have been reported in which geometries are optimized in the presence of a solvent reaction field, with subsequent determination of the electronic response using the solvent-perturbed but static geometry. It is the fast electronic response that motivates the development of ONLO materials and justifies the use of the adiabatic approximation to capture these phenomena in calculations. Tripathy

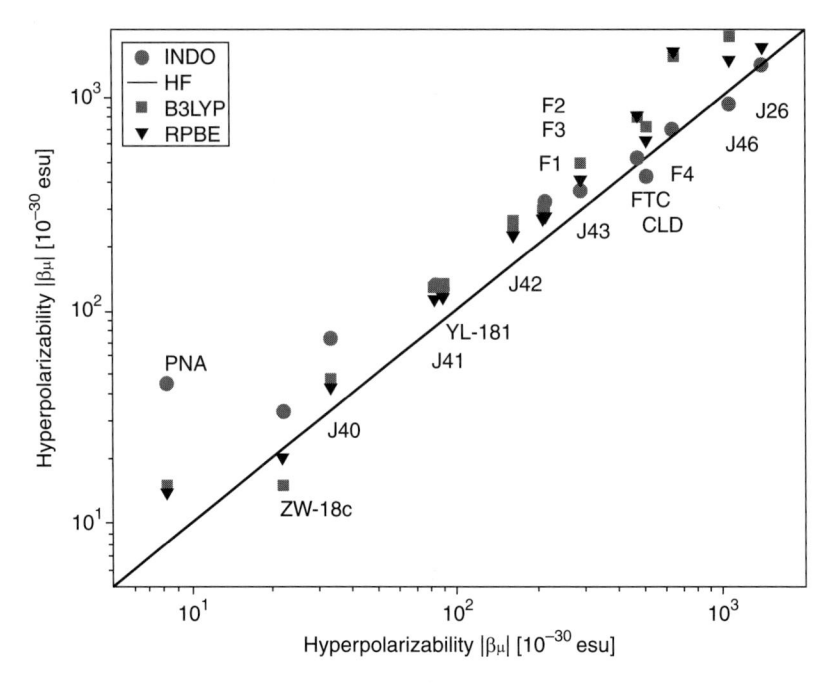

Figure 4.4 Comparison of various theoretical methods for the prediction of averaged molecular first hyperpolarizability (relevant to HRS data). Adapted from Ref. [14] with permission of the American Chemical Society.

et al. have shown that vibrations contribute only a few percent to the hyperpolarizability of Disperse Red 1, DR1 [13]. DR1 is an acronym or common name for p-nitro-aminoazobenzene (see Fig. 4.1 for the structure with D = NH_2, A = NO_2, $R_1 = R_2 = N$). Many quantum codes have implementations for external fields, enabling the user to specify the orientation and strength of several different fields as required to extract numerical derivatives of the energies or dipole vectors to determine α, β, γ, etc. The user of these codes is faced with a vast array of options for calculations; choices can be made amongst many different methods: Hartree–Fock (HF), density functional theory (DFT) with many different functionals, Møller–Plesset (MP) perturbation theory, coupled cluster (CC), etc. With any choice of method, many different basis sets can be considered. Which of these combinations is useful for a given class of problems depends upon the demands for accuracy and computational cost (usually determined by the number of electrons in the calculation). Comparisons between different methods are available (e.g., see Fig. 4.4) [14 to 16].

The structures of the chromophores considered in Fig. 4.4 are shown in Fig. 4.5.

Numerical values of μ, α, and β, computed by different methods, are given in Table 4.1 for the series of chromophores considered.

It is clear that the various methods are capable of producing trends in the variation of molecular hyperpolarizability. Specific problems related to the various methods will be discussed later in this section.

4.3 Numerical calculations

Structure	Geometry
F1: 2-[3-cyano-4-[2-(4-(dibutylamino)phenyl)vinyl]-1-(4-methoxybenzoyl)-5-oxo-1,5-dihyropyrrol-2-ylidene]-malononitrile	
F2: {3-cyano-2-(dicyanomethylene)-4-[2-(4-(dibutylamino)phenyl)vinyl]-5-oxo-2,5-dihydropyrrol-1-yl}-acetic acid ethyl ester	
F3: 2-{1-allyl-3-cyano-4-[2-(4-(dibutylamino)phenyl)vinyl]-5-oxo-1,5-dihydropyrrol-2-ylidene}-malononitrile	
F4: 2-[3-cyano-4-(2-{8'-[2-(4-(diethylamino)phenyl)vinyl]-3,4,3',4'-tetrahydro-2H,2'H-[6,6']bis[thieno[4,4-b][1,4]dioxepinyl]-8-yl}vinyl)-5,5-dimethyl-5H-furan-2-ylidene]-malononitrile	
J40: 3-phenyl-4-(2,3,6,7-tetrahydro-1H,5H-pyrido[3,2,1-ij]quinolin-9-ylmethylene)-4H-isoxazol-5-one	
J41: 3-phenyl-4-[3-(2,3,6,7-tetrahydro-1H,5H-pyrido[3,2,1-ij]quinolin-9-yl)-allyidene]-4H-isoxazol-5-one	

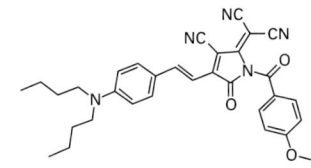

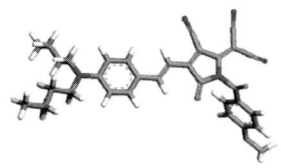

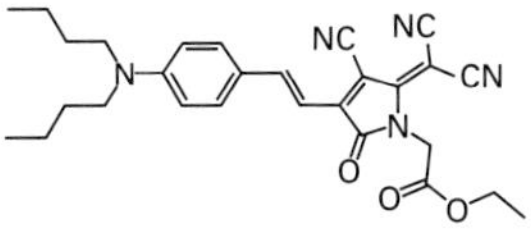

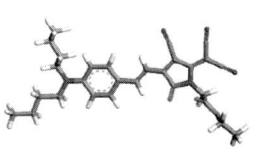

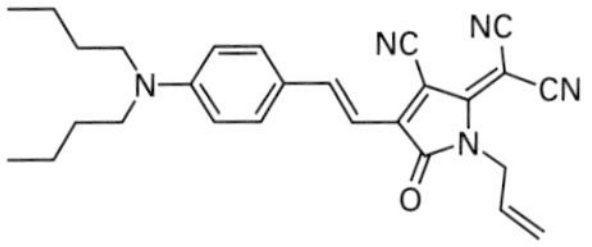

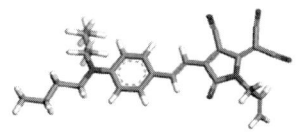

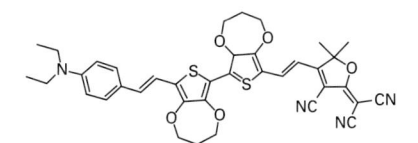

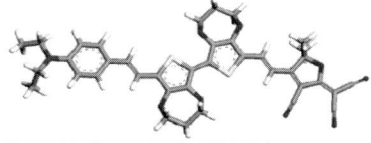

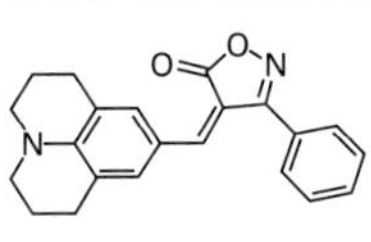

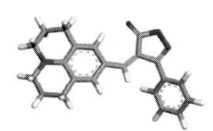

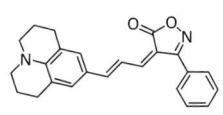

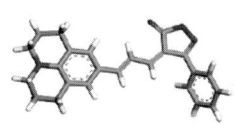

Figure 4.5 The chemical formulas, geometrics, and IUPAC names are given for the chromophores of Fig. 4.4. Adapted from Ref. [14] with permission of the American Chemical Society.

Structure	Geometry
J42: 3-phenyl-4-[5-(2,3,6,7-tetrahydro-1H,5H-pyrido[3,2,1-ij]quinolin-9-yl)-penta-2,4-dienylidene]-4H-isoazol-5-one	
J43: 3-phenyl-4-[7-(2,3,6,7-tetrahydro-1H,5H-pyrido[3,2,1-ij]quinolin-9-yl)-hepta-2,4,6-trienylidene]-4H-isoxazol-5-one	
J46: 3-phenyl-4-[13-(2,3,6,7-tetrahydro-1H,5H-pyrido[3,2,1-ij]quinolin-9-yl)-trideca-2,4,6,8,10,12-hexaenylidene]-4H-isoxazol-5-one	
J26: 1,3-dimethyl-5-{13-(2,3,6,7-tetrahydro-1H,5H-pyrido[3,2,1-ij]quinolin-9-yl)-trideca-2,4,6,8,10,12-hexaenylidene}-2-thiodihydropyrimidine-4,6-dione	
PNA: 4-nitrophenylamine or *para*-nitroaniline	
FTC: 2-[3-cyano-4-(2-{5-[2-(4-(diethylamino)phenyl)vinyl]thiopen-2-yl}vinyl-5,5-dimethyl-5H-furan-2-ylidene]-malononitrile	

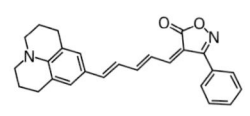

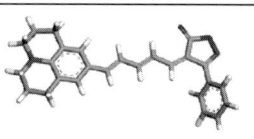

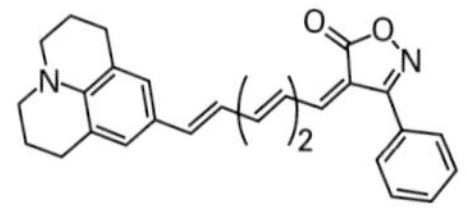

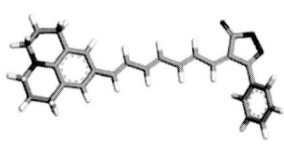

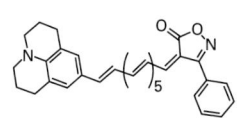

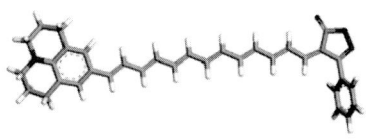

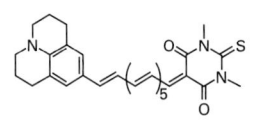

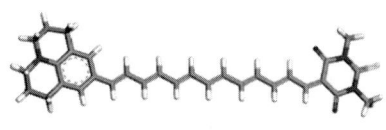

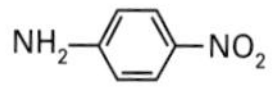

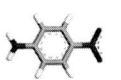

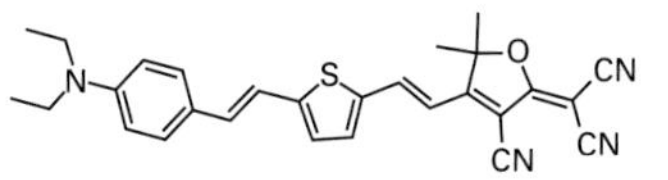

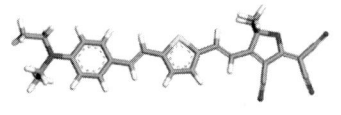

Figure 4.5 (*cont.*)

Structure	Geometry
CLD: 2-[3-cyano-4-(3-{3-[2-(4-(diethylamino)phenyl)vinyl]-5,5-dimethylcyclohex-2-enylidene}-propenyl)-5,5-dimethyl-5H-furan-2-ylidene]-molononitrile	
YL181: 2-{3-cyano-4-[2-(4-(dimethylamino)phenyl)vinyl]-5,5-dimethyl-1,5-dihydropyrrol-2-ylidene}-malononitrile	
ZW18c: 2,5-dimethyl-4-[2-(1-methylpyridinium-4-yl)vinyl]pyrazol-3-on-1-ide	

Figure 4.5 (*cont.*)

4.3.5 Analytic derivative methods

While the finite-field method is easy to grasp and efficient, analytical methods have also been developed. These "analytic derivative" methods, in which derivatives of the Hartree–Fock or the Kohn–Sham equation are developed up to sufficiently high orders, enable the functions of interest to be evaluated directly at zero field strength [17,18]. The algebraically complex equations that result have been encoded by the developers so that users interested in developing ONLO molecules do not also have to develop quantum codes. Results using both the finite-field and analytic derivative methods (using the same functional) are found to be consistent with one another to within expected numerical accuracy (routinely better than ±1% agreement between methods).

The method of choice for using one of the three methods discussed so far (sum over states, finite-field, and analytic derivatives) depends largely on the accessibility of the computer codes and the type of response function that is needed. For example, when surveying a list of structures, it may be sufficient to locate those with the most interesting properties and use only a few external field settings in a finite-field method. The design of practical chromophores for EO applications requires that optical absorptions not be too near to telecom operational frequencies. To assess the impact of resonance enhancement in these cases, a calculation that predicts the variation with optical frequency is useful.

Table 4.1 Dipole moment, polarizability, and hyperpolarizability values for the chromophores of Fig. 4.5.

(a) Dipole moments (debyes) and polarizabilities (Å³)

Molecule	μ					α			
	RPBE	INDO	HFy	B3LYP	PS-GVB	RPBE	INDO	HF	B3LYP
F1	17.3	11.5	16.4	16.7		96	50	80	95
F2	17.4	11.3	16.5	16.8		87	45	73	86
F3	17.7	11.0	16.6	17.1		84	45	70	83
F4	27.5	19.6	25.5	26.4		169	84	122	159
J40	11.2	10.7	11.0	11.1	10.7	48	28	44	50
J41	13.3	11.4	12.5	13.0	12.5	61	36	54	62
J42	16.2	13.2	14.9	15.7	15.5	78	45	66	78
J43	18.2	14.0	16.3	17.6	16.7	99	57	80	97
J46	24.0	16.0	19.6	22.7	17.7	185	99	134	177
PNA	7.5	8.0	7.5	7.5		15	13	14	15
FTC	26.0	18.2	24.6	25.2		120	66	90	114
CLD	28.4	21.7	28.0	27.9		138	73	107	133
YL181	18.1	13.9	18.1	17.8		60	35	50	59
ZW18c	10.8	12.4	12.5	11.4		37	31	36	38

(b) Scalar polarizabilities β_μ and $<\beta>$ (10^{-30} esu)

Molecule	β_μ				$<\beta>$				
	RPBE	INDO	HF	B3LYP	RPBE	INDO	HF	B3LYP	PS-GVB
F1	270	297	212	310	322	367	259	369	
F2	280	308	216	316	314	381	252	354	
F3	282	332	218	323	306	371	246	351	
F4	1630	685	655	1535	1754	827	740	1665	
J40	43	75	33	47	53	116	49	60	68
J41	111	138	83	129	129	211	112	154	120
J42	226	239	163	268	249	321	200	300	234
J43	409	366	293	493	440	471	343	537	464
J46	1491	923	1063	1891	1549	1099	1164	1980	884
J26	1706	1401	1403	2180	1713	1436	1416	2189	912
PNA	14	45	8	15	14	46	8	15	
FTC	819	522	476	783	845	553	502	818	
CLD	623	427	515	727	684	514	574	777	
YL181	117	126	89	141	132	161	106	157	
ZW18c	−20	−33	−22	−15	20	45	24	16	

4.3.6 Frequency-dependent methods

The evaluation of the frequency-dependent properties of molecules (chromophores) requires the use of time-dependent theory – either perturbation theory or analytic derivatives. Suppose that the field is given by the real form [19]

$$\mathbf{F} = \mathbf{F}_0 + \mathbf{F}_\omega \cos(\omega t) \tag{4.18}$$

where $\mathbf{F}_0$ is a static field component and ω is the frequency of the time-dependent part. When combined with Eq. (4.15b) this gives (summation convention)

$$\begin{aligned}\mu_a = \mu_a(\omega, \mathbf{F}_0, \mathbf{F}_\omega) = {}& \mu_a^0 + \alpha_{ab}(0;0)F_{0b} + \alpha_{ab}(-\omega;\omega)F_{\omega b}\cos(\omega t) \\ & + \frac{1}{2}\beta_{abc}(0;0,0)F_{0b}F_{0c} + \frac{1}{4}\beta_{abc}(0;-\omega,\omega)F_{\omega b}F_{\omega c} \\ & + \beta_{abc}(-\omega;0,\omega)F_{0b}F_{\omega c}\cos(\omega t) \\ & + \frac{1}{4}\beta_{abc}(-2\omega;\omega,\omega)F_{\omega b}F_{\omega c}\cos(2\omega t) + \cdots \end{aligned} \tag{4.19}$$

Equation (4.19) shows that several different functions can be extracted from a single frequency-dependent calculation. To clarify, the static and frequency-dependent polarizabilities are $\alpha_{ab}(0;0)$ and $\alpha_{ab}(-\omega;\omega)$, respectively. The second-order terms are all useful and readily identified by their associated field components: $\beta_{abc}(0;-\omega,\omega)$ is pertinent for optical rectification, $\beta_{abc}(-\omega,0,\omega)$ is the quantity of interest for electro-optic activity or the Pockels effect, and $\beta_{abc}(-2\omega;\omega,\omega)$ is needed for calculation of hyper-Rayleigh scattering and second-harmonic generation.

4.3.7 Hartree–Fock theory

The essential difference in philosophy between Hartree–Fock (HF) theory and density functional theory (DFT) is that HF starts with orbitals while DFT starts with electron density. The basic problem with HF is that the many-body wave function is written as a product of single-particle wave functions, and this approach fundamentally misses the correlations among the electrons. A number of orbital-based, post-Hartree–Fock (PHF) methods have been developed to address this deficiency by including high energy excitations and/or more correlation between electrons. PHF methods include Møller–Plesset (MP) and its developments MPn, configuration interaction (CI), coupled cluster (CC), etc. MP2 is often used as the first try to improve upon HF. CC methods can produce accurate β values but are restricted to small molecules (20–30 atoms). The MP2 method is less computationally demanding but still is not appropriate for the largest chromophores now being routinely considered.

The HF method was important in the early development of ONLO molecules. One of the early benchmark calculations on p-nitroaniline (PNA) was done with HF [12] (as well as MP2). A comparative study [14] of HF, INDO, and two different DFT methods (see Fig. 4.4 and Table 4.1) showed generally good agreement between HF and DFT methods for dipole moments and polarizabilities, but HF typically gives smaller hyperpolarizabilities than DFT. Larger molecules become increasingly problematic to calculate with HF, so the method is gradually being supplanted by DFT – especially popular are hybrid methods (including long-range corrected, LRC, methods) that mix some exact exchange from an HF-type calculation with the exchange-correlation functional of DFT (to be discussed more later).

4.3.8 Density functional theory

Like HF, original DFT theory has been modified to better account for electron correlation through an exchange-correlation (xc) potential. Different DFT methods correspond to different approximations to the xc potential. For example, LRC functionals mix variable amounts of xc with distance. The development of DFT methods has enabled relatively accurate calculations to be performed on quite large molecules. A general increase in the magnitude of the hyperpolarizability of a homologous series of molecules as the molecules become larger is typically observed experimentally. The scaling with the number of electrons is approximately the same for HF as for DFT, so there is no reason to prefer DFT on that account. However, fairly comprehensive studies [20,21] of the energetics of reactions have shown that DFT with generalized gradient approximation (GGA) functionals, especially with hybrid methods, perform better than HF. It is imprudent to infer that derivatives of the energy (with respect to electric field) will likewise be computed more accurately than HF on the basis of this comparison alone. Recent studies [15,22,23] have questioned the reliability of hyperpolarizabilities calculated with DFT. There are continuing developments in the controversy over wave functions vs. electron density, and it is too early to access the outcome of these. For routine work, hybrid methods compensate for the opposing shortcomings in both HF and DFT. Reference [16] provides an excellent comparison of the results of DFT calculations with different functionals (PBEO, BMK, M05, M05–2X), with MP2 calculations, and with experimental data (ratios of molecular first hyperpolarizabilities for different samples). The authors of Ref. [16] found that the BMK and especially the M05–2X functionals (functionals with the larger fraction of Hartree–Fock exchange) provide the best agreement with experimental data (comparable to that of the MP2 method). They recommend use of the M05–2X functional for the systematic prediction of molecular hyperpolarizabilities, because of the high computational cost of the MP2 method. Thus, hybrid methods seem to give fairly reasonable results, and especially when used in the context of TD-DFT (see below) and RT-TD-DFT.

4.3.9 Time-dependent DFT

The conversion from time-independent to time-dependent HF theory is relatively straightforward and involves making the molecular orbital coefficients time-dependent. For DFT the situation is different, and requires modification of the fundamentals of the Kohn–Hohenberg derivation. This was done as recently as 1984 by Runge and Gross [24]. A very fruitful approach based on the Runge–Gross theorem, but using the density matrix, was taken by Furche and Ahlrichs [25,26]. These developments have led to the TD option being available in the standard DFT and hybrid codes for calculation of excited state properties, including frequency-dependent properties. Representative results obtained with the frequency dependence implemented in Gaussian09 for a state-of-the-art high-β chromophore (YLD124) are shown in Table 4.2. These results show the importance of solvent effects as well as frequency on the hyperpolarizability. It should be noted that the permutation of tensor elements that is inherent in the

4.3 Numerical calculations

Table 4.2 Frequency-dependent hyperpolarizability of YLD124: 1310 nm

Solvent	ε	$\beta_{HRS}(-2\omega;\omega,\omega)/10^{-30}$ esu	$\beta_{zzz}(-\omega;\omega,0)/10^{-30}$ esu	$\beta_{zzz}(0;0,0)/10^{-30}10^{-30}$ esu
Vacuum	1.00	3491	702	428
CCl$_4$	2.228	3601	1567	816
CHCl$_3$	4.711	2283	2388	1086
		YLD124: 1906 nm		
CHCl$_3$	4.711	1742	1510	1086

time-independent theory is replaced by a symmetry that couples frequencies and tensor indices in the time-dependent formulation. This is generally inconsequential for the Pockels (electro-optic) effect where only one tensor element is of interest, or the hyper-Rayleigh scattering experiment, where rotational averaging mixes the tensor components.

4.3.10 Real-time TD-DFT

The direct integration of the time-dependent Schrödinger or Kohn–Sham equation is now possible with the high-performance computing power that has become available within the past few years. Application of an external time-dependent field having various strengths and orientations relative to the molecular axes (time-dependent finite-field method) enables evaluation of the real and imaginary parts of the response functions. The inverse relation between the time and frequency windows, inherent in the Fourier transform, requires lengthy runs for low-energy processes (such as those found in chromophores intended for NLO applications), making these calculations expensive [27]. As these methods continue to be developed [28 to 30], they will become indispensable tools for determination of the properties of ONLO systems.

4.3.11 Two-state model (TSM)

A two-state approximation to the SOS equations (4.7) and (4.8) is frequently used to guide insight into the critical physics that controls the magnitude of the first hyperpolarizability. For many ONLO molecules, the tensors are dominated by the components along the z-axis. Scalar measures of the electrical response functions, such as the trace of the polarizability tensor, might also be cast in a simplified scalar form. If only one excited state term ($e = 1$) is retained in Eq. (4.7) and Eq. (4.8), we have (omitting the direction subscript and all numerical factors)

$$\alpha = \frac{\mu_{10}^2}{E_{10}} \quad \text{and} \quad \beta = \frac{\mu_{10}^2 \Delta\mu_{10}}{E_{10}^2} \tag{4.20}$$

Here $E_{10} = E_1 - E_0$ and $\Delta\mu_{10} = \mu_{11} - \mu_{00}$.

An additional equation that is helpful in this analysis is the oscillator strength, $f = E_{10}\mu_{10}^2$ (in atomic units) for the lowest energy transition [31].

Experiments show that the refractive index (away from absorption) of ONLO chromophores is not unusually large. In this case, we can utilize the Clausius–Mossotti (CM) equation

$$\frac{\frac{\varepsilon}{\varepsilon_0} - 1}{\frac{\varepsilon}{\varepsilon_0} + 2} = \frac{\alpha}{v} \tag{4.21}$$

which relates the dielectric, ε, and the vacuum permittivity, ε_0, to the polarizability per unit volume, α/v, of a molecule. For a homologous series of molecules, the polarizability, α, increases in (approximate) proportion to the molar mass [18]. The validity of this assertion is the basis for well-known group additivity methods used to calculate polarizabilities. The apparent power law dependence seen for oligoacetylenes reflects the changing dielectric constants for these small molecules [32]. However, for long chains the behavior has to be asymptotic, as was shown by Tretiak et al. [33]. That is, μ_{10}^2/E_{10} can be substantially increased only by increasing the size of the molecule. Conversely, for a given molar mass, large values of the hyperpolarizability can only be obtained when the $\Delta\mu_{10}/E_{10}$ term in $\beta = \alpha(\Delta\mu_{10}/E_{10})$ is large. This is achieved when the overlap of electron densities between the HOMO and LUMO is as small as possible and/or the energy gap is made small, resulting in absorption far to the red. However, significant absorption too far to the red will lead to unacceptable optical losses. The molecular design goal is to find a happy medium where the lowest energy molecular absorption is as far to the red as possible while not so far as to lead to significant absorption at telecom wavelengths.

4.3.12 Some cautionary and additional perspective comments

Various methods have compared well with respect to the ability to predict general trends (e.g., see Fig. 4.4 and Table 4.1); however, some concerns become evident upon closer inspection. For example, theoretical methods have not been very successful in predicting the experimentally observed increase in molecular first hyperpolarizability observed in going from the heteroaromatic bridge chromophores such as YLD156 (structure given in Fig. 4.6) to chromophores in the polyene bridge structure of YLD124 (structure shown in Table 4.2). Use of TD-DFT methods with different hybrid functionals (e.g., the M05–2X functional) can predict the right trend but the magnitudes of β are not in as good agreement with experiment as observed for the B3LYP functional. Use of a single hybrid functional that yields universal desired prediction of both trends and absolute magnitudes has yet to be demonstrated.

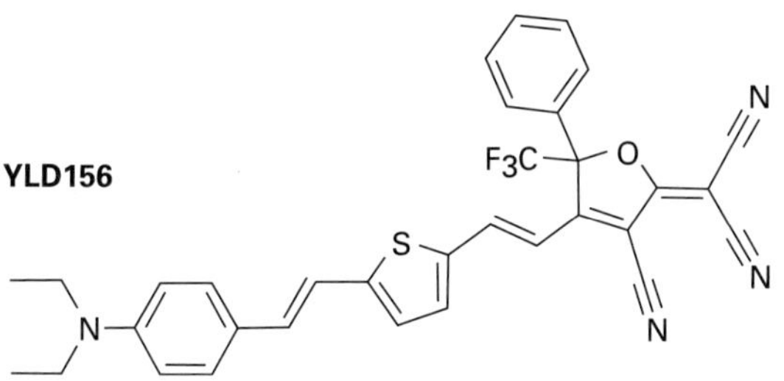

Figure 4.6 The structures of the EZ-FTC, CLD-1, and YLD156 chromophores are shown.

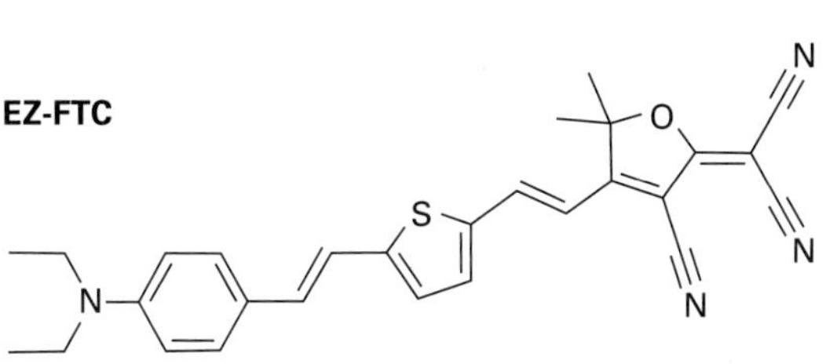

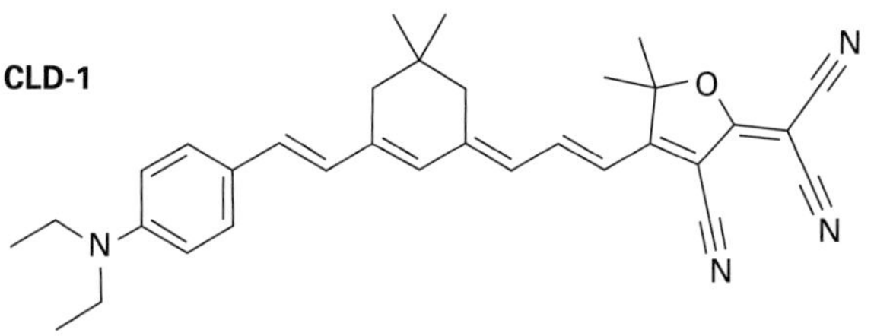

Another bothersome failure of both semi-empirical and DFT methods is that of gradient bridge chromophores where theoretically predicted and experimentally observed (by HRS) trends are not in good agreement [34,35]. A gradient bridge chromophore has the general structure strong donor–weak donor–weak acceptor–strong acceptor: for example a weak donor–weak acceptor bridge consisting of thiophene-triazole units. Such a bridge was theoretically predicted [34] to exhibit a larger β than an analogous thiophene–thiophene bridge. HRS measurements demonstrate that only modest improvement in β is realized [35].

Also, controversy continues to exist with respect to twisted bridge chromophores [36 to 43]. While EFISH measurements appear to support theoretical predictions, HRS measurements do not. Moreover, it has been impossible, to the present, to effectively translate the putative large molecular hyperpolarizabilities of twisted bridge chromophores to large macroscopic electro-optic activities, i.e., into functioning devices. Theories do suggest that twisted bridge chromophores are more complex with neutral, zwitterionic, and singlet biradical states all contributing to optical nonlinearity (Fig. 4.7).

Also, strong dipolar interactions make aggregation a serious problem, which has recently been somewhat addressed by Marks and co-workers through derivatization of the fundamental chromophore structure with bulky pendant groups. Twisted bridge chromophores remain of high interest because of their predicted exceptional optical nonlinearities (the largest predicted for any chromophore) even though these complex systems have avoided definitive characterization and utilization to the present.

Figure 4.7 The neutral ground state (NGS), singlet biradical state (SBS), and zwitterionic excited state (ZES) are shown for a twisted bridge chromophore.

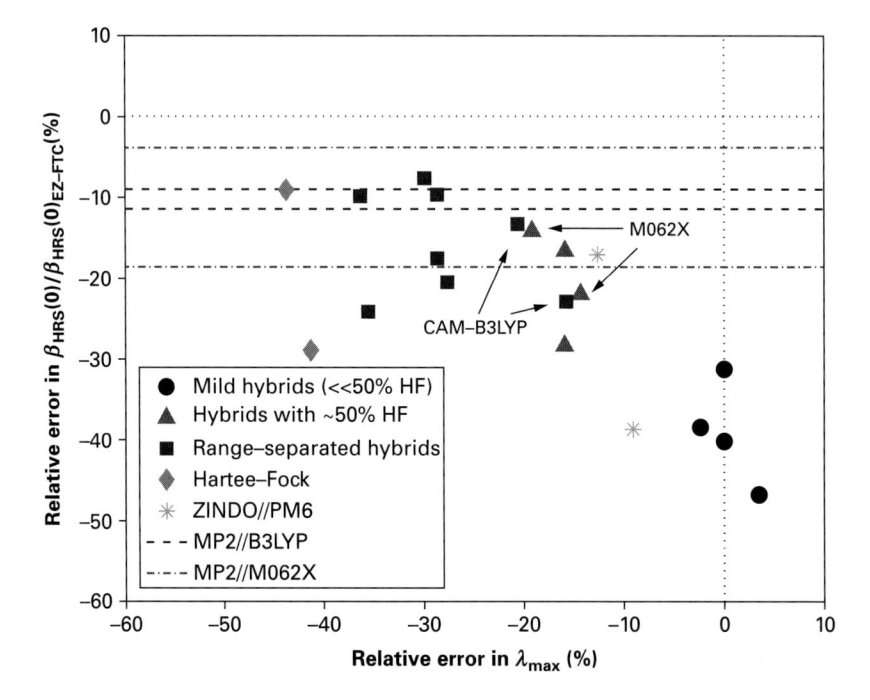

Figure 4.8 A two-dimensional representation of the relative errors (errors in the $\beta_{HRS}(0)$ ratios for CLD-1 and YLD156 versus EZ-FTC are compared with errors in calculating the lowest charge transfer excitation energy for CLD-1 and YLD156). DFT methods are divided into three families (mild hybrids, hybrids with approximately 50% HF exchange, and range-separated hybrids). Wavefunction-based methods are shown for comparison. MP2 methods are represented by lines owing to the absence of an excitation calculation. Each point or line represents a single calculation on one chromophore. Adapted from Ref. [44] with permission of the American Chemical Society.

Another bothersome problem with various computational methods is the observation that an approach that yields good reproduction of experimentally observed nonlinear optical activities (e.g., β_{HRS}) may not yield good reproduction of linear optical spectra, as illustrated in Fig. 4.8 [44]. DFT hybrid functionals with approximately 50% HF exchange and ZINDO/PM6 methods yield the best simultaneous prediction of λ_{max} and molecular first hyperpolarizability values. Efforts have been made to find a systematic approach to choosing the optimum HF exchange contribution ω. Tuning the amount of exact exchange to enforce Koopmans' theorem leads to better excitation energies but

4.3 Numerical calculations

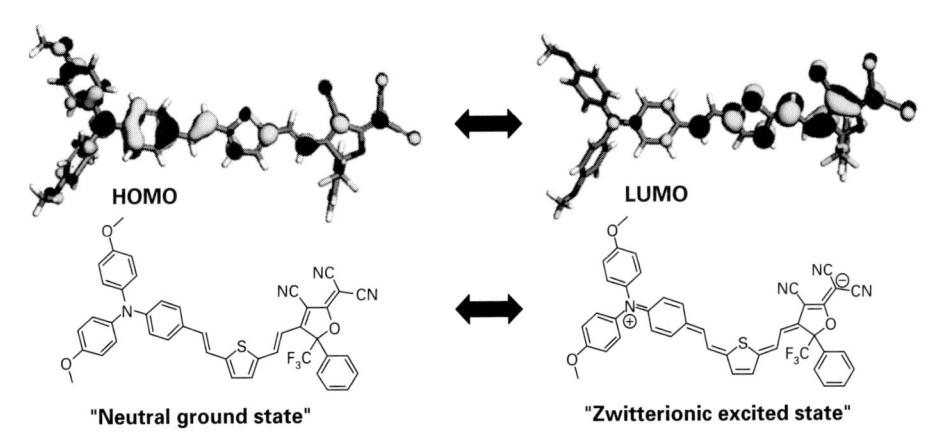

Figure 4.9 The electron density maps calculated from DFT for the HOMO and LUMO states of an electro-optic chromophore are shown together with the hypothetic limiting cases (Mulliken charge transfer states) of bond length alternation patterns for those states, which can be referred to as the neutral ground state (NGS) and the zwitterionic excited state (ZES). Actual BLA patterns for the HOMO and LUMO states will be an admixture of the two limiting forms (NGS and ZES), and application of an electric field will change that admixture.

unreliable computation of hyperpolarizability for larger chromophores. Larger values of ω are required to reproduce experimental molecular first hyperpolarizability values.

Of course, many trends predicted by theoretical calculations are observed, particularly when those trends reflect well-understood effects related to bond length alternation (BLA). Molecular first hyperpolarizability can approximately be related to the change in electron density between the ground (highest occupied molecular orbital, HOMO) and first excited state (lowest unoccupied molecular orbital, LUMO) (see Fig. 4.9). In discussion of structure/function relationships, β is often plotted versus BLA or the neutral ground state/zwitterionic excited state mixing coefficient with simple theory predicting a sinusoidal variation of β. For a BLA = 0, β is predicted to equal zero.

Introduction of groups such as highly aromatic groups, which inhibit changes in BLA upon application of an electric field, leads to reduced molecular first hyperpolarizability. This effect is seen, for example, in theoretically predicted and experimentally observed differences between ortho- and meta-X-shaped chromophores [45 to 48] as is illustrated in Fig. 4.10. The change in electron density across the chromophore is clearly much greater for the ortho-substituted phenyl ring core than for the meta-substituted phenyl.

X-shaped chromophores are of interest because of somewhat blue-shifted charge transfer absorption while retaining high molecular hyperpolarizability.

As evident from Table 4.2, theoretical methods predict the right trends for the variation of molecular first-hyperpolarizabilities with optical frequencies and material dielectric permittivity. This is illustrated in Figs. 4.11, 4.12, and 4.13. As is evident in Fig. 4.11, RTTD-DFT and some TD-DFT methods adequately reproduce the frequency dependence of the second-harmonic coefficient of p-nitroaniline measured by electric field-induced second-harmonic generation (EFISH).

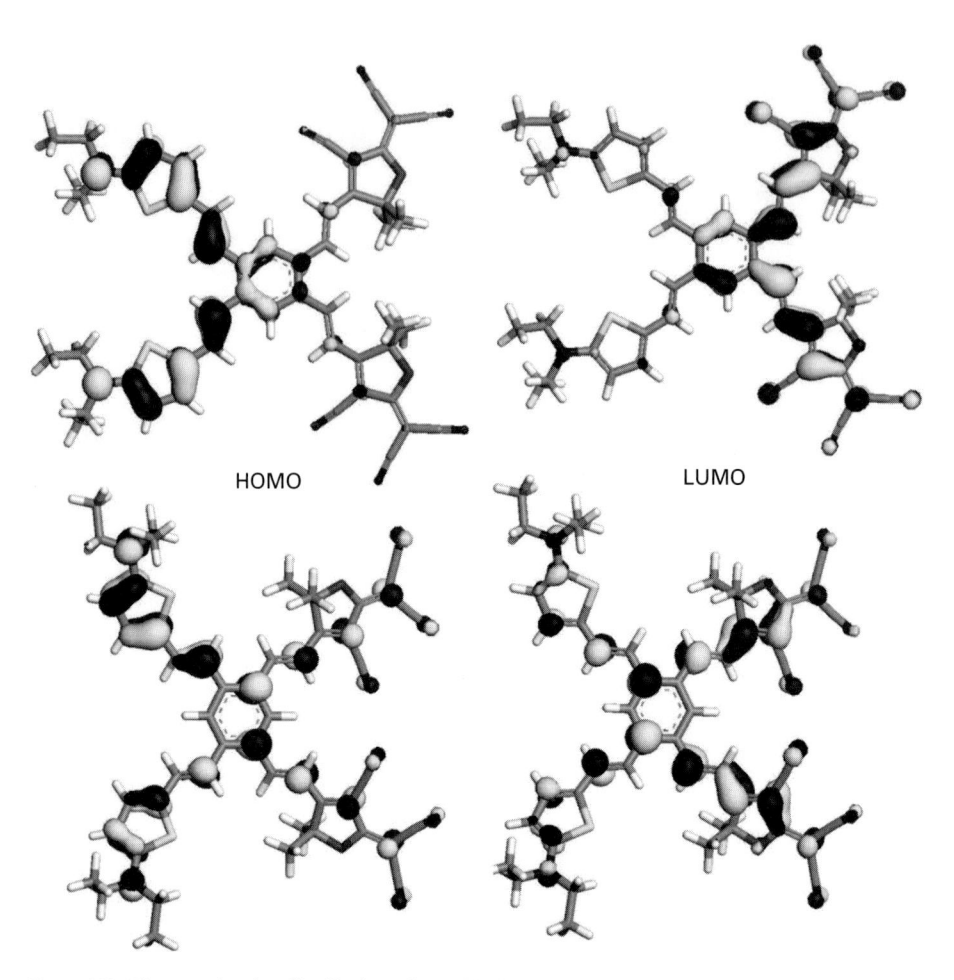

Figure 4.10 Electron density distributions for HOMO and LUMO orbitals for ortho- and meta-substituted phenyl core X-shaped chromophores.

As will be discussed in detail in Chapter 7, statistical mechanical methods can be crucial for understanding dielectric permittivity for materials with high chromophore concentrations. The combination of quantum and statistical mechanical methods can lead to very good prediction of the dielectric permittivity dependence of molecular hyperpolarizability (at least for some chromophore systems).

In summary, numerical simulations based on quantum mechanics are very useful in understanding the relationship of molecular hyperpolarizabilities to molecular structure and in some cases provide useful guidance for the design of improved chromophores. However, theoretical guidance has not reached the stage of reliability that predictions can be pursued with complete confidence. Quantum computing techniques (e.g., see March 8, 2013 issue of *Science* for a review of quantum information processing and quantum computing) may ultimately provide improved computational capabilities for electron systems with strong correlation such as those of interest here; however, as both quantum computer development and the coding for such computers is at an embryonic

4.3 Numerical calculations

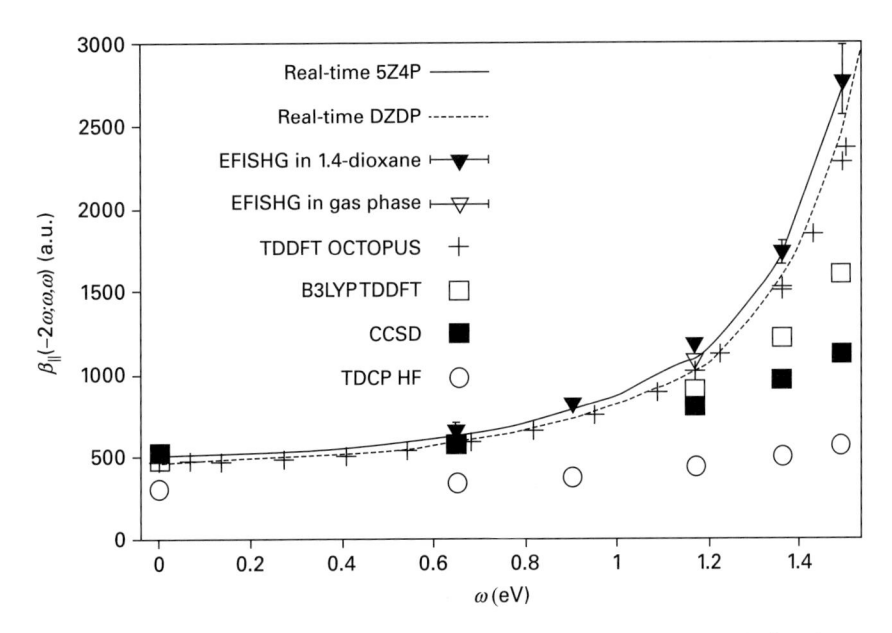

Figure 4.11 Comparison of various theoretical methods with respect to reproducing the frequency dependence of the electro-optic effect for *p*-nitroaniline (PNA).

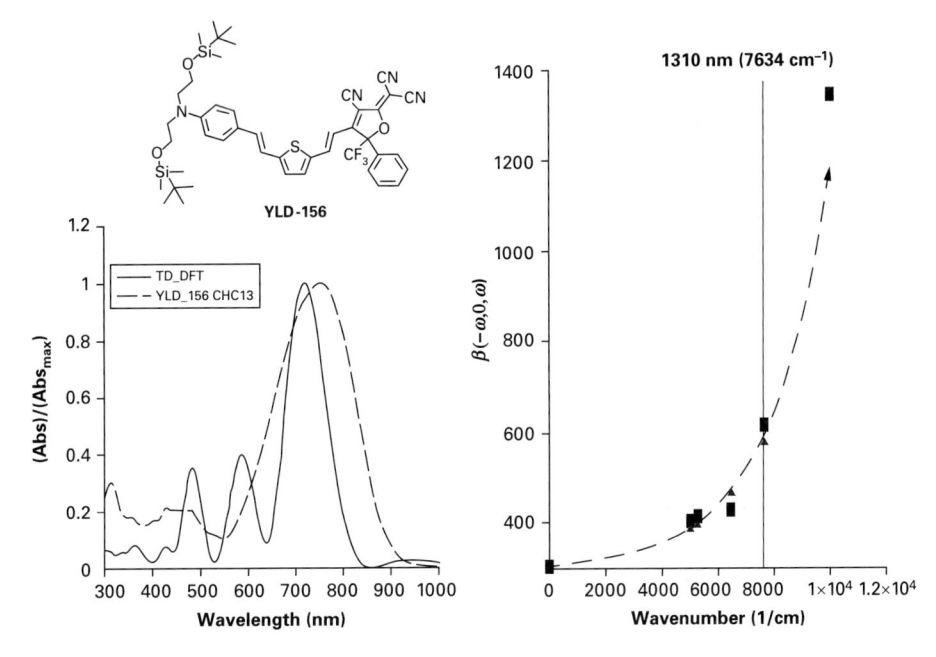

Figure 4.12 RTTD-DFT simulation of the linear absorption spectrum and low frequency (electro-optic) component of the molecular first hyperpolarizability for the YLD156 chromophore is shown. The experimental $\beta_{zzz}(-\omega,0,\omega)$ component is extracted from hyper-Raleigh scattering (HRS) data. Experimental data are indicated by a dashed line and squares while theoretical data are denoted by solid line and triangles. The dashed line on the right side indicates a best fit line to the theoretical data.

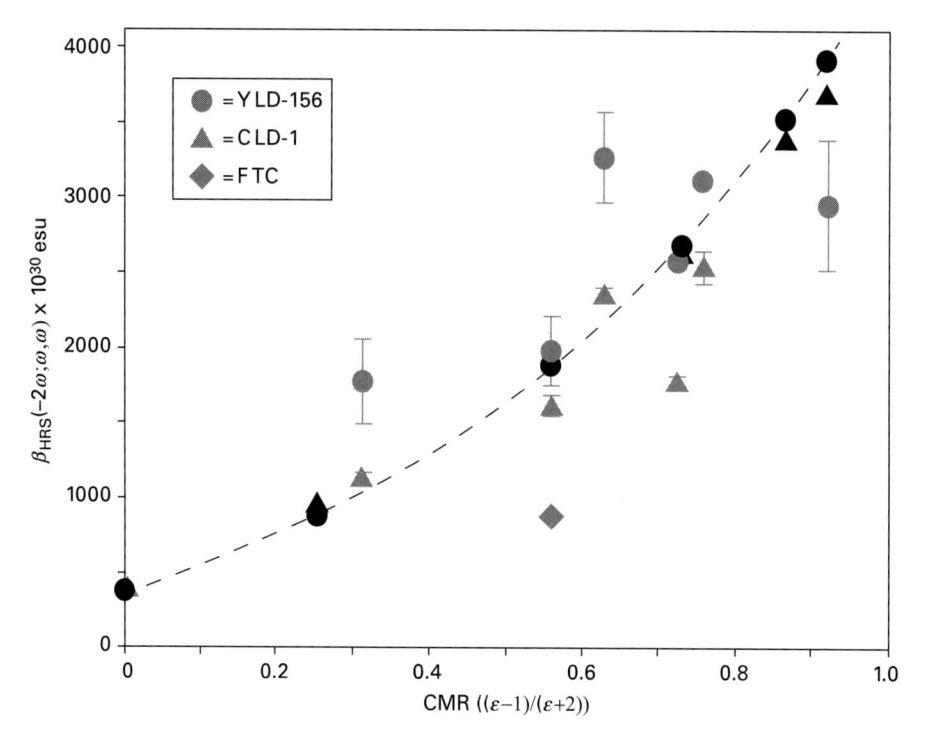

Figure 4.13 Comparison of experimental (lighter gray symbols) and theoretical (darker symbols) data for β_{HRS} versus the Clausius–Mossotti ratio (CMR) for three state-of-the-art chromophores (YLD156, CLD1, and FTC). Reproduced from Ref. [48] with permission of the American Chemical Society.

stage, it will likely be a decade or more before even preliminary calculations can be executed. A shorter-term and more likely development that could impact the utility of theoretical modeling is the potential of additional approximate wavefunction based methods, such as approximate coupled cluster methods, that circumvent the limitations of current methods. The continuing improvement of conventional computing machines may also be a factor in defining future capabilities.

4.4 Characterization methods

Characterization of molecular first hyperpolarizability has most commonly been carried out employing either hyper-Rayleigh scattering or electric field induced second-harmonic generation. A typical HRS apparatus for off-resonant measurements is shown in Fig. 4.14 and a representative EFISH apparatus is shown in Fig. 4.15.

HRS and EFISH measurements require that chromophores can rotate freely. Initially, HRS and EFISH measurements were carried out on gaseous (vapor phase) samples. The behavior of chromophores for a gaseous sample in an electric poling field reasonably well approximates that of the independent particle model. As chromophores with improved molecular first hyperpolarizability were developed, such materials typically

4.4 Characterization methods 63

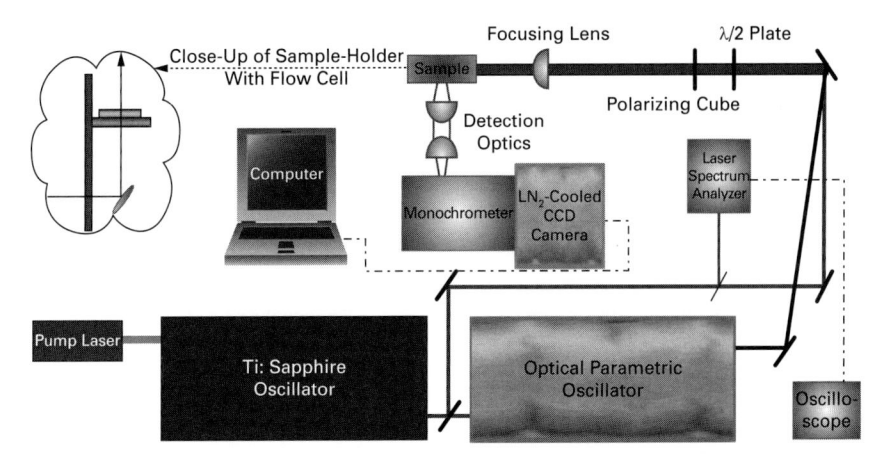

Figure 4.14 A schematic representation of a hyper-Rayleigh scattering (HRS) apparatus is shown. Such an apparatus permits femtosecond, wavelength-agile measurements. LN_2 = liquid nitrogen.

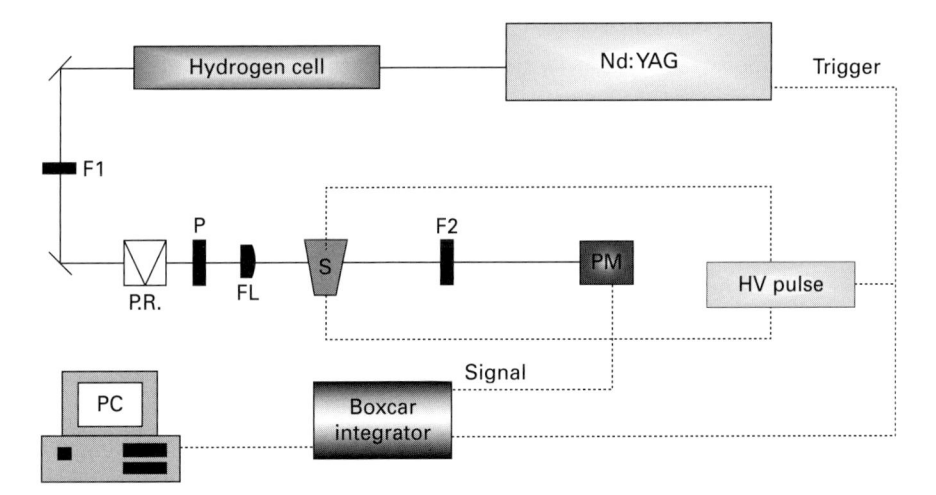

Figure 4.15 A schematic representation of an electric field induced second-harmonic (EFISH) generation apparatus is shown. F1 = combination of appropriate filters; P.R. = polarization rotator; P = polarizer aligned parallel to the applied high-voltage field axis; L = focusing lens; S = sample inside a wedged cell on translation stage; F2 = combination of a water cell, interference filter (953 nm) and neutral density filters; HV pulse = high-voltage pulse generator. Reproduced from Ref. [49] with permission of the American Chemical Society.

exhibited increased molecular weights and insufficient vapor pressure for vapor phase measurements. For such materials, measurements are typically performed by dissolving the materials in low viscosity solvents. Of course, the dielectric properties of the solvent will influence the measured hyperpolarizability as illustrated in Fig. 4.13. To extract the β value for the chromophore, HRS and EFISH measurements must be carried out as functions of concentration of the chromophore in the host solvent, and then the data must be extrapolated to yield the value for the isolated chromophores. Care must be taken to avoid aggregation effects.

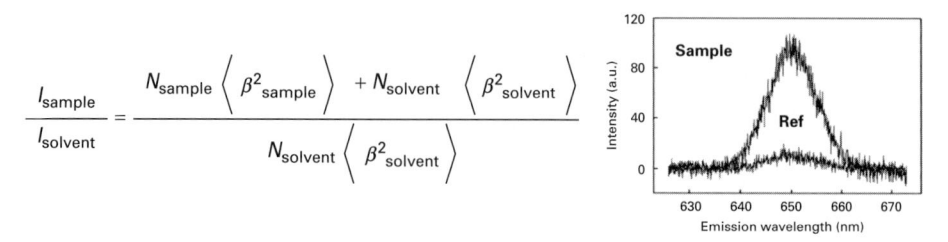

$$\frac{I_{sample}}{I_{solvent}} = \frac{N_{sample}\langle \beta^2_{sample}\rangle + N_{solvent}\langle \beta^2_{solvent}\rangle}{N_{solvent}\langle \beta^2_{solvent}\rangle}$$

Figure 4.16 Right: Experimental HRS data for a 10^{-5} M solution of a chromophore (sample of interest) in chloroform (reference sample). Left: The equation used to determine β of the sample.

Both HRS and EFISH measurements suffer from complications that reduce accuracy. Molecular first hyperpolarizabilities values are typically measured as ratios of signals to samples such as chloroform taken as a reference. For example, β values from HRS are determined as illustrated in Fig. 4.16.

Unfortunately, considerable disagreement exists as to the molecular first hyperpolarizability values of common reference samples and thus corresponding uncertainty exists in the absolute values determined for chromophores. With HRS, absolute molecular first hyperpolarizability can be determined by using an integrating sphere to collect all scattered radiation; however, such measurements are difficult and seldom executed.

HRS measures β directly while EFISH measures the product of dipole moment and molecular first hyperpolarizability, $\mu\beta$. HRS can be applied to many different types of chromophores including octupolar and ionic as well as dipolar chromophores. EFISH, on the other hand, can be applied to dipolar chromophores only but has the advantage of better signal-to-noise and can be used to determine the sign of β. HRS determines an orientationally averaged value of β, and theory must be used to gain insight into the relationship of $\beta(-2\omega,\omega,\omega)_{HRS}$ to $\beta(-\omega,0,\omega)_{zzz}$, which is the value of interest for electro-optic applications.

Both HRS and EFISH suffer from complications that contribute to uncertainty in measured values of molecular first hyperpolarizability. In addition to confusion that is associated with the two conventions for β discussed earlier (whether or not the expansion coefficient is incorporated into β), absorption at the fundamental and second-harmonic wavelengths (referred to as self-absorption) can contribute to an attenuation of β.

4.5 Synthetic methods

Although a myriad of structural modifications have been investigated in the context of improving molecular hyperpolarizability, most modern high-β (or γ) chromophores may be prepared using a relatively small number of synthetic methods. In essence, chromophores are modular materials consisting of three major subunits: electron donors, π-conjugated bridges, and electron acceptors. To achieve large β_{zzz} values, these units are combined in such a way as to create a conjugated molecular structure that is both highly electronically asymmetric and highly polarizable. The initially symmetric, polarizable, π-conjugated bridge is substituted at one terminus with an electron donor (low oxidation potential) and at the other with an electron acceptor (high electron affinity). This results in pronounced substituent-induced ground-state electron density asymmetry (see Fig. 4.9) and correspondingly large β_{zzz}. Many different donor–bridge–acceptor combinations have been explored over the past two decades. For a broader discussion, the reader is referred elsewhere [50,51]. A brief overview of the synthetic methods used to prepare 2-(3-cyano-4,5,5-trimethyl-5H-furan-2-ylidene)-malononitrile (TCF) acceptor containing materials, such as YLD156 (Fig. 4.6) and YLD124 (Table 4.2), is given here.

4.5.1 TCF acceptors

The TCF and 2-dicyanomethylene-3-cyano-4,5-dimethyl-5-trifluoromethyl-2,5-dihydrofuran (CF_3-TCF) electron acceptors have demonstrated highly desirable qualities. Theoretical and experimental analysis alike have confirmed that incorporation of TCF into chromophoric materials yields large first and second molecular hyperpolarizabilities [30,50,52]. Preparation of TCF and CF_3-TCF begins with the synthesis of the appropriate α-hydroxy ketone, A3 (Fig. 4.17).

The α-hydroxy ketone, 3-hydroxy-3-methyl-2-butanone (the dimethyl analog of A3) is commercially available, and condensation with two equivalents of malononitrile in sodium ethoxide/ethanol yields the dimethyl (R_1, R_2 = CH_3) TCF directly, at around 75% yield [53]. Modified α-hydroxy ketones may be prepared by addition of lithiated ethyl vinyl ether to the appropriate ketone precursor followed by deprotection in aqueous acid. For example, if 2,2,2-trifluoroacetophenone is used in place of reagent A1, the TCF derivative in which R_1,R_2 = CF_3/phenyl is produced. However, α-hydroxy ketone modifications can lead to significant reductions in yield in the final malononitrile condensation step [54]. Focused microwave methods have been introduced in order to combat this problem [55]. The general procedure outlined produces a racemic mixture. Although chiral TCF derivatives are potentially attractive, they have yet to be demonstrated.

4.5.2 Arylamine donors

In terms of stability and β_{zzz}, the most successful electron-donating substituents demonstrated to date fall into the class of arylamines. Three structures of this type are shown in Figure 4.18. Alkyl arylamines (Da) display excellent hyperpolarizability values.

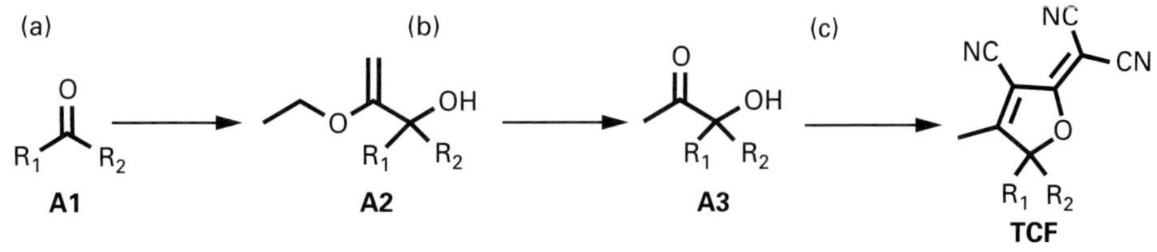

Figure 4.17 The synthesis of TCF-type electron acceptors is shown. Reagents and conditions are as follows: (a) ethyl vinyl ether, t-BuLi, −78 °C; (b) HCl, RT; (c) NaOEt, EtOH, malononitrile, microwave, 1 hr.

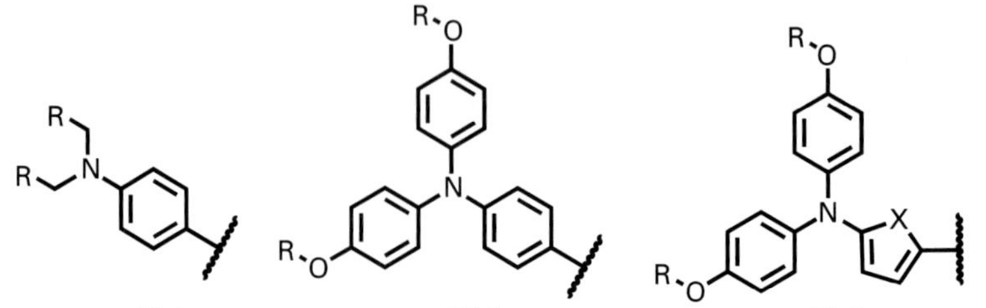

Figure 4.18 (Da) Alkyl substituted arylamine donor. (Db) Phenoxy substituted arylamine donor. (Dc) Phenoxy substituted heteroarylamine donor in which X = O, N, or S.

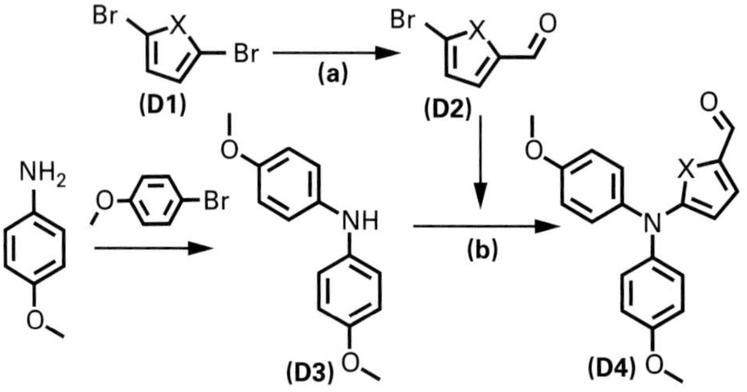

Figure 4.19 The synthesis of phenoxy substituted heteroarylamine donor, X = O, S, or N is shown. Reagents and conditions are as follows: (a) n-BuLi, THF, −78 °C then DMF, RT; (b) 2 mol% Pd2(dba)3, 4 mol% (t-Bu)3P, t-BuONa, toluene, reflux, 18 hr.

Substituted triarylamines (Db) maintain and even improve β_{zzz} values while imparting enhanced thermal stability [56, 57]. In order to further improve hyperpolarizabliity phenoxy substituted heteroarylamine donors (Dc) may be employed. This type of donor exploits the reduced aromatic stabilization energy of the heteroaryl ring (e.g., furan, thiophene, or pyrrole) to increase the energy of the HOMO.

Synthesis of the phenoxy substituted heteroarylamine donors is shown here (Fig. 4.19) in order to illustrate general synthetic methods. Formylation of the lithiated dibromo species, D1, yields the 2-formyl, 5-bromo heteroaryl ring. Palladium catalyzed cross-coupling (Buchwald/Hartwig) with the secondary amine D3 then affords the phenoxy substituted heteroarylamine donor D4 in 89% overall yield [57].

4.5.3 π-Conjugated bridges

TCF or CF_3-TCF acceptors can be directly attached to the donor via a base-catalyzed Claisen–Schmidt condensation, or as is more common, the length of the π-system may be increased by the addition of a heteroaryl or polyene bridge. Analogous to the heteroaryl moiety introduced into the donor, D4, the use of a polyene or heteroaryl bridge substituent serves to reduce or eliminate aromatic stabilization. This reduction in

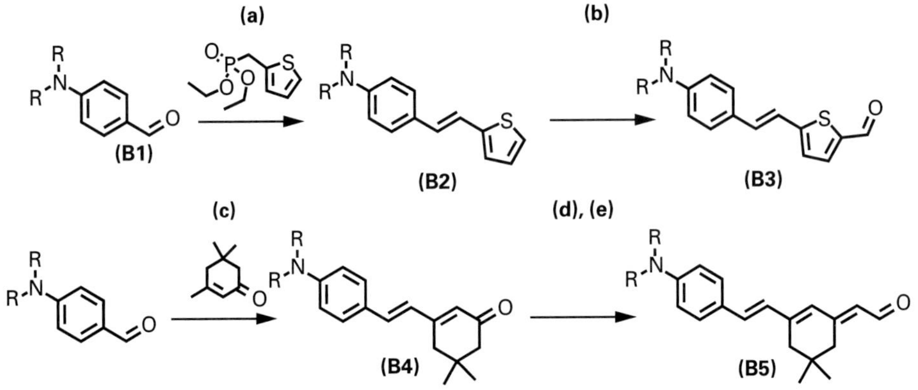

4.5 Synthetic methods

Figure 4.20 The synthesis of thienyl (B3) and polyene (B5) π-conjugated bridges is shown. Reagents and conditions are as follows: (a) t-BuOK, THF 0 °C to RT; (b) n-BuLi, THF, −78 °C, then DMF, overnight, RT; (c) EtONa, EtOH, 6 days, RT; (d) diethylcyanomethylphosphonate, n-BuLi, THF, −78 °C; (e) DIBAL-H, −20 °C.

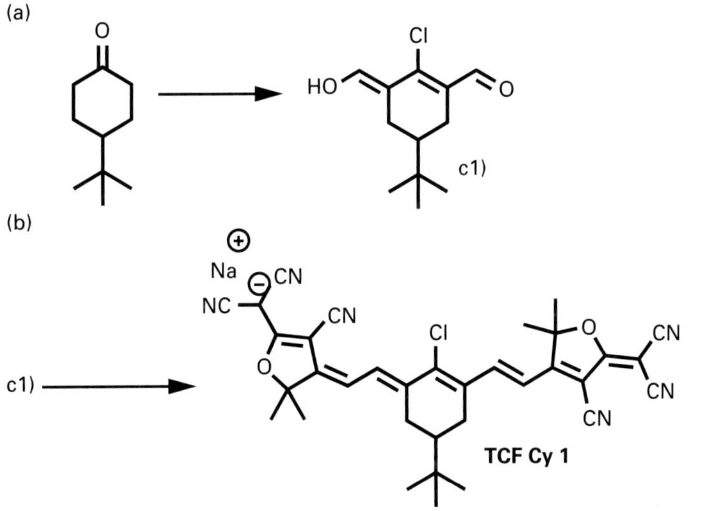

Figure 4.21 The synthesis of TCF-based cyanine dye TCF Cy 1 is shown. Reagents and Conditions are as follows: (a) POCl$_3$, DMF; (b) TCF, Ac$_2$O, AcONa.

stabilization energy reduces the barrier to electric field induced polarization, enhancing linear and nonlinear polarizabilities.

The synthetic preparation of thienyl (B3) and polyene (B5) π-conjugated bridges is illustrated in Fig. 4.20. The preparation of both B3 and B5 begins with the donor aldehyde B1. Horner–Wadsworth–Emmons (HWE), Knoevenagel, or Claisen–Schmidt olefination reactions represent strategies that are commonly employed to effect double bond formation. To prepare B3, thienyl phosphonate is used via the HWE to yield intermediate B2 at ~90%. Formylation using n-butyl lithium / DMF then affords B3 via a nearly quantitative transformation [58]. Similarly, intermediate B4 is prepared by base-catalyzed Claisen–Schmidt condensation between isophorone and B1. Aldehyde extension is then accomplished by HWE followed by DIBAL-H reduction [59]. Condensation of B3 and B5 with CF$_3$-TCF may be accomplished in high yield using a mild base catalyst such as ammonium acetate under dehydrating conditions to afford YLD156 and YLD124 respectively [60, 61].

4.5.4 TCF acceptors in high-γ cyanines

Analogous to the use of the TCF moiety in the preparation of high-β_{zzz} materials, its use has also been demonstrated in the preparation of high-γ cyanines. Cyanine dyes are well known for having very large molecular second hyperpolarizabilities γ_{ijkl} [62]. Unlike molecular first hyperpolarizability, β_{ijk}, highly symmetric ground-state electron density distributions give rise to high γ_{ijkl}.

The synthesis of TCF Cy 1 is illustrated in Fig. 4.21. The bis aldehyde (or tautomer) intermediate C1 is prepared under Vilsmier–Haak conditions from t-butyl cyclohexanone. Condensation with two equivalents of TCF under dehydrating conditions affords TCF Cy 1 [63].

References

[1] J.L. Oudar and D.S. Chemla, *J. Chem. Phys.* **66**, 2664 (1977).
[2] J. Zyss and J.L. Oudar, *Phys. Rev. A* **26**, 2028 (1982).
[3] M.G. Kuzyk, *Phys. Rev. Lett.* **85**, 1218 (2000).
[4] L. R. Dalton, S. J. Benight, L. E. Johnson, et al., *Chem. Mater.* **23**(3), 430 (2011).
[5] A. Willets, J. E. Rice, D. M. Burland, and D. P. Shelton, *J. Chem. Phys.* **97**, 7590 (1992).
[6] R. F. Shi and A. F. Garito, in *Characterization Techniques and Tabulations for Organic Nonlinear Optical Materials*, M. G. Kuzyk and C. W. Dirk, Eds., New York, Marcel Dekker (1998), p. 1.
[7] F. Meyers, S. R. Marder, B. M. Pierce, and J. L. Bredas, *J. Am. Chem. Soc.* **116** (23), 10703 (1994).
[8] D. Beljonne, J. Cornil, Z. Shuai, et al. *Phys. Rev. B (Condensed Matter)* **55**, 1505 (1997).
[9] S. Di Bella, M. A. Ratner, and T. J. Marks, *J. Am. Chem. Soc.* **114**, 5842 (1992).
[10] S. Priyadarshy, M. J. Therien, and D. N. Beratan, *J. Am. Chem. Soc.* **118**, 1504 (1996).
[11] S. R. Marder, C. B. Gorman, F. Meyers, et al. *Science* **265**, 632 (1994).
[12] F. Sim, S. Chin, M. Dupuis, and J. E. Rice, *J. Phys. Chem.* **97** (6), 1158 (1993).
[13] K. Tripathy, J. P. Moreno, M. G. Kuzyk, et al. *Chem. Phys.* **121** (16), 7932 (2004).
[14] C. M. Isborn, A. Leclercq, F. D. Vila, et al. *J. Phys. Chem. A* **111** (7) 1319 (2007).
[15] D. Jacquemin, J. M. André, and E. A. Perpète, *J. Chem. Phys.* **121** (9), 4389 (2004).
[16] K. Y. Suponitsky, Y. Liao, and A. E. Masunov, *J. Phys. Chem. A* **113**, 10994 (2009).
[17] C. E. Dykstra and P. G. Jasien, *Chem. Phys. Lett.* **109** (4), 388 (1984).
[18] G. J. B. Hurst, M. Dupuis, and E. J. Clementi, *Chem. Phys.* **89** (1), 385 (1988).
[19] J. E. Rice, R. D. Amos, S. M. Colwell, N. C. Handy, and J. J. Sanz, *Chem. Phys.* **93**, 8828 (1990).
[20] G. Fitzgerald and J. Andzelm, *J. Phys. Chem.* **95** (26), 10531 (1991).

[21] A. C. Scheiner, J. Baker, and J. J. Andzelm, *Comp. Chem.* **18** (6), 775 (1997).
[22] B. Champagne, E. A. Perpete, D. Jacquemin, *et al. J. Phys. Chem. A* **104** (20), 4755 (2000).
[23] E. A. Perpete, and D. Jacquemin, *Int. J. Quantum Chem.* **107** (11), 2066 (2007).
[24] E. Runge and E. K. U. Gross, *Phys. Rev. Lett.* **52** (12), 997 (1984).
[25] F. Furche, *J. Chem. Phys.* **114** (14), 5982 (2001).
[26] F. Furche and R. Ahlrichs, *J. Chem. Phys.* **117** (16), 7433 (2002).
[27] Y. Takimoto, F. D. Vila, and J. J. Rehr, *Chem. Phys.* **127**, 154114 (2007).
[28] O. Christiansen, P. Jorgensen, and C. Hättig, *Int. J. Quantum Chem.* **68**, 1 (1998).
[29] J. Olsen, P. Jørgensen, P. T. Helgaker, and J. J. Oddershede, *J. Phys. Chem. A* **109**, 11618 (2005).
[30] Y. Takimoto, C. M. Isborn, B. E. Eichinger, J. J. Rehr, and B. H. Robinson, *J. Phys. Chem. C* **112** (21), 8016 (2008).
[31] V. W. Geskin, C. Lambert, and J. L. Bredas, *J. Am. Chem. Soc.* **125** (50), 15651 (2003).
[32] Y. Luo, P. Norman, K. Rudd, and H. Agren, *Chem. Phys. Lett.* **285**, 160 (1998).
[33] S. Tretiak, V. Chernyak, and S. Mukamel, *Chem. Phys. Lett.* **287** (1,2), 75 (1998).
[34] E. M. Breitung, C. F. Shu, and R. J. McMahon, *J. Am. Chem. Soc.* **122**, 1154 (2000).
[35] D. M. Casmier, P. A. Sullivan, O. Clot, *et al. Proc. SPIE* **5351**, 234 (2004).
[36] I. D. L. Albert, T. J. Marks, and M. A. Ratner, *J. Am. Chem. Soc.* **119**, 3155 (1997).
[37] I. D. L. Albert, T. J. Marks, and M. A. Ratner, *J. Am. Chem. Soc.* **120**, 11174 (1998).
[38] H. Kang, A. Facchetti, C. L. Stern, *et al. Org. Lett.* **7**, 3721 (2005).
[39] H. Kang, A. Facchetti, H. Jiang, *et al. J. Am. Chem. Soc.* **129**, 3276 (2007).
[40] E. C. Brown, T. J. Marks, and M. A. Ratner, *J. Phys. Chem. B* **112**, 44 (2008).
[41] Y. Wang, D. L. Frattarelli, A. Facchetti, *et al. J. Phys. Chem. C* **112**, 8005 (2008).
[42] G. S. He, J. Zhu, A. Baev, M. Samoc, *et al. J. Am. Chem. Soc.* **133**, 6675 (2011).
[43] C. M. Isborn, E. R. Davidson, and B. H. Robinson, *J. Phys. Chem. A* **110**, 7189 (2006).
[44] L. E. Johnson, L. R. Dalton, and B. H. Robinson, *Acc. Chem. Res.* **47**, 3258 (2014).
[45] Y. Zhou, L. Shaojun, and Y. Cheng, *Synth. Met.* **137**, 1519 (2003).
[46] H. Kang, P. Zhu, Y. Yang, A. Facchetti, and T. J. Marks, *J. Am. Chem. Soc.* **126**, 15974 (2004).
[47] P. A. Sullivan, S. Bhattacharjee, B. E. Eichinger, *et al. Proc. SPIE* **5351**, 253 (2004).
[48] H. Kang, P. Zhu, Y. Yang, A. Facchetti, and T. J. Marks, *J. Am. Chem. Soc.* **126**, 15974 (2004).
[49] D. Bale, W. Liang, L. Dalton, *et al. J. Phys. Chem. B* **115**, 3505 (2011).
[50] L. R. Dalton, P. A. Sullivan, and D. H. Bale, *Chem. Rev.* **110**, 25 (2010).
[51] M. J. Cho, S. K. Lee, J.-I. Jin, *et al. Prog. Polym. Sci.* **33**, 1013 (2008).
[52] P. A. Sullivan, H. L. Rommel, Y. Takimoto *et al., J. Phys. Chem. B* **113**, 15581 (2009).
[53] G. Melikian, F. P. Rouessac, and C. Alexandre, *Synth. Commun.* **25**, 3045 (1995).
[54] M. Q. He, T. M. Leslie, and J. A. Sinicropi, *Chem. Mater.* **14**, 2393 (2002).
[55] S. Liu, M. A. Haller, H. Ma, *et al. Adv. Mater.* **15**, 603 (2003).
[56] S. M. Budy, S. Suresh, B. K. Spraul, and D. W. Smith, *J. Phys. Chem. C,* **112**, 8099 (2008).
[57] J. A. Davies, A. Elangovan, P. A. Sullivan *et al. J. Am. Chem. Soc.* **130**, 10565 (2008).
[58] P. A. Sullivan, A. J. P. Akelaitis, S. K. Lee, *et al. Chem. Mater.* **18**, 344 (2006).
[59] C. Zhang, L. R. Dalton, M. C. Oh, *et al. Chem. Mater.* **13**, 3043 (2001).
[60] Y. Liao, C. A. Anderson, P. A. Sullivan, *et al. Chem. Mater.* **18**, 1062 (2006).
[61] T. Baehr-Jones, M. Hochberg, G. Wang, *et al. Opt. Express.* **13**, 5216 (2005).
[62] J. M. Hales, J. Matichak, S. Barlow, *et al. Science*, **327**, 1485 (2010).
[63] P. A. Bouit, E. D. Piazza, S. R. Rigaut, *et al. Org. Lett.* **10**, 4159 (2008).

5 Acentric self-assembled films

To obtain macroscopic quadratic nonlinear optical responses in bulk materials, acentric ordering of the constituted molecules is required [1 to 10]. Usual nonlinear optical chromophores consisting of long π-conjugated bridges between electron donor and electron acceptor groups exhibit large ground-state dipole moments μ_g. The highly polar chromophores easily form centrosymmetric arrangements, owing to strong intermolecular dipole–dipole aggregation and in some cases electrostatic ion–dipole interactions [1 to 3; 11 to 13] (see Chapter 7 for a more quantitative discussion). Therefore, many supramolecular approaches for obtaining acentric molecular ordering in bulk materials have been developed: for example self-assembled films, acentric crystals, and poled polymers [1 to 5]. In contrast to poled polymers where acentric order is introduced by application of an external electric poling field, the constituent molecules in acentric crystalline materials align under the influence of specific intermolecular electrostatic interactions. This chapter focuses on generation of organic electro-optic thin-film materials through exploitation of sequential synthesis/self-assembly techniques based on Langmuir–Blodgett and Merrifield approaches. Three techniques are introduced for achieving thin films characterized by large order parameters [2,9]: (1) polar Langmuir–Blodgett films; (2) acentric solution-deposited films; and (3) acentric vapor-deposited films. Sequential synthesis/self-assembly techniques are based not only on utilization of covalent bonds but also upon exploitation of intermolecular electrostatic non(covalent bond)-bonding interactions such hydrogen bonds, van der Waals forces, ionic, and π–π stacking forces.

5.1 Polar Langmuir–Blodgett films

Langmuir–Blodgett (LB) films are well known, and the LB technique is widely used for preparation of highly oriented thin films of organic materials [3,5,14]. Figure 5.1 illustrates the Langmuir–Blodgett sequential synthesis technique: (1) spreading of molecules in dilute solution on hydrophilic liquid surface, in most cases water; (2) compression of the spreading molecules on the liquid surface by movable walls, which control the surface pressure–area isotherm; (3) deposition of the first layer onto the substrate; and (4) repeating the deposition to fabricate multilayers [3,14].

For fabricating LB films, amphiphilic molecules (or polymers) having both hydrophilic and hydrophobic groups are required. Table 5.1 lists the influence of hydrophilic

5.1 Polar Langmuir–Blodgett films

Table 5.1 The influence of hydrophilic groups of amphiphiles on the LB film formation and stability. Reprinted with permission from [14]

Very weak (no film)	Weak (unstable film)	Strong (stable film with C16 chain)	Very strong (C16 compounds dissolve)
Hydrocarbon	-CH$_2$OCH$_3$	-CH$_2$OH	-SO$_3$-
-CH$_2$I	-C$_6$H$_4$OCH$_3$	-CO$_2$H	-OSO$_3$-
-CH$_2$Br	-CO$_2$CH$_3$	-CN	-C$_6$H$_4$SO$_4$-
-CH$_2$Cl		-CONH$_2$	-NR$_4^+$
-NO$_2$		-CH=NOH	
		-C$_6$H$_4$OH	
		-CH$_2$COCH$_3$	
		-NHCONH$_2$	
		-NHCOCH$_3$	

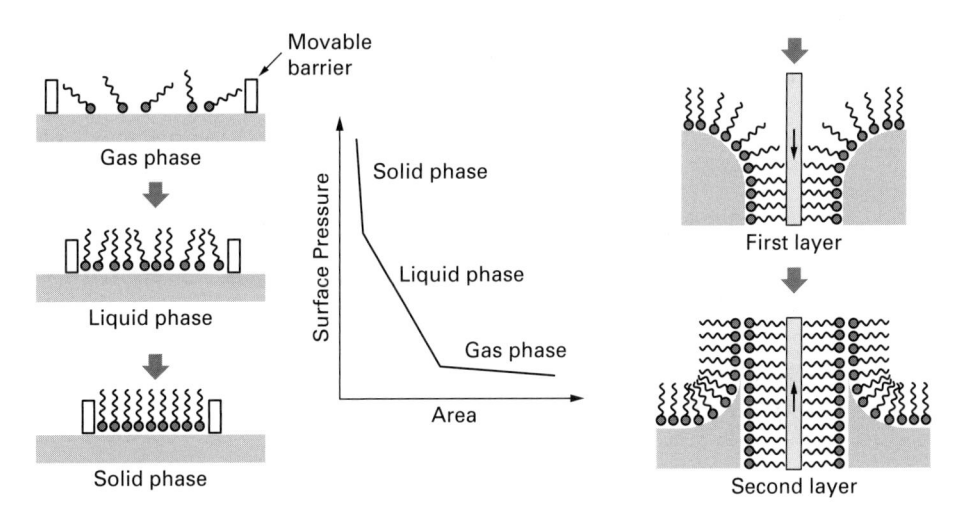

Figure 5.1 Process of Langmuir–Blodgett film technique: spreading, compression, and deposition of a Y-type multilayer (see Fig. 5.3). Solid circles and lines represent hydrophilic heads and hydrophobic tails of amphiphiles. The graph shows the surface pressure (π) versus area per molecule (A) isotherm during the LB process.

groups of amphiphiles on the LB film formation and stability with C16 compounds [14]. For many molecules and polymers used for fabrication of LB films, the hydrophilic groups are positioned at one end of the molecules and the hydrophobic groups at the other. Many π-conjugated molecules and polymers have been designed for LB film formation [5,15 to 39]. Typical examples of chemical structures of molecules and polymers for LB films for nonlinear optical applications are shown in Fig. 5.2 [16,18,27,35]. Owing to the high hydrophilicity of salts and easy chemical modifications, hemicyanine dyes with long alkyl tails have been widely investigated for both molecular and polymeric LB films [16 to 26].

Figure 5.2 Typical examples of the chemical structure of nonlinear optical molecules and polymers for LB films [16,18,27,35].

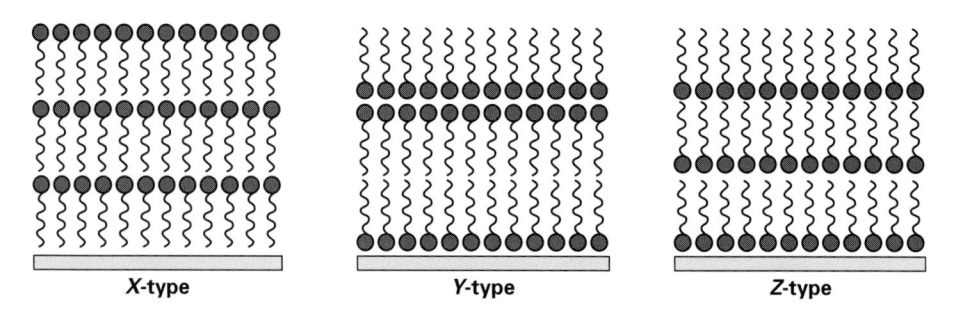

Figure 5.3 Schematic diagram of acentric X- and Z-type LB films and centrosymmetric Y-type LB films.

The LB film technique gives a monolayer film with ideally perfect alignment of the molecules and low defects [3]; however, in reality, dipolar interactions will cause chromophores to tilt away from the normal to the deposition surface, and defects will propagate with subsequent sequential steps. For macroscopic second-order optical nonlinearity, thicker LB films (i.e., multilayers of 1–3 µm thickness) are typically required. Figure 5.3 shows a schematic representation of multilayer LB films: X-type, Y-type, and Z-type [3]. As shown in Fig. 5.1, by transferring the second layer onto the first monolayer, centrosymmetric Y-type LB films are often obtained. The strong interlayer interactions between similar hydrophilicities of the first and the second layer, such as hydrophobic–hydrophobic groups and hydrophilic–hydrophilic groups, allow stable film formation. The X- and Z-type LB films, with acentric ordering of molecules, are suitable for quadratic nonlinear optical properties (e.g., electro-optic modulation

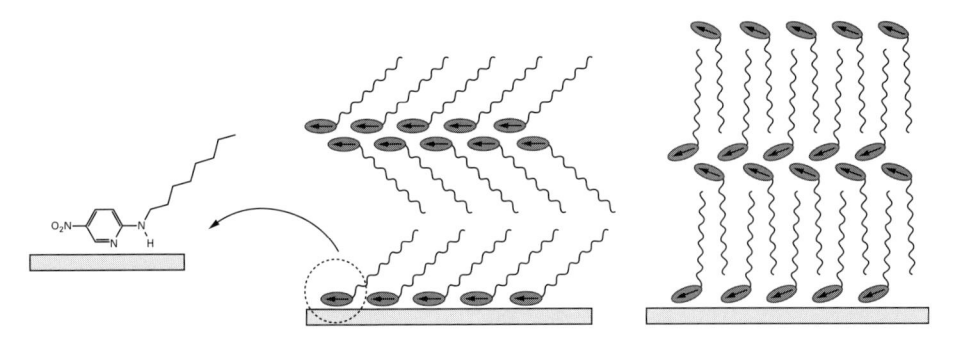

Figure 5.4 Schematic illustration of the alignment of DCANP molecules in tilted and intercalated Y-type LB layers [27 to 29].

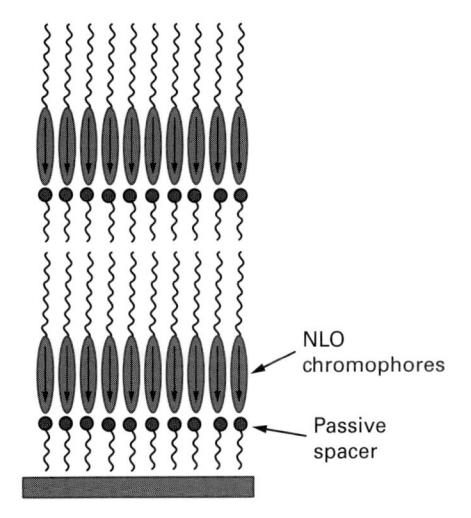

Figure 5.5 Schematic representation of acentric LB films alternating with nonlinear optical (NLO) chromophores and passive spacers. Adapted from [23].

and second-harmonic generation). However, as these X- and Z-type LB films are based on weak interlayer interactions and on opposite hydrophilicities (i.e., interaction between hydrophobic and hydrophilic groups), they often exhibit instability, large defect populations, and film inhomogeneity [3,15,40]. Many approaches have been developed to overcome the centrosymmetric arrangement of chromophores in the Y-type LB films and the low quality of the X- and Z-type LB films [15].

Figure 5.4 presents an example of modified Y-type LB films with acentric molecular ordering [27 to 29]. In the LB monolayer, DCANP (2-(docosylamino)-5-nitropyridin) molecules are aligned parallel to the hydrophilic substrate. The polar axis of the monolayer is also parallel to the substrate, because the first hyperpolarizability is in the same direction as the long axis of the molecules. Moreover, the polar axis is maintained during multilayer formation [27 to 29].

Another example of modified Y-type LB multilayers presents acentric alternating films, as illustrated in Fig. 5.5 [23 to 41]. In such LB films, two different layers

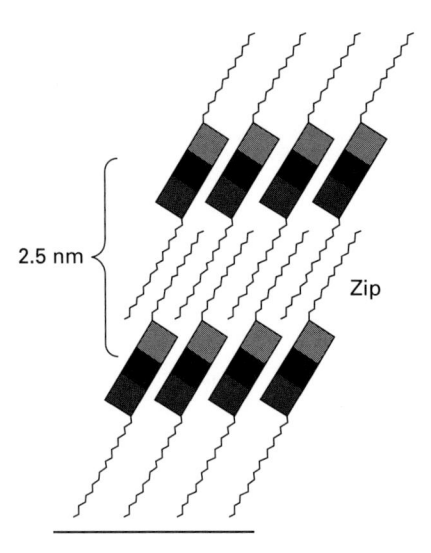

Figure 5.6 Schematic representation of an example of modified Z-type LB multilayers. Reprinted with permission from [15].

alternate: one layer consists of nonlinear optical chromophores and the other layer consists of passive spacers without optical nonlinearity. The highly polar nonlinear optical chromophores of one layer form strong interlayer interactions with hydrophilic groups of passive spacers such as fatty acids.

Figure 5.6 shows an example of the modified Z-type LB multilayers [15,42]. The amphiphilic molecules have two hydrophobic alkyl tails at each end. Each layer is interlocked with adjacent layers, and the polar axis of the LB films is normal to the plane [15,42].

One of the successful examples of preventing dipole–dipole aggregations of highly polar nonlinear optical chromophores in LB films is mixing of two different amphiphilic molecules; zwitterionic chromophores having a negative hyperpolarizability ($-\beta$) and neutral ground-state chromophores having a positive hyperpolarizability ($+\beta$) with respect to the ground-state dipole moment of the molecule [43]. This mixing, eliminating the electrostatic effect between the molecules, has been used previously in a polymeric system [44]. Figure 5.7 shows a schematic diagram of a LB film with mixed zwitterionic and neutral ground-state chromophores. The resulting monolayers are densely packed with a high order parameter and are thermodynamically stable [43].

As discussed above, for nonlinear optical applications, the LB technique possesses the following advantages: (1) relatively simple technique for preparing mono- and multilayers with high density of nonlinear optical chromophores and with high order parameter (see Fig. 5.7); (2) easy control of film thickness with sub-nm scale; and (3) control of the ordering of multilayers [3,5]. Despite their advantages, LB films still show some limitations. For nonlinear optical applications, environmentally stable films with suitable thickness for specific applications are required, for example 0.1 μm to few μm thickness for integrated photonic devices [2], and 0.1 mm to few mm thickness for

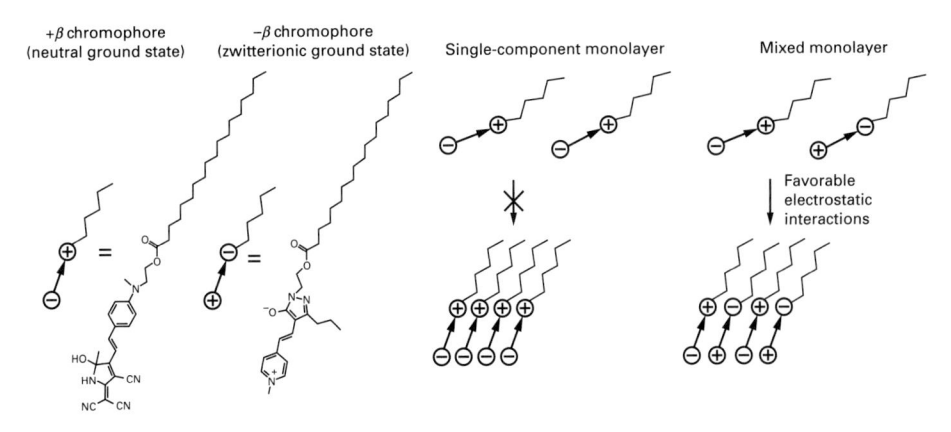

Figure 5.7 LB films containing a $+\beta$ neutral ground-state amphiphilic chromophore and a $-\beta$ zwitterionic ground-state amphiphilic chromophore. Reprinted with permission from [43].

THz-wave applications [45 to 47]. Preparing thick multilayers, even when the sequential deposition steps are automated, is a time-consuming process. Also, the LB films often exhibit low environmental and thermal stabilities because intermolecular interactions within and between the layers are less stable than covalent bonds [40].

5.2 Acentric solution-deposited films

In order to overcome the limitations of LB techniques based on weak secondary bonds, film preparation methods based on self-assembly with stronger intermolecular interactions have been developed. In solution, there are mainly two kinds of self-assembly methods based on (1) covalent bonds and (2) electrostatic forces.

5.2.1 Covalent bond based self-assembly

Self-assembled mono- and multilayers (SAMs) mainly utilize thiol and silane compounds, which form specific interactions with surface groups of substrate [48 to 58]. The thiol group strongly adsorbs on metallic surfaces, typically gold. For nonlinear optical applications, a thiol compound containing nonlinear optical chromophores [48 to 50] is used as shown in Fig. 5.8. The SAM technique exploiting a thiol linkage to form a surface monolayer is useful for modifying surfaces for crystal growth and other applications [59 to 61].

Another well-known SAM technique for nonlinear optical applications is silanization [51 to 58]. Figure 5.9 shows a schematic illustration of the SAM preparation method based on silanization [53 to 55]. The coupling agent possesses two chemical groups at the ends of the molecule to form covalent bonds. One group can react with the surface group of the substrate and the other can react with the chromophore precursor.

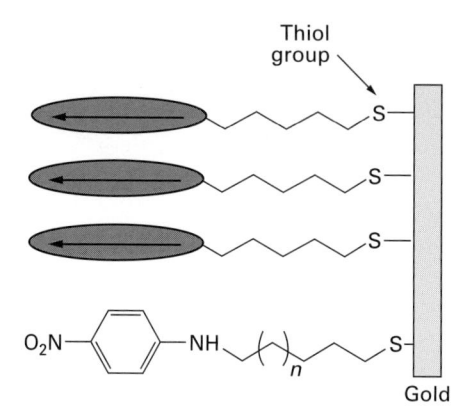

Figure 5.8 SAMs of thiol compound containing nonlinear optical chromophores. Reprinted with permission from [48].

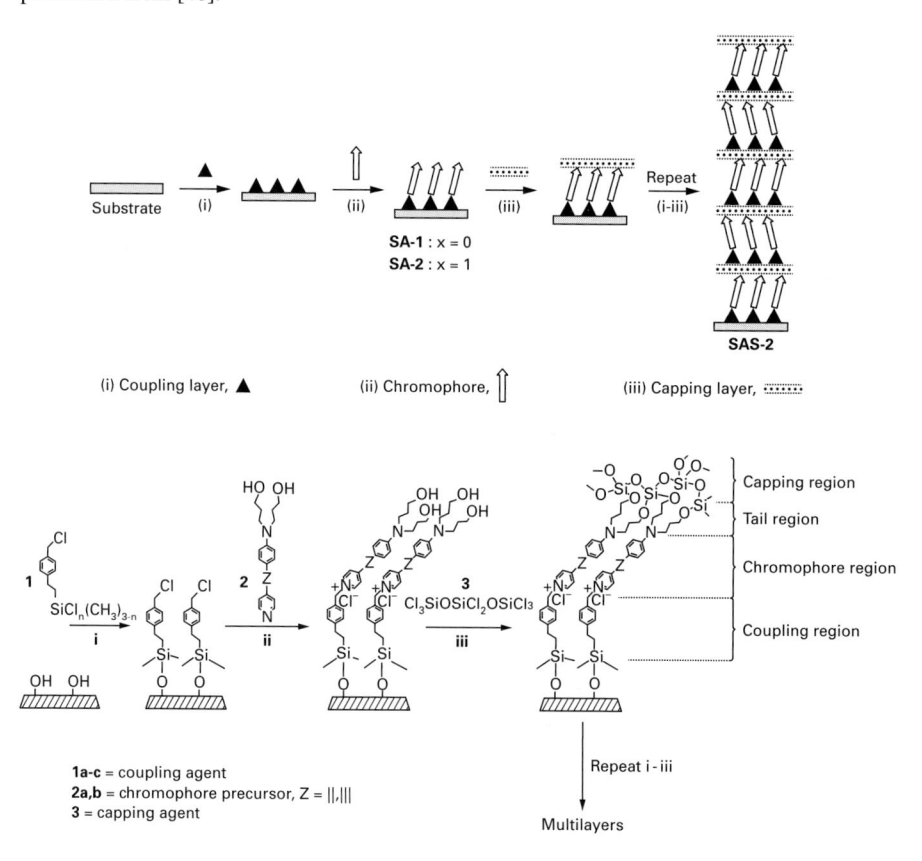

Figure 5.9 SAM technique with silanization. Reprinted with permission from [53,55].

As illustrated in Fig. 5.9, one terminal group of the coupling agent, the silane group $(SiCl_n(CH_3)_{3-n})$ reacts with hydrophilic substrates and forms a coupling layer. The other terminal group, an alkyl halide group, reacts with the chromophore precursor to form an acentric chromophore layer, in which the chromophore precursor is modified

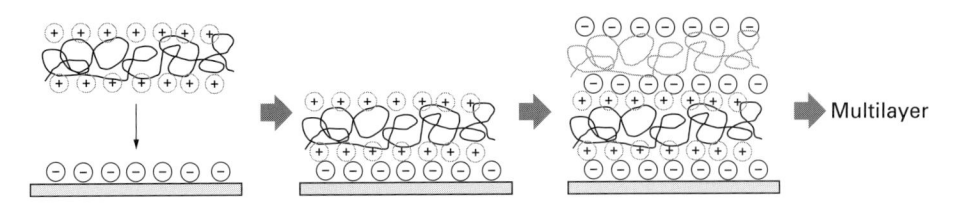

Figure 5.10 Electrostatic self-assembly (ESA) technique.

to a general chromophore with a stronger electron acceptor (from pyridine to pyridinium) [53,55]. The reactive group at the end of the chromophore, such as the OH group in Fig. 5.9, reacts with a capping layer having similar surface groups to the substrate. By repeating this process, acentric multilayers can be obtained [53,55]. The silane coupling agents can be changed to any functional group reacting with chemical groups on the surface.

5.2.2 Electrostatic self-assembly

The electrostatic self-assembly (ESA) technique is based on electrostatic forces between positive or negative charged polymeric materials and substrates [62 to 65]. The general preparation of an ESA film is illustrated in Fig. 5.10. The substrate, having a charged surface, is immersed into a solution containing an oppositely charged material. A self-assembled layer is formed through the electrostatic interactions between the positive charges of the polymers and the negative charges of the substrate, as shown in Fig. 5.10. Conversely, a positively charged substrate can interact with a negatively charged polymer to form a self-assembled layer. Interactions between positive and negative charges are typically weaker than covalent bonds, but many such interactions are sufficient to yield high-stability layers. During formation of an ESA film, the polymer chains can fold, yielding a layer whose thickness is uniform but depends on the experimental conditions and the characteristics of polymers. The positively charged ESA film in Fig. 5.10 can be immersed again into another solution containing a negatively charged polymer. Repeating immersions in solutions containing positively and negatively charged polymers yields a multilayer film [62 to 65]. This ESA technique can be applied to any charged substrate.

For second-order nonlinear optical response, a charged polymer having nonlinear optical chromophores is needed. Figure 5.11 illustrates an example of charged polymers having nonlinear optical hemicyanine chromophores, and their use in ESA film formation. Wholly aromatic polyamide and polyester (POST) polymers possess a non-charged main chain and charged side-chains incorporating nonlinear optical hemicyanine groups. When the negatively charged glass substrate is immersed into the polymer solution, an ESA film can be obtained. In the ESA film, the hemicyanine chromophores are densely packed with a high order parameter and exhibit large macroscopic optical nonlinearities [66,67].

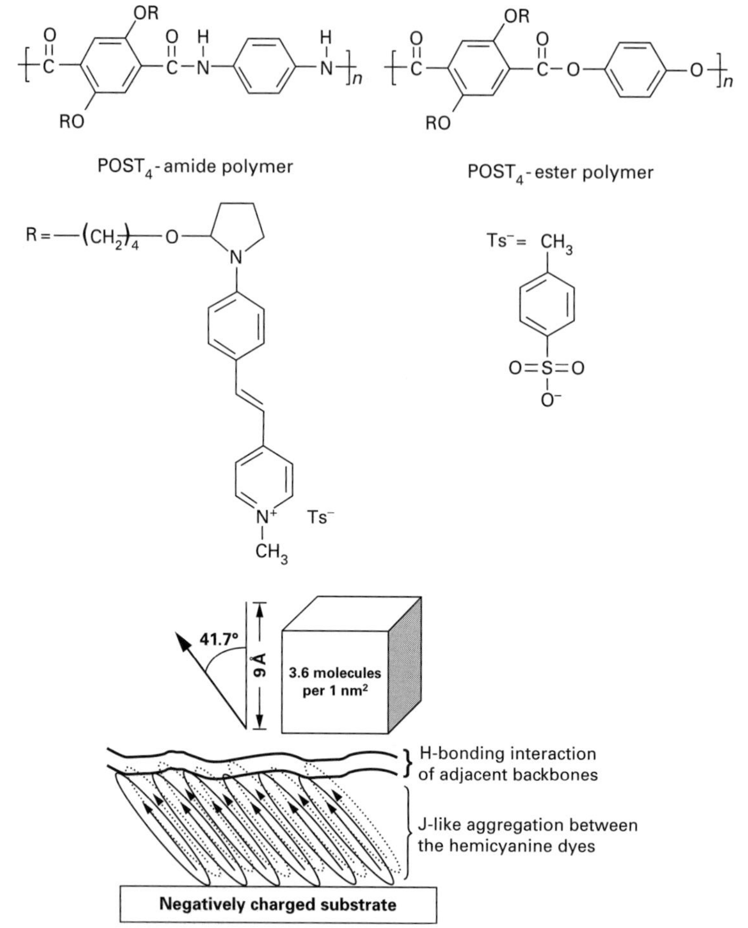

Figure 5.11 Self-assembled monolayer based on ESA technique. Reprinted with permission from [67].

5.3 Acentric vapor-deposited films

5.3.1 Volatile non-ionic organic materials

Oblique-incidence organic molecular beam deposition. Organic molecular beam deposition (OMBD) is an attractive technique for obtaining an organic thin film on a substrate using physical vapor deposition [68 to 81]. In the OMBD technique, organic thin films are formed by sublimation of desired organic molecules from the effusion cell onto the substrates at a certain temperature, usually under ultrahigh vacuum conditions [68 to 71]. The characteristics of grown films such as growth rate, film thickness, and surface quality can be precisely controlled and optimized by changing the temperature of the effusion cell and the substrate, and by the vapor pressure in the chamber, which can be monitored with *in-situ* instruments and spectroscopy such as highly sensitive

5.3 Acentric vapor-deposited films

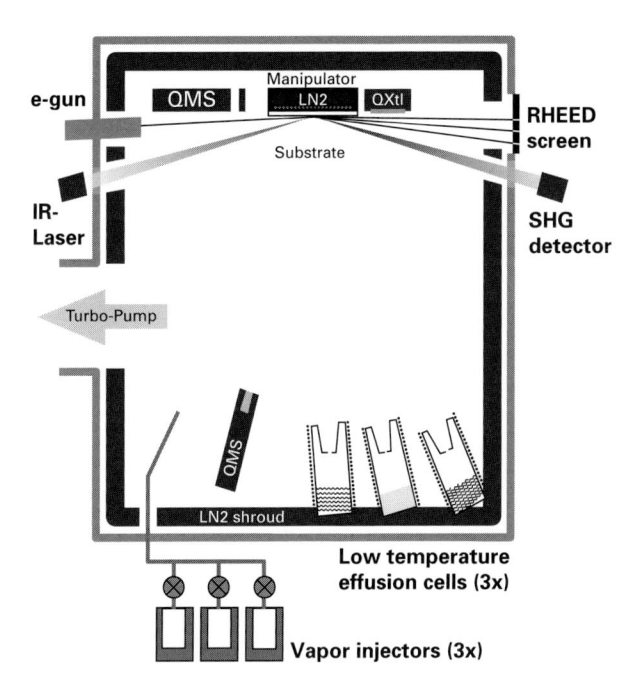

Figure 5.12 OMBD set-up. QMS: quadrupole mass spectrometer; QXtl: quartz crystal; RHEED: reflective high-energy electron diffraction; SHG: second-harmonic generation; LN2, liquid nitrogen. Reprinted with permission from [68].

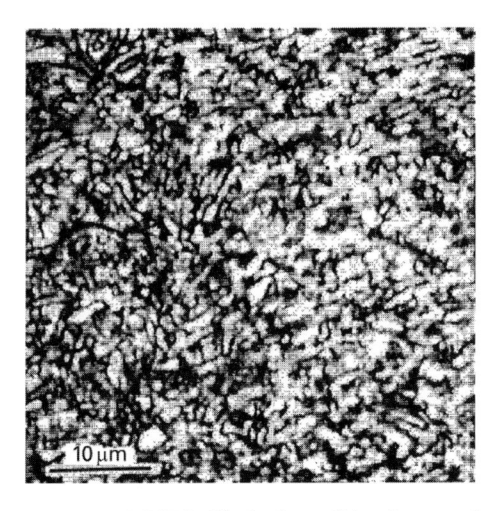

Figure 5.13 MNBA (4′-nitrobenzyliden-3-acetamino-4-methoxyaniline) crystallites grown on lattice-mismatched silicon substrates. Reprinted with permission from [70].

Figure 5.14 The chemical structure of P2BA and P3BA and their acentric hydrogen bonds. Reprinted with permission from [71].

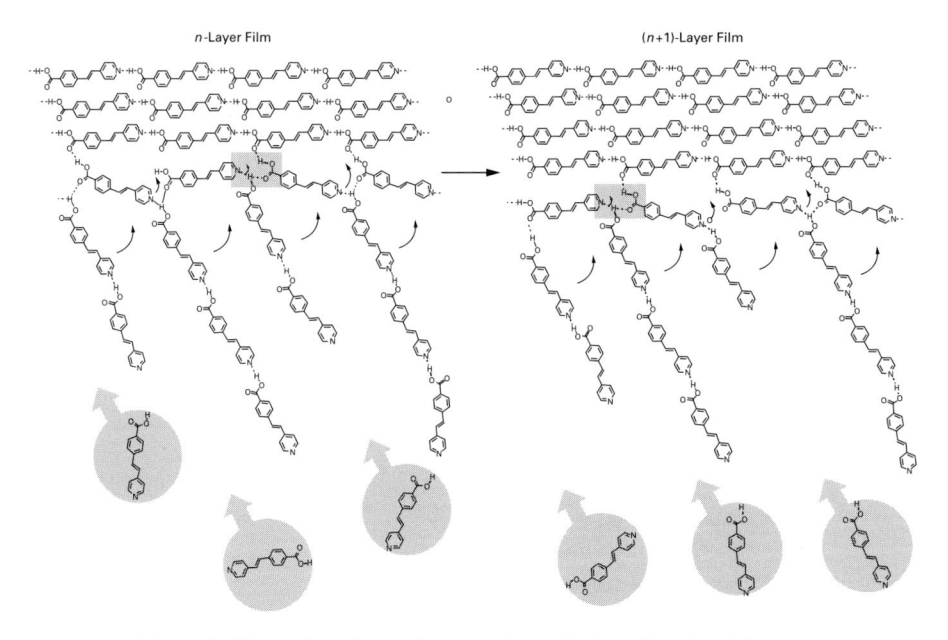

Figure 5.15 Schematic illustration of growth mechanism of P2BA films in oblique-incidence OMBD. Reprinted with permission from [69].

thickness monitors and ellipsometers (see Fig. 5.12) [68 to 71]. In contrast to the acentric-film solution deposition techniques as described in Section 5.2, organic thin films grown by OMBD techniques in ultrahigh vacuum condition possess many advantages such as avoiding solvent inclusion, fast growth rate with a large enough thickness for practical applications, and easy optimization of the film characteristics with *in-situ* monitoring [71].

Owing to the requirement for lattice matching between organic materials and the substrates in epitaxial growth [70,82], organic thin films grown on lattice-mismatched substrates are usually polycrystalline without long-range ordering of the crystals,

5.3 Acentric vapor-deposited films

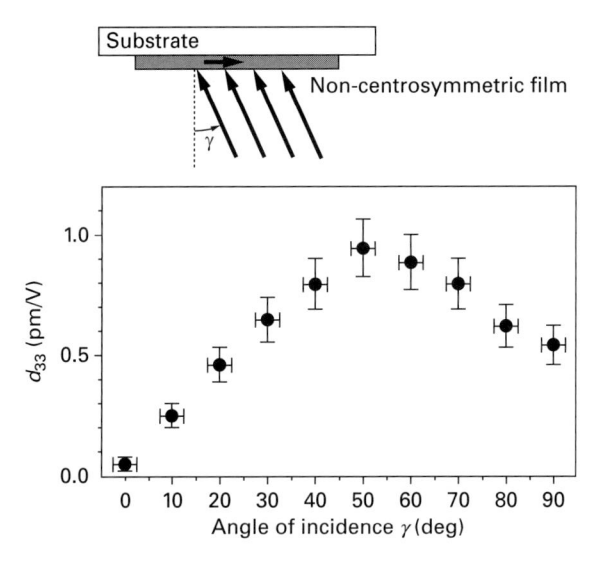

Figure 5.16 Macroscopic nonlinear optical susceptibility as a function of the incident angle of the molecular beams. Reprinted with permission from [78].

and exhibit rough surfaces (see Fig. 5.13). This is one of the limitations of the OMBD techniques [68,70].

In order to overcome the difficulties of OMBD, an oblique-incidence OMBD technique with head-to-tail intermolecular hydrogen bonds has been developed [68 to 72,77,78]. In OMBD techniques, the incident angle of the molecular beam to the substrate is an important factor in determining the direction of molecular ordering in the film [68 to 71,77,78]. Figure 5.14 shows the chemical structures of dipolar molecules 4-[trans-(pyridin-4-ylvinyl)]benzoic acid (P2BA) and 4-(pyridin-4-ylethynyl)benzoic acid (P3BA) [71]. Both P2BA and P3BA molecules form strong head-to-tail intermolecular hydrogen bonds between COOH⋯N on the pyridine group, which leads to acentric polymer-like chains by self-assembly [69,71,72,77]. An illustration of the growth mechanism based on strong hydrogen bonds is shown in Fig. 5.15 [69]. The P2BA films grown in oblique-incidence OMBD show a strong dependence of the nonlinear optical susceptibility on the incident angle of molecular beams (see Fig. 5.16) [78].

Specific interactions on substrates. Although growth of organic thin films with the oblique-incidence OMBD method permits control over the orientation of the molecules on the lattice-mismatched substrate, films grown this way exhibit mainly in-plane orientation of the constituent molecules [68 to 72,77,78,83]. To overcome these limitations of oblique-incidence OMBD, strong intermolecular hydrogen bonds between the constituent molecules and the surface of substrates are introduced [83 to 86].

Acentric organic thin films with almost perfect ordering of nonlinear optical chromophores on amorphous substrates have been demonstrated by physical vapor

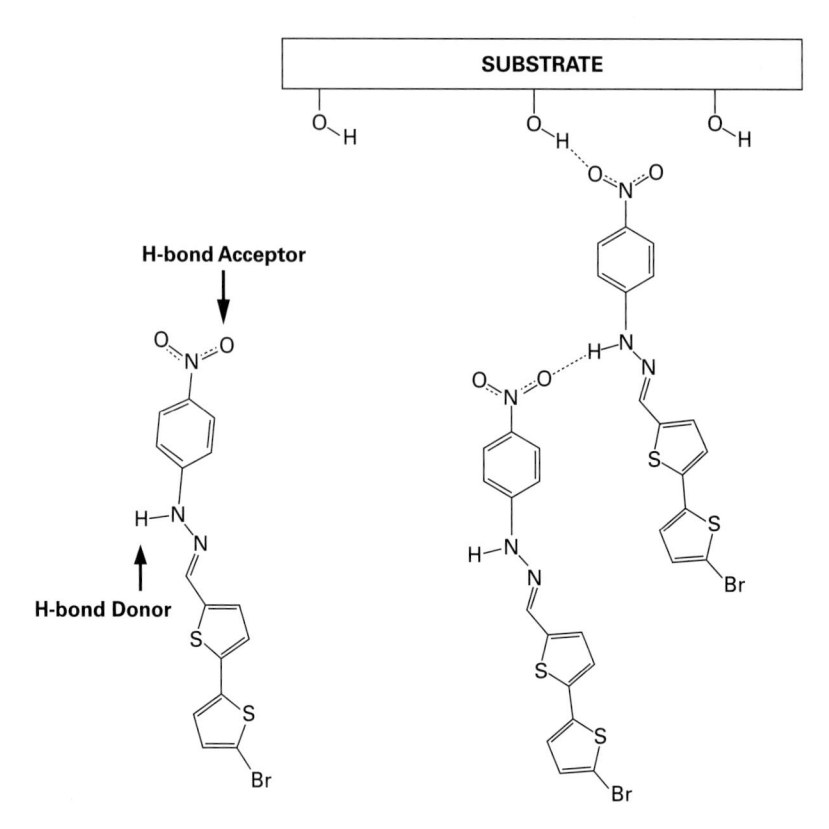

Figure 5.17 Intermolecular interactions between BBT-NPH molecules and the substrates are shown. Reprinted with permission from [84].

deposition [84]. Figure 5.17 shows the chemical structure of hydrazone derivative BBT-NPH (5´-bromo-2,2´-bithiophene-5-carbaldehyde)-4-nitrophenylhydrazone). The BBT-NPH is a derivative of highly efficient nonlinear optical crystals based on 4-nitrophenylhydrazone, which form strong hydrogen bonds between the hydrazone and the nitro groups (i.e., N-H···O-N groups) in the crystalline state [87 to 90]. In BBT-NPH films grown on a hydrophilic glass substrate by physical vapor deposition, BBT-NPH molecules are aligned perpendicular to the substrate owing to strong hydrogen bonds between the nitro group of BBT-NPH and the OH group of the substrate surface, as shown in Fig. 5.17 [84].

Recently, Marks and coworkers have also reported on organic thin films based on hydrogen-bond-directed orientation by vapor growth technique by using surface modification with silanization [83]. The schematic diagram of their process is given in Fig. 5.18. The DTPT (5-{4-[2-(4,6-diamino-[1,3,5]triazin-2-yl)-vinyl]-benzylidene}-pyrimidine-2,4,6-trione) chromophore can form head-to-tail triple hydrogen bonds between the pyrimidine-2,4,6-trione group and 4,6-diamino-1,3,5-trizine-2-yl

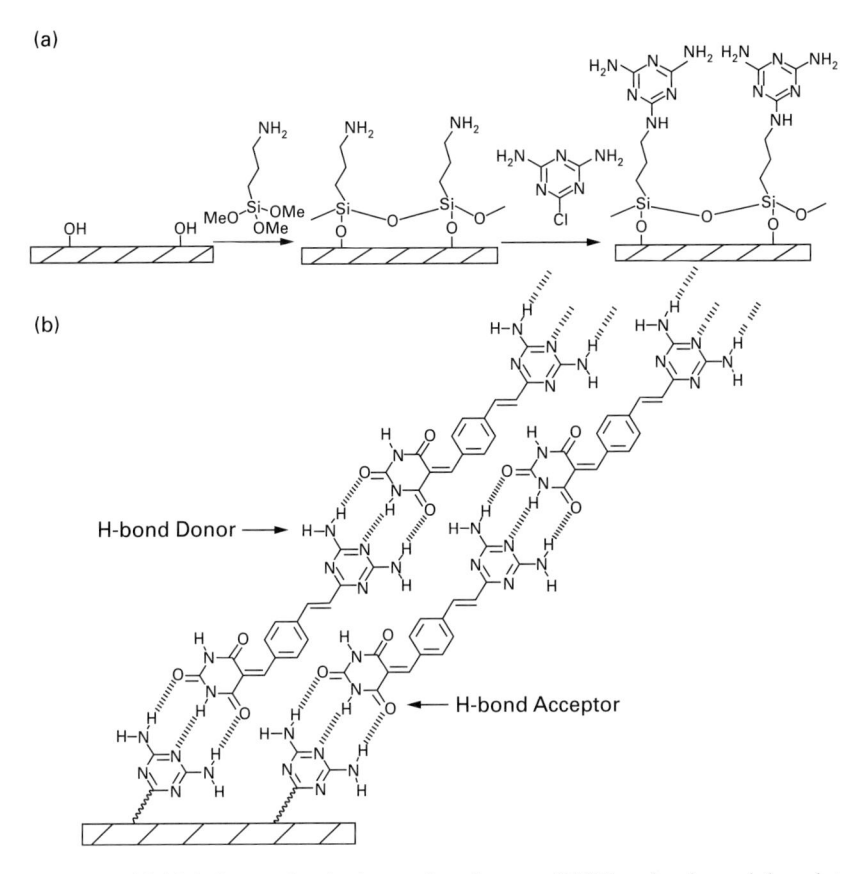

Figure 5.18 Multiple intermolecular interactions between DTPT molecules and the substrates are illustrated. Reprinted with permission from [84].

group. By silanization, the substrate surface is functionalized with the 4,6-diamino-1,3,5,-trizine-2-yl group. This functional group is identical to the hydrogen donor part of the DTPT chromophores and forms intermolecular triple hydrogen bonds with chromophores as shown in Fig. 5.18 [83]. Another successful example of vapor growth with surface modification by silanization is illustrated in Fig. 5.19 [86]. These acentric films grown on a self-assembled monolayer utilizing chromophores exhibit large macroscopic nonlinearities with good long-term stability and optical quality.

5.3.2 Non-volatile ionic organic materials: physical vapor deposition with chemical reactions

For physical vapor deposition, it is essential to use materials exhibiting a suitable vapor pressure and to achieve materials with a high thermal stability. One of the benchmark

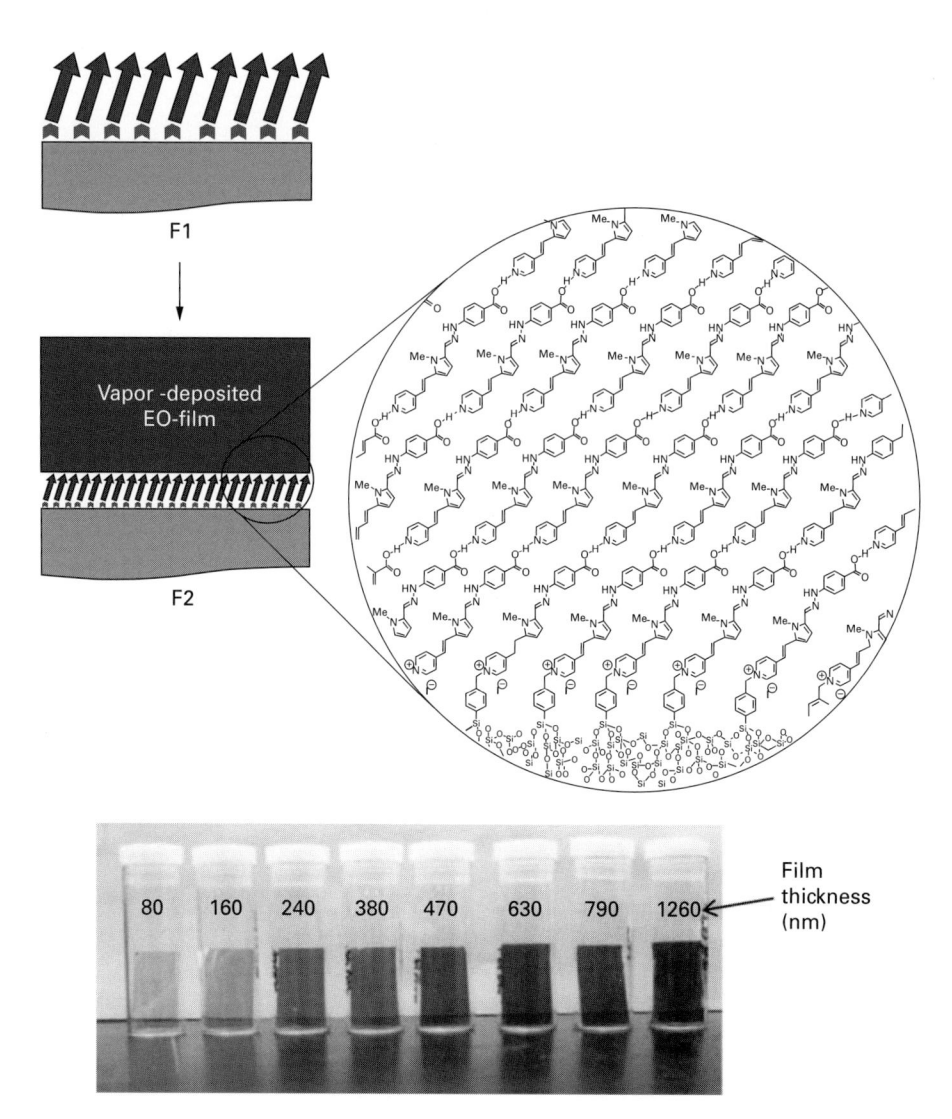

Figure 5.19 Hydrogen-bonded films grown by physical vapor deposition on a self-assembled monolayer incorporating chromophores. Adapted with permission from [84].

nonlinear optical materials is ionic DAST crystal and its derivatives [91,92]. However, the salt-type DAST derivatives decompose before evaporation can take place. To eliminate thermal degradation by direct vacuum evaporation of DAST, Forrest and co-workers used vapor deposition with chemical reaction [73,93]. As shown in Fig. 5.20, the volatile precursors, DASI (N,N-dimethylamino-N′-methylstilbazolium iodide) and MT (methyl tosylate) are evaporated in vacuum and then form DAST at the substrate. The resulting DAST thin films exhibit a polycrystalline morphology with a rough surface [73,93].

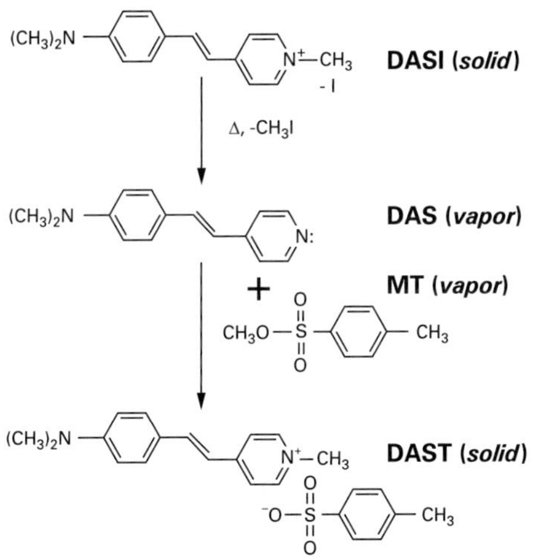

Figure 5.20 The physical vapor deposition of ionic DAST. Reprinted with permission from [73].

References

[1] Ch. Bosshard, M. Bösch, I. Liakatas, M. Jäger, and P. Günter, in *Nonlinear Optical Effects and Materials*, P. Günter, Ed., Berlin, Springer-Verlag (2000), Ch. 3, pp. 163–299.
[2] M. Jazbinsek, O. P. Kwon, Ch. Bosshard, and P. Günter, in *Handbook of Organic Electronics and Photonics*, S. H. Nalwa, Ed., Los Angeles, American Scientific Publishers (2008), Ch. 1, pp. 1–32.
[3] Ch. Bosshard, K. Sutter, Ph. Prêtre, *et al. Organic Nonlinear Optical Materials; Advances in Nonlinear Optics*, Langhorne, PA, Gordon & Breach Science Publishers (1995), Vol. 1.
[4] *Introduction to Organic Electronic and Optoelectronic Materials and Devices*, S. S. Sun and L. R. Dalton, Eds., Boca Raton, FL, CRC Press (2008).
[5] H. S. Nalwa, T. Watanabe, and S. Miyata, in *Nonlinear Optics of Organic Molecules and Polymers*, H. S. Nalwa and S. Miyata, Eds., Boca Raton, FL, CRC Press (1997), Ch. 4, pp. 89–350.
[6] D. S. Chemla and J. Zyss, in *Nonlinear Optical Properties of Organic Molecules and Crystals*, Orlando, FL, Academic Press (1987), Vol. 1.
[7] L. R. Dalton, P. A. Sullivan, and D. H. Bale, *Chem. Rev.*, **110**, 25 (2010).
[8] O. P. Kwon, M. Jazbinsek, H. Yun, *et al. Cryst. Eng. Comm.*, **11**, 1541 (2009).
[9] S. J. Kwon, O. P. Kwon, J. I. Seo, *et al. J. Phys. Chem. C*, **112**, 7846 (2008).
[10] O. P. Kwon, M. Jazbinsek, H. Yun, *et al. Cryst. Growth Des.*, **8**, 4021 (2008).
[11] O. P. Kwon, B. Ruiz, A. Choubey, *et al. Chem. Mater.*, **18**, 4049 (2006).
[12] O. P. Kwon, S. J. Kwon, M. Jazbinsek, V. Gramlich, and P. Günter, *Adv. Funct. Mater.*, **17**, 1750 (2007).
[13] Yi Liao, S. Bhattacharjee, K. A. Firestone, *et al. J. Am. Chem. Soc.*, **128**, 6847 (2006).
[14] A. Ulman, in *An Introduction to Ultrathin Organic Films*, San Diego, CA, Academic Press (1991), Ch. 2, pp. 101–236.
[15] G. J. Ashwell, *J. Mater. Chem.*, **9**, 1991 (1999).
[16] D. Lupo, W. Prass, U. Scheunemann, *et al. J. Opt. Soc. Am. B*, **5**, 300 (1988).

[17] I. R. Girling, N. A. Cade, P. V. Kolinsky, and C. M. Montgomery, *Electronics Letters*, **21**, 169 (1985).
[18] P. Stroeve, D. D. Saperstein, and J. F. Rabolt, *J. Chem. Phys.*, **92**, 6958 (1990).
[19] J. S. Schildkraut, T. L. Penner, C. S. Wiland, and A. Ulman, *Opt. Lett.*, **13**, 134 (1988).
[20] R. Advincula, E. Aust, W. Meyer, W. Steffen, and W. Knoll, *Polym. Adv. Technol.*, **7**, 571 (1996).
[21] G. J. Ashwell, T. W. Walker, and P. Leeson, *Langmuir*, **14**, 1525 (1998).
[22] F. Gao, C. Wang, H. Zeng, and S. Ma, *Colloids Surf. A*, **7**, 321 (2008).
[23] G. H. Cross, I. R. Girling, I. R. Peterson, N. A. Cade, and J. D. Earls, *J. Opt. Soc. Am. B*, **4**, 962 (1987).
[24] D. Locatelli, S. Quici, S. Righetto, *et al.*, *Prog. Solid State Chem.*, **33**, 223 (2005).
[25] G. J. Ashwell, R. Ranjan, A. J. Whittam, and D. S. Gandolfo, *J. Mater. Chem.*, **10**, 63 (2000).
[26] I. R. Girling, N. A. Cade, P. V. Konlinsky, *et al.*, *J. Opt. Soc. Am. B*, **4**, 950 (1987).
[27] G. Decher, B. Tieke, C. Bosshard, and P. Günter, *Ferroelectrics*, **91**, 193 (1989).
[28] W. M. K. P. Wijekoon, B. Asgharian, M. Casstevens, *et al.*, *Langmuir*, **8**, 135 (1992).
[29] W. M. K. P. Wijekoon, S. P. Karna, G. B. Talapatra, and P. N. Prasad, *J. Opt. Soc. Am. B*, **10**, 213 (1993).
[30] I. Ledoux, D. Josse, P. Vidakovic, *et al.*, *Europhys. Lett.*, **3**, 803 (1987).
[31] T. F. Heinz, H. W. K. Tom, and Y. R. Shen, *Phys. Rev. A*, **28**, 1883 (1983).
[32] T. F. Heinz, C. K. Chen, D. Ricard, and Y. R. Shen, *Phys. Rev. Lett.*, **48**, 478 (1982).
[33] G. Panambur, Y. Zhang, A. Yesayan, *et al.*, *Langmuir*, **20**, 3606 (2004).
[34] Z. Guo, Y. Li, Z. Ge, *et al.*, *Appl. Phys.*, **71**, 545 (2000).
[35] D. West, D. Dunne, P. Hodge, N. B. McKeown, and Z. Ali-Adib, *Thin Solid Films*, **323**, 227 (1998).
[36] K. Kajikawa, T. Anzai, H. Takezoe, *et al.*, *Appl. Phys Lett.*, **62**, 2161 (1993).
[37] J. P. Cresswell, M. C. Petty, I. Ferguson, *et al.*, *Adv. Mater. Opt. Electron.*, **6**, 33 (1996).
[38] W. M. K. P. Wijekoon, S. K. Wijaya, *et al.*, *J. Am. Chem Soc.*, **118**, 4480 (1996).
[39] L. S. Miller, D. J. Walton, P. J. W. Stone, A. M. McRoberts, and R. S. Sethi, *J. Mater. Sci. Mater. Electron.*, **5**, 75 (1994).
[40] A. Ulman, *Adv. Mater.*, **3**, 298 (1991).
[41] I. R. Girling, P. V. Kolinsky, N. A. Cade, J. D. Earls, and I. R. Peterson, *Opt. Commun.*, **55**, 289 (1985).
[42] G. J. Ashwell, R. Hamilton, B. J. Wood, I. R. Gentle, and D. Zhou, *J. Mater. Chem.*, **11**, 2966 (2001).
[43] M. Halter, Y. Liao, R. M. Plocinik, *et al.*, *Chem. Mater.*, **20**, 1778 (2008).
[44] Y. Liao, S. Bhattacharjee, K. A. Firestone, *et al.*, *J. Am. Chem. Soc.*, 2006, **128**, 6847–6853.
[45] O. P. Kwon, S. J. Kwon, M. Stillhart, *et al.*, *Cryst. Growth Des.*, **7**, 2517 (2007).
[46] A. Schneider, M. Neis, M. Stillhart, *et al.*, *J. Opt. Soc. Am. B*, **9**, 1822 (2006).
[47] O. P. Kwon, S. J. Kwon, M. Jazbinsek, *et al.*, *Adv. Funct. Mater.*, **18**, 3242 (2008).
[48] O. Dannenberger, J. J. Wolff, and M. Buck, *Langmuir*, **14**, 4679 (1998).
[49] R. Naraoka, G. Kaise, K. Kajikawa, *et al.*, *Chem. Phys. Lett.*, **362**, 26 (2002).
[50] K. Tsuboi, K. Seki, Y. Ouchi, K. Fujita, and K. Kajikawa, *Jpn. J. Appl. Phys.*, **42**, 607 (2003).
[51] D. Li, M. A. Ratner, and T. J. Marks, *J. Am. Chem. Soc.*, **112**, 7389 (1990).
[52] W. Lin, W. Lin, G. K. Wong, and T. J. Marks, *J. Am. Chem. Soc.*, **118**, 8034 (1996).
[53] S. B. Roscoe, A. K. Kakkar, and T. J. Marks, *Langmuir*, **12**, 4218 (1996).

[54] P. Zhu, M. E. van der Boom, H. Kang, *et al.*, *Chem. Mater.*, **14**, 4982 (2002).
[55] A. Facchetti, A. Abbotto, L. Beverina, *et al.*, *Chem. Mater.*, **15**, 1064 (2003).
[56] A. Facchetti, A. Abbotto, L. Beverina, *et al.*, *Chem. Mater.*, **14**, 4996 (2002).
[57] J. U. Choi, C. B. Lim, J. H. Kim, *et al.*, *Synth. Metals*, **71**, 1729 (1995).
[58] W. Silwa, *Chem. Hetero. Comp.*, **34**, 56 (1998).
[59] A. L. Briseno, J. Aizenberg, Y. J. Han, *et al.*, *J. Am. Chem. Soc.*, **127**, 12164 (2005).
[60] R. Hiremath, J. A. Basile, S. W. Varney, and J. A. Swift, *J. Am. Chem. Soc.*, **127**, 18321 (2005).
[61] R. Hiremath, S. W. Varney, and J. A. Swift, *Chem. Mater.*, **16**, 4948 (2004).
[62] J. C. Huie, *Smart Mater. Struct.*, **12**, 264 (2003).
[63] P. T. Hammond, *Curr. Opin. Coll. Interface Sci.*, **4**, 430 (2000).
[64] S. H. Lee, S. Balasubramanian, D. Y. Kim, *et al.*, *Macromolecules*, **33**, 6534 (2000).
[65] S. S. Lee, K. B. Lee, and J. D. Hong, *Langmuir*, **19**, 7592 (2003).
[66] O. P. Kwon, J. H. Im, J. H. Kim, and S. H. Lee, *Macromolecules*, **33**, 9310 (2000).
[67] J. H. Im, O. P. Kwon, J. H. Kim, S. H. Lee, *Macromolecules*, **33**, 9606 (2000).
[68] R. U. A. Khan, O. P. Kwon, A. Tapponnier, A. N. Rashid, and P. Günter, *Adv. Funct. Mater.*, **16**, 180 (2006).
[69] C. Cai, B. Müller, J. Weckesser, *et al.*, *Adv. Mater.*, **11**, 750 (1999).
[70] R. Schlesser, T. Dietrich, Z. Sitar, *et al.*, *J. Appl. Phys.*, **78**, 4943 (1995).
[71] C. Cai, M. M. Bösch, B. Müller, *et al.*, *Adv. Mater.*, **11**, 745 (1999).
[72] J. V. Barth, J. Weckesser, C. Cai, *et al.*, *Angew. Chem. Int. Ed.*, **39**, 1230 (2000).
[73] S. R. Forrest, *Chem. Rev.*, **97**, 1793 (1997).
[74] C. Cai, M. M. Bösch, Y. Tao, *et al.*, *J. Am. Chem. Soc.*, **120**, 8563 (1998).
[75] B. Esembeson, M. L. Scimeca, T. Michinobu, F. Diederich, and I. Biaggio, *Adv. Mater.*, **20**, 4584 (2008).
[76] B. Müller, M. Jäger, Y. Tao, *et al.*, *Opt. Mater.*, **12**, 345 (1999).
[77] B. Müller, C. Cai, A. Knüdig, *et al.*, *Appl. Phys. Lett.*, **74**, 3110 (1999).
[78] A. Tapponnier, E. Delvigne, I. Biaggio, and P. Günter, *J. Opt. Soc. Am. B*, **21**, 685 (2004).
[79] A. Yamashita and T. Hayashi, *Adv. Mater.*, **8**, 791 (1996).
[80] W. Kowalsky, T. Benstem, A. Bohler, *et al.*, *Phys. Chem. Chem. Phys.*, **1**, 1719 (1999).
[81] F. Schreiber, *Phys. Status Solidi A*, **201**, 1037 (2004).
[82] T. Yamashiki, S. Fukuda, K. Tsuda, and T. Gotoh, *Appl. Phys. Lett.*, **83**, 605 (2003).
[83] P. Zhu, H. Kang, A. Facchetti, *et al.*, *J. Am. Chem. Soc.*, **125**, 11496 (2003).
[84] A. N. Rashid, C. Erny and P. Günter, *Adv. Mater.*, **15**, 2024 (2003).
[85] D. Frattarelli, M. Schiavo, A. Facchetti, M. A. Ratner, and T. J. Marks, *J. Am. Chem. Soc.*, **131**, 12595 (2009).
[86] A. Facchetti, E. Annoni, L. Beverina, *et al.*, *Nat. Mater.*, **3**, 910 (2004).
[87] Ch. Serbutoviez, Ch. Bosshard, G. Knopfle, *et al.*, *Chem. Mater.*, **7**, 1198 (1995).
[88] M. S. Wong, V. Gramlich, Ch. Boshard, and P. Günter, *J. Mater. Chem.*, **7**, 2021 (1997).
[89] O. P. Kwon, M. Jazbinsek, H. Yun, *et al.*, *Cryst. Growth Des.*, **8**, 4021 (2008).
[90] O. P. Kwon, M. Jazbinsek, J. I. Seo, *et al.*, *J. Phys. Chem. C*, **113**, 15405 (2009).
[91] S. R. Marder, J. W. Perry, and W. P. Schaefer, *Science*, **245**, 626 (1989).
[92] F. Pan, M. S. Wong, Ch. Bosshard and P. Günter, *Adv. Mater.*, **8**, 592 (1996).
[93] M.A. Baldo, M. Deutsch, P.E. Burrows, *et al.*, *Adv. Mater.*, **10**, 1505 (1998).

6 Crystalline materials

6.1 Non-centrosymmetric organic crystalline packing: approaches

In the past, there have been significant advances in understanding and optimizing classical π-conjugated donor–acceptor chromophores with large first-order molecular hyperpolarizabilities in the area of organic nonlinear optics [1 to 11]. However, there are only a few chromophores with very large molecular hyperpolarizabilities that form potentially useful crystalline materials.

The basic requirement for any second-order nonlinear optical material is that it crystallizes in a non-centrosymmetric space group, so that the microscopic effects favorably add up to a macroscopic effect. For crystalline materials, the suitable packing of molecules, which yields optimized macroscopic effects, can be evaluated by the oriented-gas model, as discussed in Chapter 4 (see Fig. 4.3). However, controlled crystallization of large organic molecules with desired optical properties is a challenging topic. A major problem in achieving this is that most organic molecules will crystallize in a centrosymmetric space group, which is usually attributed to dipolar interactions that favor anti-parallel chromophore alignment.

Crystal growth is the prototype of self-assembly in nature. The molecules pack into the crystalline lattice that corresponds to a minimum in the potential energy, which will depend on many parameters, such as the geometry of the molecules (close-packing principle), and the electrostatic (Coulomb) including ionic, dipolar (e.g., hydrogen-bond), and Van der Waals interactions. A correct determination of the crystalline structure requires a more fundamental quantum mechanical approach that is computationally very difficult. Prediction of the crystal structure based on π-conjugated donor–acceptor chromophores has not yet been possible. However, there have been several approaches identified to obtain non-centrosymmetric nonlinear optical organic crystals, which are listed below and illustrated in Fig. 6.1. For more details, see e.g. Refs. [8,12,13].

- **Use of molecular asymmetry:** When crystallizing, molecules tend to form simpler shapes, which gives rise to dimers and then high-order aggregates following the close-packing principle in the solid state [14]. The high probability of anti-parallel aggregation of achiral chromophores due to such a close-packing driving force can be reduced by changing the shape of the molecules. By decreasing the symmetry of a molecule, dimerization and subsequent aggregation may no longer be advantageous, which increases the probability of acentric

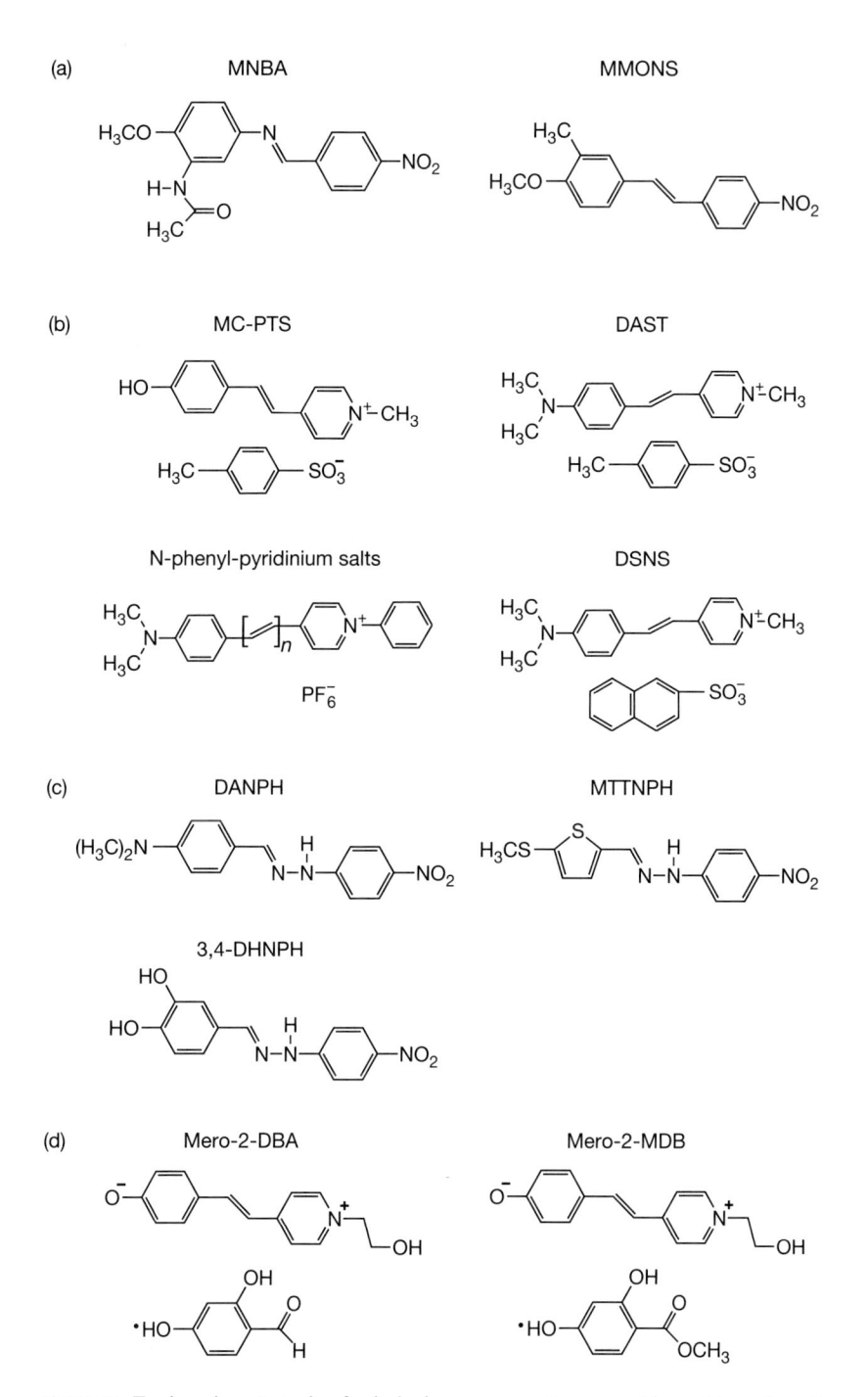

Figure 6.1 Engineering strategies for inducing a non-centrosymmetric packing of nonlinear chromophores and examples: (a) use of molecular asymmetry, (b) use of strong Coulomb interactions, (c) use of non-rod-shaped π-conjugated cores, and (d) supramolecular synthetic approach.

crystallization. This symmetry reduction can be accomplished by either the introduction of molecular (structural) asymmetry or the incorporation of steric (bulky) substituents into the chromophore. These two approaches have been widely and successfully applied to benzenoid chromophores. An introduction of a substituent at the 3-position of 4-nitrobenzylidene 4-donor-substituted-aniline can induce a favorable non-centrosymmetric packing for large optical nonlinearities [15]. This led to the discovery of 4′-nitrobenzylidene-3-acetamino-4-methoxyaniline, MNBA (Fig. 6.1(a)), which shows a large SHG efficiency that is 230 times that of the urea standard. Another example using this approach is 3-methyl-4-methoxy-4′-nitrostilbene, MMONS (Fig. 6.1(b)), which shows a SHG powder efficiency of 1250 times that of urea [16]. More recently, this approach has been adapted to configurationally locked polyene (CLP) molecules [17], often combined with additional supramolecular interactions such as hydrogen bonds to achieve favorable non-centrosymmetric packing. The most promising CLP crystals are presented in Section 6.5.

- **Use of strong Coulomb interactions** can help to override the weak dipole–dipole interactions to result in a higher probability of non-centrosymmetric packing. The validity of this concept was proved for the case of 4-dimethylamino-N-methylstilbazolium salts by Meredith [18]. This led to the discovery of 4-dimethylamino-N-methylstilbazolium methylsulfate, DMSM, which shows a SHG efficiency 220 times that of urea [18]. The 4-toluenesulfonate anion was found to be an effective counter-ion to induce the non-centrosymmetric packing of stilbazolium chromophores, which led to the development of 4-hydroxy-N-methylstilbazolium 4-toluene-sulfonate, MC-PTS [19] (Fig. 6.1(b)). Marder and coworkers adopted this strategy to perform an extensive investigation varying the counter-ions of various stilbazolium chromophores including 2-N-methylstilbazolium and 4-N-methylstilbazolium cations [20,21]. They found that whereas rod-shaped 4-N-methylstilbazolium cations can often be forced to crystallize non-centrosymmetrically, this is not true for non-rod-shaped 2-N-methylstilbazolium cations. DAST, 4-N,N-dimethylamino-4′-N′-methylstilbazolium tosylate, was shown to exhibit a very large powder SHG efficiency (1000 times urea) at 1.9 μm. Currently, DAST is still known as the best organic nonlinear optical crystal, and its optical and nonlinear optical properties are presented in more detail in Section 6.3 together with other promising DAST derivatives developed more recently [10, 22 to 25] (see Fig. 6.1(b)).

- **Use of non-rod-shaped π-conjugated cores:** In contrast to donor–acceptor disubstituted stilbene derivatives, hydrazone derivatives generally adopt a bent, non-rod-shaped conformation in the solid state because of the non-rigid nitrogen–nitrogen single bond –CH=N–NH–. Donor-substituted (hetero)-aromatic aldehyde-4-nitrophenylhydrazones show a relatively very high tendency to form a non-centrosymmetric packing, often combined with a high second-order optical nonlinearity [26 to 28] (Fig. 6.1(c)), at least two orders of magnitudes greater than that of urea. The best example in this class is 4-dimethylaminobenzaldehyde-4-nitrophenyl-hydrazone, DANPH (Fig. 6.1(c)), which exhibits a very strong SHG signal, comparable to that of DAST. Another potential candidate is 5-(methylthio)thiophenecarboxaldehyde-4-nitrophenylhydrazone, MTTNPH, which also shows the same order of powder efficiency as DANPH. Also promising are isomorphous crystals 2,4- and 3,4-dihydroxybenzaldehyde-4-nitrophenylhydrazone (DHNPH) with an excellent alignment of the chromophores in the crystal lattice and a molecular hyperpolarizability comparable to DANPH [27,28].

- **Supramolecular synthetic approach:** This involves molecular or ionic aggregates or assemblies to favor the desirable crystallographic packing. This approach offers more design feasibility, as one or both molecules can be tailor-made or can be modified to fit one another to acquire the desirable molecular properties in the solid state. Furthermore, the physical properties such as melting point and solubility as well as the crystal properties such as crystallinity and ease of crystal growth of the co-crystals can usually be improved compared with those of their starting components. Etter and Frankenbach first demonstrated the induction of a net dipole moment with a complimentary host–guest pair of 4-aminobenzoic acid and 3,5-dinitrobenzoic acid; however, the SHG signal generated by this co-crystal is of the order of the urea standard [29]. Co-crystals formed from the merocyanine dyes and the class I phenolic derivatives (Fig. 6.1(c)) have also shown a high tendency of forming acentric co-crystals [30 to 32] and additionally a very large macroscopic second-order optical nonlinearity at least two orders of magnitude higher than that of urea. Mero-2-DBA and Mero-2-MDB crystals (Fig. 6.1(c)) show very interesting variations of macroscopic optical and non-linear optical properties depending on the position of the proton in the hydrogen bond; more details are given in Section 6.5.

6.2 Examples of organic electro-optic crystals

There have been many organic electro-optic crystals developed during the past decades. In Table 6.1, we list only some of the most investigated ones. The crystals are organized with respect to the cut-off wavelength λ_c. For more examples, see Refs. [8] and [12].

6.3 Ionic crystals: stilbazolium salts

6.3.1 DAST

DAST is the best-known and most widely investigated organic electro-optic crystal. It was first reported in 1989 by Marder *et al.* [20] and Nakanishi *et al.* [84], and was developed based on the stilbazolium salt approach introduced by Meredith [18]. DAST is still recognized as the state-of-the-art organic nonlinear optical crystal. High optical quality and large size crystals, grown from methanol solution by the slow

Table 6.1 Examples of molecular crystals that have been investigated for their nonlinear optical and/or electro-optic response. λ_c is the cut-off wavelength in the bulk, d is the nonlinear optical coefficient, r is the electro-optic coefficient, and T_m is the melting point. The values of the nonlinear optical coefficients d were adjusted to the same reference value (d_{11}(α-quartz) = 0.3 pm/V (1064 nm) and 0.28 pm/V (1907 nm)). For a more extended overview of organic crystals see e.g. Refs. [8] and [12]. This table is a short version of Table 3.3 in Ref. [8], but includes some of the new materials developed in the past few years.

Material	Point group	λ_c	d, r (pm/V)	T_m (°C)	Ref.
\multicolumn{6}{c}{Cut-off wavelength $\lambda_c \leq 450$ nm}
Urea	$\bar{4}2m$	200	d_{14}(480–640 nm) =1.0 d_{14} (1064 nm)=1.1 r_{41} (633 nm)=1.9 r_{63} (633 nm)=0.8	133–135	[33–37]
DMACB (dimethylaminocyanobiphenyl)	m	420	d_{11} (1064 nm)=276 r_{11} (1064 nm)=55	–	[38,39]
\multicolumn{6}{c}{Cut-off wavelength 450 nm $\leq \lambda_c \leq$ 550 nm}
MNA (2-methyl-4-nitroaniline)	m	480	d_{11} (1064 nm)=150 d_{12} (1064 nm)=23 r_{11} (633 nm)=67	133	[16, 40–46]
DAN (4-(N,N-dimethylamino)-3-acetamidonitrobenzene)	2	485	d_{23} (1064 nm)=38 r_{32} (633 nm)=13	166	[47, 48]
mNA (meta-nitroaniline)	$mm2$	500	d_{31} (1064 nm)=11.7 d_{33} (1064 nm)=12.3 r_{33} (633 nm)=16.7	112	[49–51]
POM (3-methyl-4-nitropyridine-1-oxide)	222	460	d_{14} (1064 nm)=6.0 r_{41} (633 nm)=3.6 r_{52} (633 nm)=5.2 r_{63} (633 nm)=2.6	136	[52–54]
NPP (N-(4-nitrophenyl)-(L)-prolinol)	2	500	d_{21} (1064 nm)=51 d_{22} (1064 nm)=16.8	116	[55]

Table 6.1 (*cont.*)

Material	Point group	λ_c	d, r (pm/V)	T_m (°C)	Ref.
PNP (2-(N-prolinol)-5-nitropyridine)	2	490	d_{21} (1064 nm)=51 d_{22} (1064 nm)=16.2 r_{22} (514 nm)=28.3 r_{22} (633 nm)=12.8	83	[40,56,57]
BNA (N-benzyl-2-methyl-4-nitroaniline)	mm2	500	d_{33} (1064 nm)=234 d_{32} (1064 nm)=15.6	102	[58–60]
COANP (2-cyclooctylamino-5-nitropyridine)	mm2	490	d_{32} (1064 nm)=32 r_{33} (514 nm)=28 r_{33} (633 nm)=15	73	[40,56,61,62]
(-)MBANP ((-)-2-(α-methylbenzylamino)-5-nitropyridine)	2	450	d_{22} (1064 nm)=36 r_{eff} (488 nm)=31.4 r_{eff} (514 nm)=26.6 r_{eff} (633 nm)=18.2	83	[63–66]
AANP (2-adamantylamino-5-nitropyridine)	mm2	460	d_{31} (1064 nm)=48 d_{33} (1064 nm)=36	167	[67,68]
MMONS (3-methyl-4-methoxy-4′-nitrostilbene)	mm2	515	d_{33} (1064 nm)=112 d_{24} (1064 nm)=43 r_{33} (633 nm)=40	–	[69]
MNBA (4′-nitrobenzylidene-3-acetamino-4-methoxyaniline)	m	520	d_{11} (1064 nm)=131 d_{11} (532 nm)=50 r_{11} (633 nm)=29	–	[70]

Table 6.1 (*cont.*)

Material	Point group	λ_c	d, r (pm/V)	T_m (°C)	Ref.
MC-PTS (4-hydroxy-N-methylstilbazolium-4-toluene sulfonate)	1	510	d_{11} (1064 nm)=314	–	[19,71,72]

Cut-off wavelength $\lambda_c \geq 550$ nm

Material	Point group	λ_c	d, r (pm/V)	T_m (°C)	Ref.
DAST (4-N,N-dimethylamino-4′-N′-methylstilbazolium tosylate)	m	700	d_{11} (1318 nm)=1010 d_{11} (1542 nm)=290 d_{26} (1542 nm)=39 d_{11} (1907 nm)=210 d_{12} (1907 nm)=32 d_{26} (1907 nm)=25 r_{11} (720 nm)=92 r_{11} (1313 nm)=53 r_{11} (1535 nm)=47	256	[73–75]
DSTMS (4′-dimethylamino-N-methyl-4-stilbazolium 2,4,trimethylbenzenesulfonate)	m	690	d_{11} (1907 nm)=210 d_{12} (1907 nm)=31 d_{26} (1907 nm)=35	258	[76,77]
DAPSH (trans-4-dimethylamino-N-phenyl-stilbazolium hexafluorophosphate)	m	700	d_{11} (1907 nm)=290	–	[24,25,78]
Mero-2-MDB (4-{2-[1[(2-hydroxyethyl)-4-pyridylidene]-ethylidene}-cyclo-hexa-2,5-dien-1-one-methyl-2,4-dihydroxybenzoate)	m	Phase II: 680 Phase I: 615	Phase II: d_{11} (1318 nm)=267 r_{11} (1313 nm)=34 Phase I: d_{11} (1318 nm)=108 r_{11} (1313 nm)=24	185	[31,32]

6.3 Ionic crystals

Table 6.1 (*cont.*)

Material	Point group	λ_c	d, r (pm/V)	T_m (°C)	Ref.
DANPH (4-dimethylaminobenzaldehyde-4-nitrophenylhydrazone)	m	670	d_{12} (1542 nm)=200 d_{11} (1542 nm)=150	186	[79]
DAT2 (2-(3-(2-(4-dimethylaminophenyl) vinyl)-5,5-dimethylcyclohex-2-enylidene) malononitrile)	2	700	r_{12} (1550 nm)=7.4 r_{22} (1550 nm)=6.7	235	[17,80,81]
OH1 (2-(3-(4-hydroxystyryl)-5,5-dimethylcyclohex-2-enylidene) malononitrile)	$mm2$	600	d_{33} (1907 nm)=120 r_{33} (633 nm)=109 r_{33} (1319 nm)=52 r_{23} (1319 nm)=30 r_{13} (1319 nm)=6.8	212	[82,83]

cooling method [85,86], have permitted accurate determination of the dielectric, linear, and nonlinear optical properties of DAST. The reasons for the growing interest in obtaining high-quality DAST crystals include the high second-order nonlinear optical and the electro-optic coefficients, which are respectively ten times and twice as large as those of the inorganic standard LiNbO$_3$. Its extraordinarily high nonlinearities in combination with a low dielectric constant allow for high-speed electro-optic applications and broadband THz-wave generation.

The structural, dielectric, optical, electro-optic, and nonlinear optical properties of DAST most relevant for photonic applications, including frequency conversion and THz-wave generation, have been recently reviewed [87]. Here we summarize the most important material parameters relevant for electro-optics.

- **Crystal structure and linear optical properties of DAST:** DAST is an organic salt that consists of a positively charged stilbazolium cation and a negatively charged tosylate anion as shown in Fig. 6.2(a). The stilbazolium cation is one of the most efficient NLO active chromophores that pack in an acentric structure,

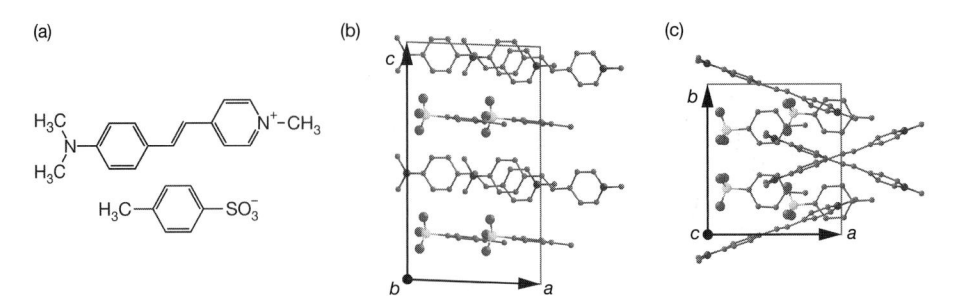

Figure 6.2 (a) Molecular units of the ionic DAST crystal. The positively charged, NLO active chromophore methyl-stilbazolium and the negatively charged tosylate. (b,c) X-ray structure of the ionic DAST crystal with the point group symmetry m showing molecules from one unit cell, projected along the crystallographic axes b and c. Hydrogen atoms have been omitted for clarity. The ac crystallographic plane is the (glide) mirror plane. The charge transfer axis of the chromophores makes an angle of about 20° with respect to the polar a axis.

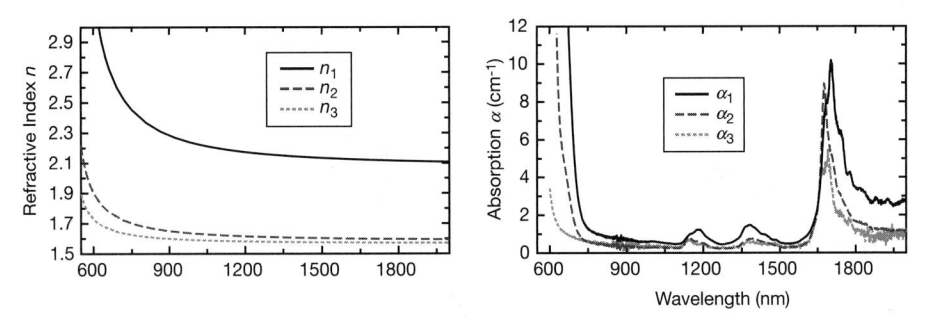

Figure 6.3 Refractive indices n_1, n_2, n_3 and absorption coefficients α_1, α_2, α_3 of DAST, represented by full, dashed, and dotted curves, respectively [73,74,85] (from [87], reprinted with permission from the IEEE).

whereas the counter-ion tosylate is used to promote non-centrosymmetric crystallization [20,21]. The structure of DAST is shown in Fig. 6.2(b,c). DAST crystals are composed of consecutive layers of stilbazolium and tosylate molecules with the space (point) group symmetry Cc (m); the crystallographic a axis makes an angle of $\beta = 92.2°$ with the crystallographic c axis [85]. The chromophores are packed with their main charge transfer axis oriented at about $\theta = 20°$ with respect to the polar a axis, resulting in a high order parameter of $\cos^3 \theta = 0.83$.

There is a slight difference between the crystallographic and the dielectric axes for the optical waves (~5°), which can be neglected for most practical applications [73,87].

Figure 6.3 shows the refractive indices n_1, n_2, and n_3 as a function of the wavelength, measured with an interferometric technique [73,74], as well as the absorption coefficients calculated from the measured transmission spectra taking into account multiple reflections at the crystal surfaces [85]. DAST crystals are highly anisotropic with a

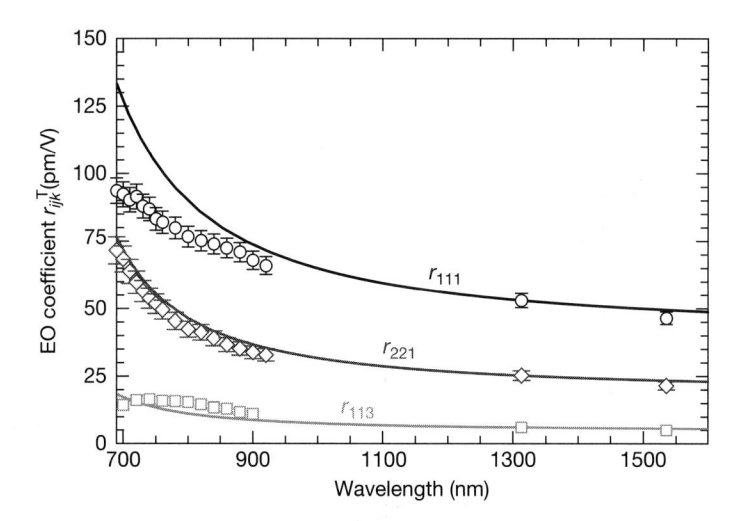

Figure 6.4 Electro-optic coefficient of DAST: Dispersion of the free electro-optic coefficient of DAST shown for r_{111}, r_{221}, and r_{113} [74] (from [87], reprinted with permission from the IEEE).

refractive index difference $n_1 - n_2 > 0.5$ in the visible and infrared wavelength range. They show small absorption bands at 1700 nm, 1400 nm, and 1100 nm, which correspond to overtones of the C–H stretching vibrations [88]. For applications in telecommunications, DAST crystals are well-suited with a material absorption that is smaller than 1 cm^{-1} at 1.3 μm and 1.55 μm wavelengths.

The dielectric constants of DAST in the low frequency range, below acoustic and optical lattice vibrations (see Fig. 3.1), were determined as $\varepsilon_1^T = 5.2 \pm 0.4$, $\varepsilon_2^T = 4.1 \pm 0.4$, and $\varepsilon_3^T = 3.0 \pm 0.3$ [74], and are considerably lower than those of inorganic EO materials, e.g., LiNbO$_3$ with $\varepsilon_1^T = 85 \pm 1$ and $\varepsilon_3^T = 28 \pm 1$ [89], or KNbO$_3$ with $\varepsilon_1^T = 154 \pm 5$, $\varepsilon_2^T = 985 \pm 20$, and $\varepsilon_3^T = 44 \pm 2$ [90].

- **Electro-optic properties of DAST:** For the electro-optic modulators based on the linear electro-optic (Pockels) effect, materials with high EO figures-of-merit (leading to low driving voltages) are required. DAST has a favorable acentric orientation of the chromophores (see Fig. 6.2) with an order parameter of $\langle \cos^3 \theta_{kz} \rangle = 0.83$, which is close to the optimum for EO applications.

The low-frequency (unclamped) electro-optic r_{ijk}^T coefficients of DAST were measured by using an interferometric method in the spectral range of 700–1535 nm [74]. DAST features large electro-optic coefficients, e.g., $r_{111} = 77 \pm 8$ pm/V at 800 nm and $r_{111} = 47 \pm 8$ pm/V at 1535 nm. The measured dispersion of the free electro-optic coefficients r_{111}, r_{221}, and r_{113} of DAST bulk crystals is shown in Fig. 6.4. The other coefficients r_{333}, r_{331}, and r_{223} are all smaller than 1 pm/V. The experimentally measured dispersion was modeled by a theoretical dispersion calculated according to the Sellmeier function and the two-level model, and is represented by solid curves in Fig. 6.4 [74]. The deviation at shorter wavelengths stems from resonance effects when approaching the absorption edge.

Owing to the large electro-optic coefficients and refractive indices of DAST, its EO figure-of-merit is high, $n^3 r = 455 \pm 80$ pm/V at wavelength $\lambda = 1535$ nm, and therefore the reduced half-wave voltage $v = \lambda/(n^3 r)$ compares favorably with inorganic single crystals and many poled electro-optic polymers.

The acoustic phonon contribution to the linear electro-optic effect in DAST was measured by applying a step voltage to the sample [91] as discussed in Chapter 3. A very low contribution from acoustic phonons ($r^a = -1 \pm 0.1$ pm/V at 1535 nm) to r^T_{111} was observed (see Fig. 3.2). The electronic contribution to the electro-optic effect r^e can be calculated from the measured nonlinear optical susceptibility $\chi^{(2)}_{111}(-2\omega, \omega, \omega)$ $= 2d_{111}(\omega)$ using the oriented-gas and two-level models [3, 92 to 94]. $r^e = -36 \pm 2$ pm/V at 1535 nm, which is about 75% of the measured unclamped coefficient [88]. Therefore, DAST is a very favorable material for broadband electro-optic switches, since the electro-optic response is large and follows the applied voltage almost perfectly from DC to at least 1 GHz, and most likely up to much higher frequencies below the optical phonon resonances [91].

The electro-optic coefficients measured in thin films of DAST close to the resonance at 720–750 nm are larger than those measured in bulk crystals by a factor of 5, reaching values of $r_{111} = 530$–445 pm/V [95 to 97].

Electro-absorption properties of DAST were also investigated [98]. High-speed in-line intensity modulation [99] and electro-optic sensing [100,101] were demonstrated with DAST crystals, confirming its great potential for EO applications.

6.3.2 Stilbazolium salt derivatives

Owing to the exciting nonlinear optical properties of DAST, together with a fairly good possibility to grow high-quality bulk crystals of DAST, there are still efforts going on to develop new molecules for organic salt crystals based on stilbazolium salts. Even very minor chemical-structure modifications of the non-active part of DAST, the counter-ion tosylate, may considerably change the crystal structure and material properties [21,23,76,102 to 107]. There are several reasons that motivate this kind of material development. The most obvious one is to achieve better molecular packing of the stilbazolium chromophores to result in higher macroscopic nonlinearities. Another motivation is to separate and align the molecules in such a way as to enhance and not reduce the molecular (microscopic) nonlinearities. The intermolecular interactions leading to these effects are not yet completely understood, but could potentially lead to much higher macroscopic nonlinearities. For example, for DAST it has been shown that the molecular nonlinearity in the solid state is only about 20% of the value in solution, attributed to the influence of intermolecular interactions [108]. The third reason is probably the most important from the applications point of view: a different counter-ion can significantly change the processing possibilities of such crystals, including bulk and thin-film growth, as well as material structuring.

Figure 6.5 lists some examples of stilbazolium salt derivatives reported in Ref. [23] and obtained by changing the counter-ion. In this study, three series of stilbazolium derivatives with various sizes of the counter-anions were systematically investigated

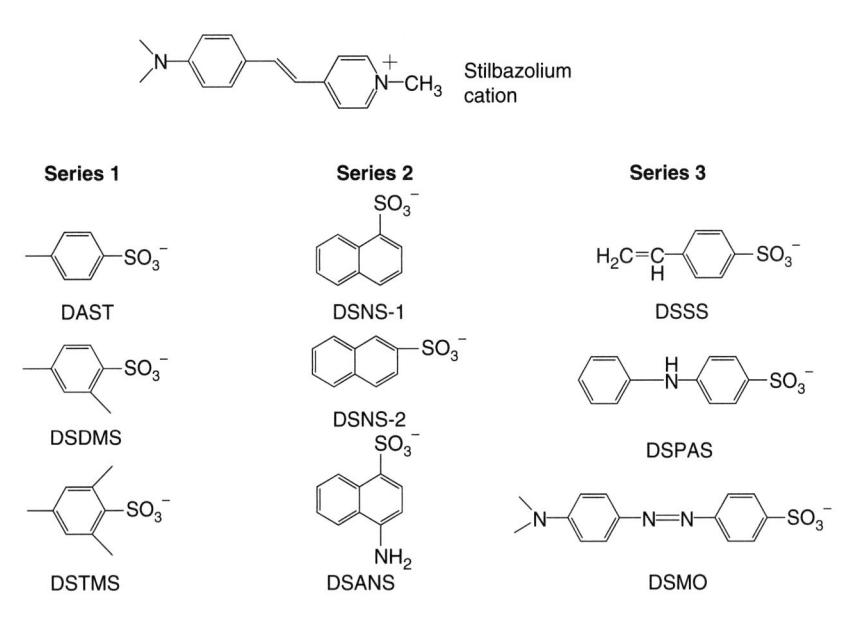

Figure 6.5 Counter-anions promote different crystalline packing of stilbazolium chromophores, which may considerably change the physical and optical properties. The examples in this figure were reported in Ref. [23] and are categorized in three series, with relatively small counter-ions (Series 1), mid-size (Series 2), and large-size (Series 3).

to understand the effect of counter-anions on the crystal structure and NLO activity of these compounds.

The potential for nonlinear optical applications of new crystalline materials is usually evaluated using the so-called Kurtz and Perry powder second-harmonic generation test [109], which was utilized for the compounds presented in Fig. 6.5 performed at the (non-resonant) fundamental wavelength of 1907 nm relative to the DAST powder. Out of the eight compounds, six exhibited second-harmonic signals, i.e. a non-centrosymmetric structure, out of which five showed strong NLO activity comparable to DAST (see Table 6.2). The NLO activity of DSNS-2 is 50% larger than that of DAST and is among the largest powder SHG efficiencies measured so far [103].

For three of the DAST-derivative crystals (DSDMS, DSNS-1, DSNS-2), optimal, perfectly parallel packing of the stilbazolium chromophores with the order parameter $\langle \cos^3 \theta_{kz} \rangle = 1$ has been obtained. Contrary to expectations, DSNS-2 shows a powder test efficiency of more than twice that of DSDMS and DSNS-1, attributed to the influence of the intermolecular interactions on the microscopic nonlinearity of chromophores in the solid state [23].

Stilbazolium salts also exhibit very different crystal processing characteristics. Slow cooling and slow evaporation solution growth techniques were used to grow single crystals of the salts. The compounds of series 1 nucleated much more easily than those of series 2 and 3. Bulk crystals of series 1 with dimensions of the order of 1 cm could be grown from methanol. Compounds of series 2 can be grown in single

Table 6.2 Some properties of DAST and its derivatives from Fig. 6.5: space and point symmetry group, order parameter $\langle \cos^3\theta_{kz} \rangle$, and second-harmonic powder test efficiency at 1907 nm fundamental wavelength relative to that of DAST.

Series	Compound	Space (point) group	$\langle \cos^3\theta_{kz} \rangle$	Powder NLO activity
1	DAST	Cc (m)	0.83	1.0
	DSDMS	$P1$ (1)	1.00	0.7
	DSTMS	Cc (m)	0.78	1.0
2	DSNS-1	$P1$ (1)	1.00	0.7
	DSNS-2	$P1$ (1)	1.00	1.5
	DSANS	$P2_1/c$ (2/m)	0	0.0
3	DSSS	-	-	1.0
	DSPAS	-	-	0.1
	DSMO	-	-	0.0

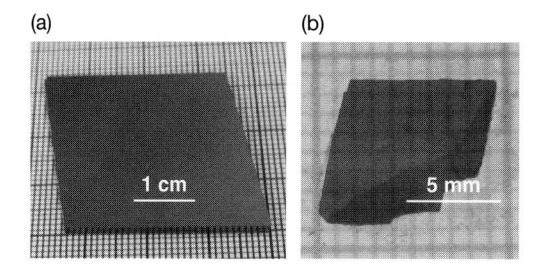

Figure 6.6 (a) DSTMS bulk crystal (size $33 \times 33 \times 2$ mm³); the large surface is the (001) face. (b) Large-area DSTMS thin-film single crystal with a thickness of about 20 µm, grown by the capillary method as described in Ref. [76].

crystalline form up to 1–2 mm length; they are all considerably smaller than those of series 1. In series 3, however, no single crystals could be grown. The results suggest that, with bulky counter-anions, the possibilities for growth of high-quality large-size bulk crystals tend to be suppressed [23].

DSTMS (4-N,N-dimethylamino-4′-N′-methyl-stilbazolium 2,4,6-trimethylbenzenesulfonate) shows the best crystal-growth characteristics among the investigated stilbazolium salt derivatives. The growth of high-quality crystals from methanol solution is relatively easy and faster than for DAST, which was partially attributed to the enhanced solubility [76]. Measurements of the linear and nonlinear optical properties of DSTMS have shown that it is a very attractive alternative to DAST, since its second-order optical nonlinearities are the same or even higher [77], which is a consequence of the isomorphous crystal structure of these two compounds. DSTMS also shows an improved THz-generation spectrum with a larger bandwidth and smaller absorption in the THz range than DAST [110]. Thin-film growth of DSTMS was also investigated with very promising first results, as shown in Fig. 6.6.

Another promising DAST derivative is DASC with *p*-chlorobenzenesulfonate counter-ion, which also shows an isomorphous crystal structure and therefore similar optical/NLO properties to DAST and DSTMS, but diferent physical properties such as

6.4 Supramolecular hydrogen bonded crystals

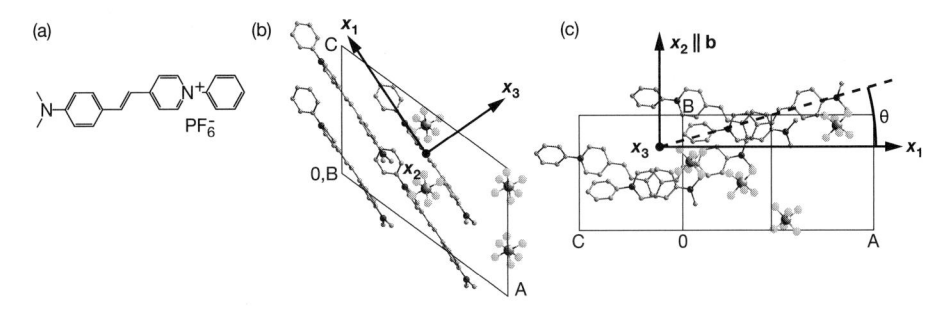

Figure 6.7 Molecular diagram (a) and the crystal structure of DAPSH with the point group symmetry m projected along the crystallographic axis b (b) and the dielectric axis x_3 (c); hydrogen atoms have been omitted for clarity. The polar axis x_4 points approximately along the $a-c$ crystallographic vector in the mirror plane that is normal to the b axis (b). The angle between the long axes of the molecules z and the polar axis x_1 is about $\theta_{1z} = 15°$.

the melting point and solubility [106]. Several other DAST derivatives with the optimal parallel chromophore orientation have also been identified [105], which proves that this approach is very successful for obtaining highly NLO organic crystals for various applications.

Based on the favorable results with N-methyl stilbazolium chromophore of DAST, Coe and others developed a new series of N-aryl pyridinium analogs with considerably higher molecular polarizabilities [10,24,78,111]. Several new promising salts with comparable powder test efficiencies to DAST, or even higher, were thus obtained [24]. Out of these compounds, trans-4-dimethylamino-N-phenyl-stilbazolium hexafluorophosphate (DAPSH) presents best physical properties for the solution growth of crystals [112]. Growth from solution in acetone or acetonitrile yields non-centrosymmetric crystals with the monoclinic space group Cc (point group m) and a very favorable chromophore orientation with $\theta_{kz} = 15°$, leading to a very high order parameter of $\langle \cos^3 \theta_{kz} \rangle = 0.9$ (see Fig. 6.7).

The optical and NLO properties of polished DAPSH bulk crystals were measured in the wavelength range from 0.7 to 2 µm. A very large anisotropy of the refractive index of $\Delta n = 1.17 \pm 0.06$ at $\lambda = 0.83$ µm and $\Delta n = 0.83 \pm 0.04$ at $\lambda = 1.55$ µm was measured, which is due to the highly ordered alignment of the NLO chromophores in DAPSH. DAPSH has also a very high diagonal second-order susceptibility tensor element $\chi^{(2)}_{111} = 580 \pm 80$ pm/V at $\lambda = 1.907$ µm, considerably higher than that of DAST ($\chi^{(2)}_{111} = 420 \pm 110$ pm/V), and presents the highest non-resonant second-order susceptibility measured so far in organic crystals [25].

6.4 Supramolecular hydrogen bonded crystals

Supramolecular hydrogen bonded co-crystals, formed by the merocyanine dyes (Mero-1 and Mero-2) and the class I phenolic derivatives, in which the electron acceptor is *para*-related to the phenolic functionality together with a substituent either in the

Mero-1: R=CH$_3$
Mero-2: R=CH$_2$CH$_2$OH

Class I phenolic derivatives

Figure 6.8 Chemical structures of Mero-1, Mero-2, and class I phenolic derivatives.

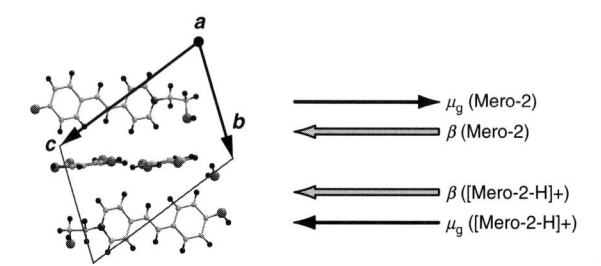

Figure 6.9 Crystal packing of Mero-2-DBA crystal projected along the *a* axis with the directions of the ground-state dipole moments and the main directions of the first-order hyperpolarizabilities.

ortho- or *meta-*position (Fig. 6.8), have shown the highest tendency of forming acentric co-crystals. In addition, a relatively large fraction of acentric co-crystals (25%) based on Mero-2 and the class I phenolic derivatives exhibit strong second-harmonic signals that are at least two orders of magnitude larger than that of urea. Their packing motifs can be distinctively divided into two groups.

The type I co-crystal is generally characterized by anionic and cationic assemblies or arrays. An interesting example in this class is the co-crystal Mero-2-DBA (DBA = 2,4-dihydroxybenzaldehyde; see Fig. 6.1(d)) [30]. Mero-2-DBA contains a water molecule and packs non-centrosymmetrically with space group *P*1 and point group 1. The anionic assembly is constructed by the co-aggregation of two DBA molecules in which one of the molecules gives up a proton and bonds to another by a hydrogen bond. Additionally, Mero-2 acquires a proton and co-aggregates in anti-parallel fashion with another Mero-2 by a short hydrogen bond constituting a cationic assembly. Therefore, although the net dipole moment almost vanishes in this arrangement, the Mero-2-DBA co-crystal exhibits a large second-harmonic signal in the powder test. This can be explained by the asymmetric position of the hydrogen bonded proton between the two Mero-2 dyes, which results in a positive reinforcement of molecular hyperpolarizabilities within the cationic assembly, since Mero-2 has a negative sign of the hyperpolarizability and the protonated form of Mero-2 ([Mero-2-H]$^+$) has a positive sign of the hyperpolarizability (see Fig. 6.9). As a consequence, the co-crystal Mero-2-DBA is a potential candidate for linear electro-optic effects because of its perfectly parallel alignment of molecular hyperpolarizabilities in the solid state.

6.4 Supramolecular hydrogen bonded crystals

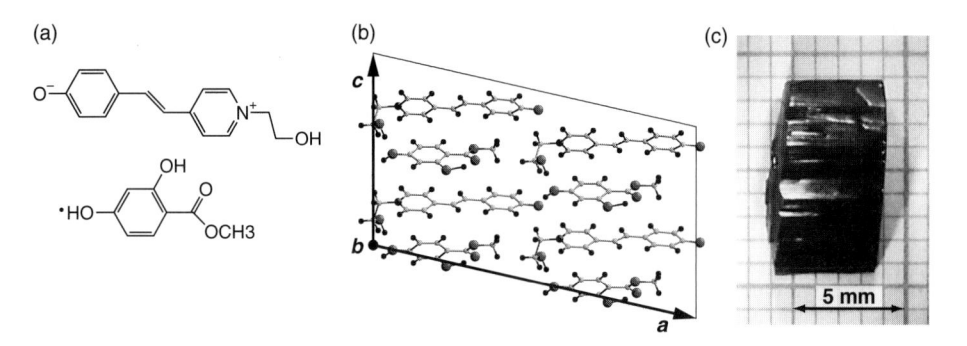

Figure 6.10 (a) Molecular structure, (b) acentric layer structure projected along the crystallographic axis *b*, and (c) an as-grown co-crystal Mero-2-MDB.

Type II co-crystals are formed by linear molecular aggregates. One of the representative examples in this class is the co-crystal Mero-2-DAP, which exhibits a very strong SHG signal that is three orders of magnitude larger than that of urea [31]. The molecular aggregate is assembled by the highly electronegative oxygen of Mero-2 and the acidic proton of the phenolic derivative through a short hydrogen bond. These rod-like aggregates connect laterally by hydrogen bonds, resulting in a staircase-like polar chain. These polar chains align in a parallel fashion constituting a two-dimensional acentric layer, which is found to be the common and key feature of all the highly non-centrosymmetric co-crystals in this class. Since the charge-transfer axis of Mero-2 is inclined by an angle of about 70° to the polar direction of the crystal, this co-crystal is a candidate for phase-matched nonlinear optical effects. In addition, in this system the orientation of the merocyanine dye can be changed and tuned within the crystal lattice by a careful selection of a guest molecule–phenolic derivative, provided that the linear molecular aggregate and the acentric layer packing motifs are maintained. Although Mero-2 by itself exists only in an amorphous state, both types of co-crystals formed show greatly improved crystalline and physical properties compared with their constituents.

Another interesting type II crystal, shown in Fig. 6.10, is Mero-2-MDB, which is optimized for electro-optic applications owing to the almost parallel alignment of the nonlinear optical chromophores.

Three phases of the co-crystal Mero-2-MDB were found. Beside a centrosymmetric Mero-2-MDB-c form, orange-red Mero-2-MDB-a and dark-red Mero-2-MDB-b non-centrosymmetric crystals have an identical crystal structure with a space group symmetry *Cc* (point group *m*), as determined by the X-ray diffraction, but show considerably different linear and nonlinear optical properties. For example, Mero-2-MDB-a has a nonlinear optical coefficient $d_{111} = 108 \pm 10$ pm/V, while the same coefficient for Mero-2-MDB-b is higher by more than a factor of two, $d_{111} = 267 \pm 25$ pm/V, both measured at 1318 nm fundamental wavelength [32]. This interesting property in the Mero-2-MDB co-crystals was attributed to a different proton location of the short hydrogen bond O–H–O in the aggregation between Mero-2 and MDB, which

6.5 Molecular crystals: configurationally locked polyene crystals

The chromophores for highly nonlinear optical crystals in general exhibit a limited temperature stability. In case of DAST (and similarly for its derivatives), the chromophores start to decompose at about the melting temperature, which is for DAST at 256 °C. Therefore, the processing possibilities of stilbazolium salts are for most practical situations limited to solution-based techniques. On the other hand, melt growth is very attractive for several reasons, e.g. higher growth rates, higher purity without solvent inclusion problems, and very attractive waveguide processing possibilities, as will be discussed in Chapter 9. Until recently, the short π-conjugated chromophores with relatively low melting temperatures (T_m < 150 °C), but also relatively low first hyperpolarizabilities, such as mNA, NPP, MAP, COANP (see Table 6.1), were the only organic nonlinear optical crystals obtainable by melt growth techniques. Unfortunately, the electro-optic figure-of-merit $n^3 r$ of these crystals may be one order of magnitude smaller than that of the best stilbazolium salts. This is the so-called nonlinearity–thermal stability tradeoff, which limits high-temperature processing of highly nonlinear molecules. Therefore, to design organic EO materials with a broad spectrum of processing possibilities, the challenge is to simultaneously achieve a high thermal stability and nonlinearity in one compound.

To obtain a large first-order hyperpolarizability, the long conjugated polyene chromophores provide one of the most effective pathways for efficient charge delocalization between donor and acceptor groups without loss of the aromaticity [7, 113]. However, similarly as for stilbazolium salts, the thermal and photochemical instability for conventional, "non-locked" polyene (CNP, Fig. 6.11) chromophores is not sufficient for the melt processing owing to the nonlinearity–thermal stability tradeoff. More recently, several chromophores with the polyene chain incorporated into a ring system have been

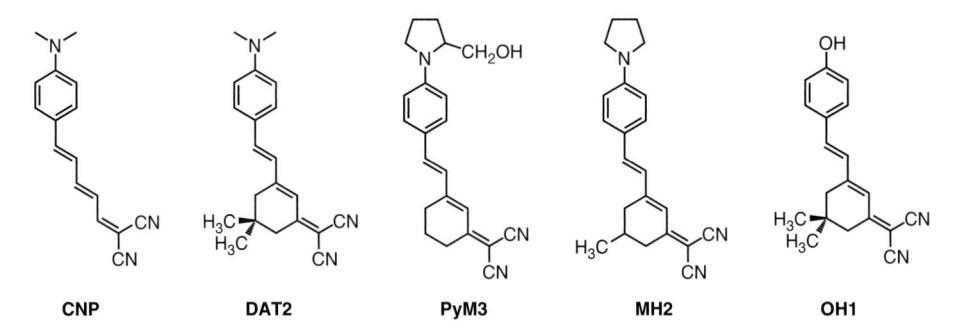

Figure 6.11 Chemical structures of the conventional "non-locked" polyene (CNP) and various configurationally locked polyene (CLP) chromophores with dicyanomethylidene electron acceptor and various electron donor groups: DAT2 [17], PyM3 [120], MH2 [122], and OH1 [82].

developed to improve thermal and photochemical stability [114 to 119]. To circumvent the nonlinearity–thermal stability tradeoff of organic crystalline materials, different series of configurationally locked polyene (CLP) chromophores have been recently developed [17,82,120 to 122]. These consist of a π-conjugated hexatriene bridge between dialkylamino, methoxy, prolinol or phenolic electron donors and a dicyanomethylidene [>C=C(CN)$_2$] electron acceptor; the non-π-conjugated part in the hexatriene bridge is connected with one or two methyl groups or hydrogen atoms. Several of these chromophores crystallize in a non-centrosymmetric structure with a high powder test efficiency, which has the same order of magnitude as that of DAST [17,82,120 to 122]. Here, we more specifically describe two of these crystals: DAT2 and OH1 (see Fig. 6.11) that have already demonstrated particular promise for integrated electro-optic applications.

6.5.1 Configurationally locked polyene DAT2 crystals

The configurationally locked polyene DAT2 (2-(3-(2-(4-dimethylaminophenyl)vinyl)-5,5-dimethylcyclohex-2-enylidene)malononitrile) chromophore consists of a phenylhexatriene bridge between a dimethylamino electron donor group and a dicyanomethylidene electron acceptor, leading to a first hyperpolarizability $\beta = 1100 \times 10^{-40}$ m^4/V at 1.9 μm, which is among the largest microscopic nonlinearities of chromophores that crystallize in a non-centrosymmetric crystalline arrangement [17]. DAT2 crystals present a monoclinic space-group symmetry $P2_1$ (point group symmetry 2), the crystal structure is shown in Fig. 6.12. The twofold symmetry axis is along the polar b axis. The main charge transfer axis of the chromophore is tilted by an angle of about 56° with respect to the largest as-grown crystal face, which is parallel to the ab-crystallographic plane. Figure 6.12 shows the orientation of the molecules, as well as the orientation of the dielectric system with respect to the unit cell in the ac-plane. More details on the crystal orientation and the linear optical properties can be found in Ref. [80].

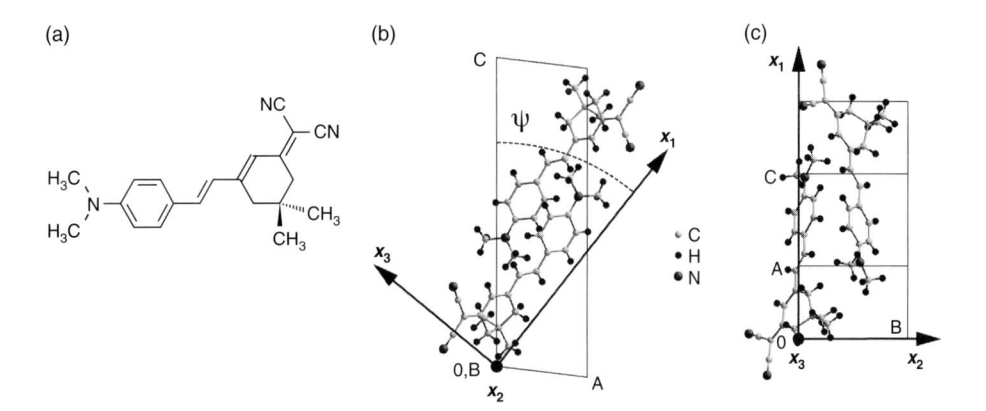

Figure 6.12 (a) DAT2 chemical structure and its crystal structure with point group symmetry 2 projected along the polar axis b (b) and along the dielectric axis x_3 (c). The dielectric axes x_1, x_2, and x_3 were determined in the optical frequency range in Ref. [80]; $\psi = 39 \pm 3°$.

The thermal stability of DAT2 chromophores has been investigated using differential scanning calorimetry (DSC) and thermogravimetric analysis (TGA) [123]. DAT2 chromophores do not show any evidence of decomposition up to at least 300 °C in the DSC measurement. The starting temperature of melting transition is at 216 °C. Resulting from the thermal stability and also favorable sublimation properties, single-crystalline DAT2 crystals were grown from both melt and vapor [17,80,123], in addition to solution growth [17,124].

Unfortunately, the crystal packing of DAT2 chromophores is not optimal with a low order parameter $\langle \cos^3 \theta_{zz} \rangle < 0.01$, which leads to relatively low electro-optic tensor elements; the largest being r_{112} of about 7 pm/V at 1.55 µm, as estimated from waveguide-modulation experiments [80], which may be still sufficient for several applications. The big advantage of DAT2 is the excellent range of possibilities for thin-film processing – from solution, vapor, and melt – which also gives great motivation for developing derivative materials with better chromophore alignment in the crystalline state. Indeed, there are now several promising examples of CLP chromophores with a more favorable crystalline packing [82,120 to 122], including the OH1 chromophore discussed in the following section.

6.5.2 Phenolic configurationally locked polyene OH1 crystals

The phenolic configurationally locked polyene crystal OH1 (2-(3-(4-hydroxystyryl)-5,5-dimethylcyclohex-2-enylidene)malononitrile) has been developed recently and has very promising properties for applications in nonlinear optics. Its chemical structure and the crystal packing diagram are shown in Fig. 6.13. The OH1 molecule and its crystal structure were first reported by Lemke [125] and Kolev et al. [126], while its potential for applications in electro-optics and in THz generation and detection has been only recently demonstrated [82]. The OH1 chromophore consists of the configurationally locked π-conjugated hexatriene bridge linked between phenolic electron donor and dicyanomethylidene [>C=C(CN)$_2$] electron acceptor, which can also act as

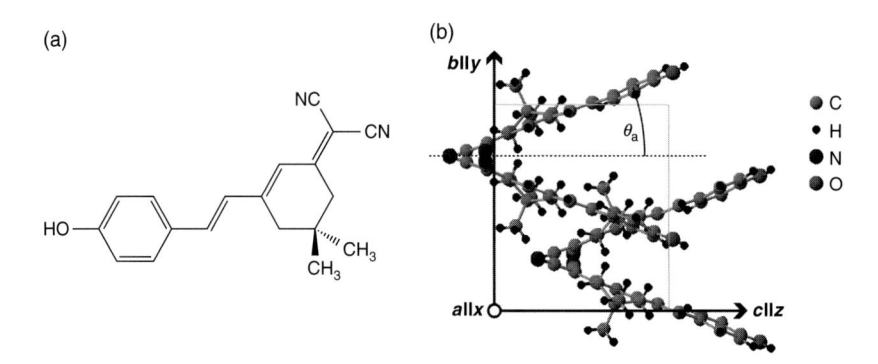

Figure 6.13 (a) OH1 molecule and (b) its crystal-packing diagram (one unit cell) projected along the a axis. The crystal structure is orthorhombic $Pna2_1$ (point group $mm2$) and the molecules make an angle of $\theta_a \cong 20°$ with respect to the polar c axis in this plane.

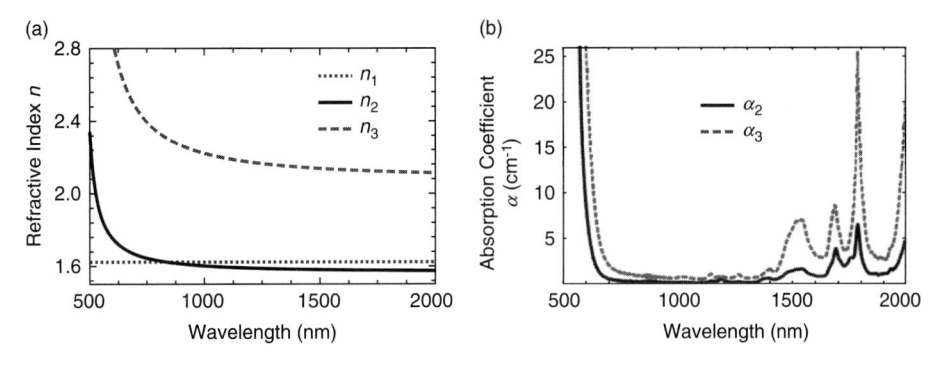

Figure 6.14 (a) Refractive indices n_1, n_2, and n_3 along the main dielectric axes x, y, and z, respectively. Note the large birefringence $\Delta n > 0.5$ between the refractive index along the polar axis z and perpendicular axes x or y, which is due to the favorable chromophore alignment along the z axis (see Fig. 6.13(b)). (b) Absorption coefficients α_2 and α_3 along the main dielectric axes y and z, respectively.

hydrogen bond donor and acceptor site, respectively. The melting temperature T_m is at about 212 °C and the thermal weight-loss temperature at about 325 °C, which involves sublimation and/or decomposition [82]. The high thermal stability of OH1 crystals is an advantage for applying melt-based crystal processing, which would be difficult for ionic salts such as DAST that decompose at the melting temperature.

The main supramolecular interactions of the OH1 crystals are strong hydrogen bonds of C=N···H−O. With the help of these two hydrogen bonding sites at opposite ends, nearly planar OH1 molecules build a hydrogen-bonded polar polymer-like chain in the crystalline solid. The polar polymer-like chains make an angle of about 20° with the polar crystallographic c axis in the bc plane, as shown in Fig. 6.13(b). OH1 crystals feature a relatively high point-group symmetry compared with other organic crystals. They belong to the orthorhombic crystal system, for which the dielectric axes coincide with the crystallographic axes. This is of advantage since it allows much simpler crystal preparation and orientation for optical characterization and applications. Most of the known highly nonlinear optical crystals feature lower crystal symmetry, most often crystallizing in monoclinic crystal systems (e.g., DAST, DSTMS, DAPSH, DAT2).

The direction of the maximum first hyperpolarizability β_{max} of the OH1 molecules is aligned at a small angle of $\theta_{3z} = 28°$ with the polar c axis, i.e., the order parameter of the OH1 crystal is $\langle \cos^3 \theta_{3z} \rangle \cong 0.7$, which results in a large macroscopic second-order optical nonlinearity.

The refractive indices of OH1 have been measured by two independent methods: by interferometric measurements with bulk crystals and by measuring transmission spectra (fringes) of thin films of OH1 based on multiple coherent reflections in the film [83].

The absorption was measured for light polarized along the polar axis z and along the y axis (Fig. 6.14(b)). The absorption is low over a broad wavelength range between 800 and 1400 nm.

Electro-optic properties of OH1: The electro-optic coefficients r_{113}, r_{223}, and r_{333} have been determined by the phase modulation technique using a Mach–Zehnder

Table 6.3 Electro-optic coefficients of OH1.

λ (nm)	632.8	785	1064	1319
r_{113} (pm/V)	13.8±0.7	10.1±0.6	7.6±0.3	6.8±0.3
r_{223} (pm/V)	90±6	52±3	35±2	30±2
r_{333} (pm/V)	109±4	75±7	56±2	52±7

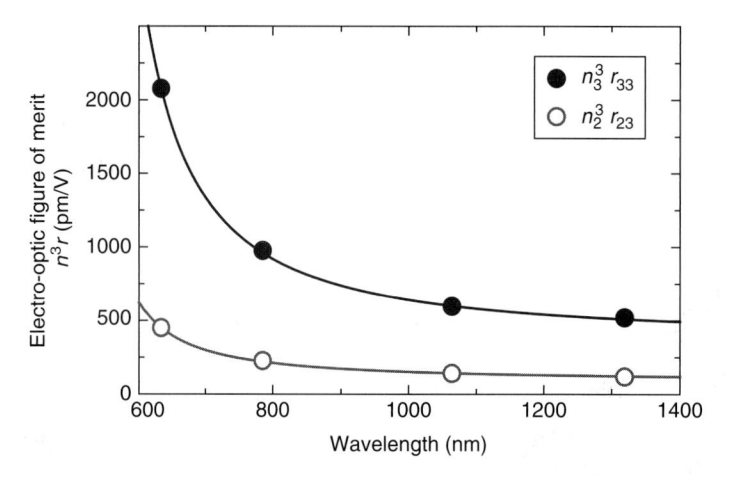

Figure 6.15 Wavelength dispersion of the electro-optic figure-of-merit $n^3 r$ measured for the electric field applied along the polar x_3 axis and polarization direction along the x_2 (empty circles) and x_3 (full circles) axis.

interferometer [83]. Owing to the alignment of the charge-transfer axis of the OH1 molecules in the crystal, the coefficients r_{ii3} can be expected to be the largest ones. The results of our measurements at different wavelengths are shown in Table 6.3. The coefficients $r_{333} = 75 \pm 8$ and 52±7 pm/V have been measured at wavelengths of 785 and 1319 nm, respectively. These values are, within experimental errors, the same as the electro-optic coefficients $r_{111} = 77 \pm 8$ and 53±6 pm/V at 800 and 1313 nm, respectively, measured in DAST [87]. The dispersion of the electro-optic figures-of-merit $n_2^3 r_{223}$ and $n_3^3 r_{333}$ is shown in Fig. 6.15. The electro-optic figure-of-merit for OH1 is $n_3^3 r_{333} = 970 \pm 100$ pm/V and 2070 ± 80 pm/V at 785 and 632.8 nm, respectively, which is among the largest measured in organic materials.

6.6 Crystal growth techniques

6.6.1 Solution, melt, and vapor crystal growth

Crystallization of organic materials is based on solution growth, melt growth, or vapor growth. To produce bulk, thin-film, or one-dimensional wire-like crystals, different and in some cases rather complex growth techniques are required. The choice of an appropriate technique depends on different material properties, as well as the desired crystalline form:

(i) The production of three-dimensional bulk, two-dimensional thin platelet, epitaxial and one-dimensional wire crystals;
(ii) The melting temperature, the solubility, and the vapor pressure;
(iii) The long-term thermal and chemical stability;
(iv) The purity of starting materials;
(v) The formation of solution or solvent inclusions and other defects.

Solution growth techniques, even for the bulk crystals, are most common, but limited to rather low growth rates (in many cases ∼0.1–1 mm/day) and in some cases solvent or solution inclusion problems. Melt growth techniques on the other hand do not possess these restrictions and exhibit relatively fast growth rates compared with solution growth techniques. Growth from the solution represents in many cases the only possibility to obtain single crystals, owing to insufficient thermal stability of the organic molecules to apply melt growth techniques. The vapor growth method is one of the simplest techniques to obtain crystals of high purity and lattice perfection. However, as a consequence of the physical vapor deposition process at a low vapor pressure, it often exhibits low growth rates in the range 0.1–0.01 mm/hr for high crystalline quality, which is mostly not sufficient for obtaining bulk crystals of a reasonable size, but can be an optimal technique in cases where thin films or wires are desired.

6.6.2 Growth of bulk crystals from solution

For most of the highly nonlinear organic materials, such as DAST, DSTMS, DAPSH, Mero-2-MDB, and OH1, best-quality bulk crystals are grown from solution. Most commonly, slow temperature-lowering techniques or slow isothermal evaporation techniques are used, which can be combined with temperature gradients at the growth position. Also, the growth by spontaneous nucleation requires different optimization compared with the seeded growth. The optimization of the growth procedure from solution depends on the thermodynamic properties of the solvent/solute system. The most important parameters are the solubility and the variation of the solubility as a function of temperature, which relate to the changes of entropy and enthalpy during dissolution. For inducing nucleation and optimizing the temperature during the growth, an important parameter is the metastable-zone range of the solvent/solute system. Different solvents will lead to different growth parameters and yields, as well as different morphology of the final crystals and in some cases even a different crystal structure/polymorph.

An example of the solution-growth process optimization for obtaining high-optical-quality bulk crystals of DAST can be found in Ref. [86]. By optimizing the cooling ramp during the growth, a constant growth rate was achieved, which improved the quality of the crystals, as well as reducing the growth time by a factor of 3; high optical-quality bulk crystals of DAST with side lengths of up to 3 cm and thickness of up to 1 cm-like required about one month for their growth. Compared with DAST,

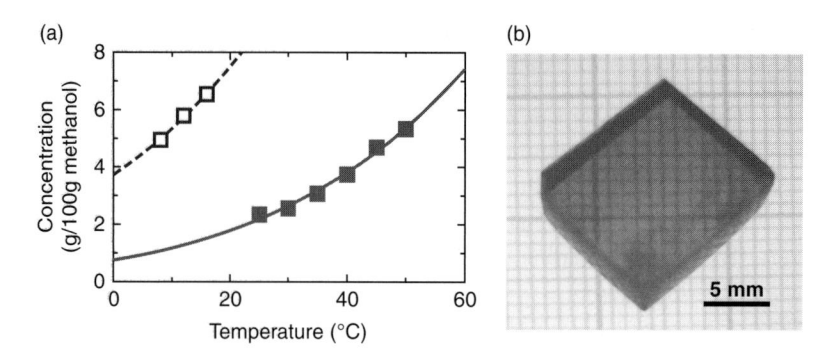

Figure 6.16 (a) OH1 solubility curve (filled squares and solid line) and the metastable-zone boundary (empty squares, dotted line) measured in methanol solvent [127]. (b) Photograph of a larger OH1 bulk crystal (4 mm thickness) grown from solution by the slow-cooling method.

growth of DSTMS crystals and OH1 crystals from solution may be faster and easier, which is due to the favorable thermodynamic properties of these materials [76,127]. The solubility and the metastable-zone range of OH1 in methanol as a function of temperature are shown in Fig. 6.16(a).

The metastable-zone width is a very important parameter for inducing both the nucleation and growth of high optical quality and large single crystals. The metastable-zone width for OH1 in methanol is very large, about 40 °C, as shown in Fig. 6.16(a). For comparison, for DAST crystals in methanol the metastable zone is only about 7 °C wide in the same temperature range. The very large metastable-zone width is related to strong hydrogen-bonding interactions between the OH1 molecules and the methanol molecules, and plays an essential role in the growth of bulk and in particular thin films, as discussed below.

6.6.3 Organic single-crystalline thin films

High-quality single-crystalline thin films of highly nonlinear optical materials are essential for the fabrication of integrated photonic devices. If starting from bulk crystals, then complicated, expensive, and time-consuming cutting, polishing, and structuring procedures are required to fabricate waveguiding devices. Obviously thin films may be much more compatible with simpler and cheaper waveguiding structures for applications such as electro-optic modulators.

For integrated optics, one is interested in fabrication techniques for organic nonlinear optical crystalline films with a thickness in the range of 0.2–10 μm and a large area. There have been several different approaches investigated for the fabrication of single crystalline films, using solution, melt, or vapor growth techniques; examples are listed in Table 6.4. More detail can be found in the respective references; an overview of different approaches for thin-film fabrication is reviewed in more detail in Ref. [128]. However, although there exist several examples and approaches for thin-film fabrication, there are only a few reports on the successful development of optical waveguiding

Table 6.4 Growth methods for thin organic single-crystalline films with second-order nonlinear optical activity

Material	Thickness (µm)	Single crystal area	Ref.
(a) Mechanical method			
Cut and polish			
DAST	20–25	~20 mm² before polishing	[99]
Etching			
DAST	~18	0.5–2 mm waveguide length	[129]
(b) Epitaxial growth method			
Solution epitaxial growth			
DAST	10–35	4 mm²	[130]
OH1	0.1–4	>2 cm²	[131]
Organic molecular beam epitaxy			
MNBA	1–4	40 mm²	[132, 133]
(c) Capillary method between two plates			
Solution capillary method with shear			
DAST	0.1–18	~1 cm²	[95]
NPP	3	1 cm²	[134]
Solution capillary method without shear			
DAST	20	12 mm²	[130]
DSTMS	30	30 mm²	[76]
DAT2	1–40	10 mm²	[124]
Melt capillary method			
MNA	5–10	2–3 mm sides	[135]
MNA	1	1 cm²	[136]
AANP	6–14	Not given	[137]
COANP	5–20	Few mm sides	[3, 138]
NPP	10	0.25 mm²	[139]
DAT2	0.025–1	7 mm waveguide length	[80]
(d) Planar solution growth method			
Two-dimensional ΔT method			
DAST	-	-	[130]
Traveling cell method			
DAST	20–40	4 mm²	[130]
Undercooled flow cell method			
DAST	40	11 mm²	[130]
(e) Vapor growth method			
DAT2	0.2–5	15 mm²	[123]
DAST	1–5	0.2 mm diameter	[140]

in reliable and stable crystalline thin films of materials with high macroscopic nonlinearity. This is because for many of the examples in Table 6.4, the single-crystalline area is too small for subsequent waveguide structuring, or the fabricated films cannot be attached with high optical quality and without breaking to an appropriate substrate, or in some cases thin crystals are inaccessible for further structuring, such as when crystals are grown between two glass plates and cannot be removed without damage. Here we present one recent example, in which thin-film single crystals of OH1 were grown

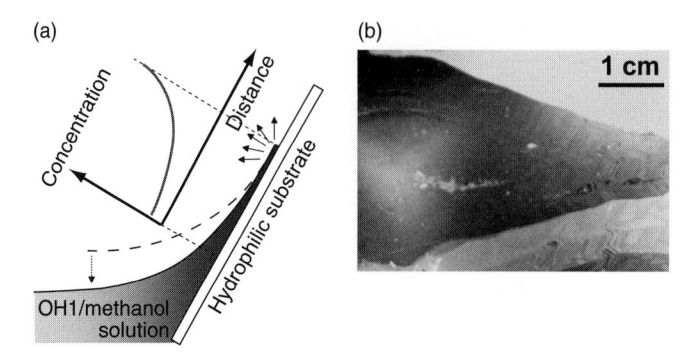

Figure 6.17 (a) Schematic of evaporation-induced local supersaturation with surface interactions (ELSSI) principle. (b) Photograph of the OH1 film grown on an OH-modified substrate from above between crossed polarizers; large single-crystalline domains can be observed [131].

epitaxially on various substrates. These crystals are promising for further waveguide processing, as well as for a cladding modulation concept using silicon nanowire waveguides, both of which have already been demonstrated (see Chapter 9).

6.6.3.1 Evaporation-induced local supersaturation growth

A new method for thin-film deposition, called evaporation-induced local supersaturation with surface interactions (ELSSI), was recently developed [131]. This method allows for relatively large-area (>2 cm^2) single-crystalline thin films (0.1–4 μm thickness) on various substrates, including glass, gold, chromium, and silicon. The adhesion of the crystal to the substrate is very strong, allowing mechanical polishing of the top surface without removing the film, and also lithographic processes to fabricate different microstructures. Thin films can also be deposited directly on the structured substrates, which is particularly attractive for silicon-on-insulator nanowire optical waveguides. The optimal growth of thin organic films on substrates was first developed for OH1 single crystals. Hydrophilic substrates allow good attachment of the OH1 molecules from OH1/methanol solutions [131]. Using this method, the substrates with modified surfaces are immersed in the OH1/methanol solutions at a constant temperature ($\sim$40 °C), which is then left for slow evaporation. OH1 crystals nucleate and grow on the modified substrates. A resulting OH1 crystal grown directly on an amorphous glass substrate is shown in Fig. 6.17(b) [131].

The ELSSI principle is illustrated in Fig. 6.17(a). For an appropriately treated substrate, the surface tension is very low, which results in strong wetting by the solution. In this thin wetting layer, the evaporation is enhanced, which leads to locally high supersaturation and will eventually result in a nucleation. The solution concentration used here is considerably lower than saturated (about one order of magnitude lower), so that the nucleation can only occur within the thin wetting layer. After nucleation, OH1 crystals continuously grow by following the evaporation direction on the substrate without additional nucleation, indicating that the concentration of the solution in the wetting layer remains within the solution metastable zone. Owing to

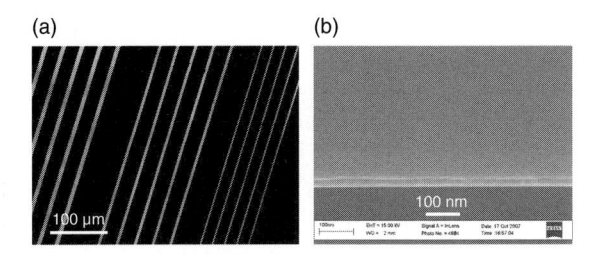

Figure 6.18 (a) Transmission microscope image of approximately 25-nm-thick DAT2 crystalline stripes grown from the melt, as seen between crossed polarizers. (b) Scanning electron micrograph of the corresponding end-facet (from Figi et al. [80]).

the very large metastable zone of OH1 (see Fig. 6.16), additional nucleation points usually do not occur, and very large-area single crystalline domains may result.

The photograph of Fig. 6.17(b) was taken between crossed polarizers, and therefore areas of the same shade represent single-crystalline domains. These domains reach lengths of over 4 cm and thicknesses in the range of 0.1–4 µm with good optical quality, suitable for the fabrication of photonic devices.

For applications, one of the important factors is the orientation of the films. X-ray diffraction and optical transmission investigations [131] have shown that the *a* axis is perpendicular to the surface of the films and that thin films grown this way are oriented with the polar axis *c* close to the evaporation/growth direction with a small variation for different single-crystalline domains ($\sim 20°$). This can be related to the fact that the growth rate of OH1 is the fastest along the polar axis direction [127].

6.6.3.2 Single-crystalline micro- and nano-wires

One of the most attractive solutions for integrated photonics devices is direct growth of the desired micro- and nanostructures in the desired position. This can be done by first structuring standard inorganic templates made of materials such as glass, silicon and electrode materials, with void structures at positions where active organic crystalline materials are desired. This method was recently demonstrated by using melt-processable materials, namely the configurationally locked polyene DAT2 [80] and the small chromophore COANP [141]; both materials were chosen because of their favorable growth characteristics from the melt as well as the tendency for thin-film formation (see Table 6.4). By this method, single-crystalline wires several mm long with a thickness ranging from several µm down to below 30 nm have been obtained, as shown in Fig. 6.18. More details on fabrication are given in Chapter 9.

References

[1] L.R. Dalton, P.A. Sullivan, and D.H. Bale, *Chem. Rev.*, **110**, 25 (2010).
[2] J. Zyss, *Molecular Nonlinear Optics: Materials, Physics, Devices*, Boston, MA, Academic Press (1994).

[3] C. Bosshard, K. Sutter, P. Prêtre, *et al.*, *Organic Nonlinear Optical Materials*, Amsterdam, Gordon & Breach Science Publishers (1995).
[4] J. Zyss and J.F. Nicoud, *Curr. Opin. Solid State Mater. Sci.*, **1**, 533 (1996).
[5] C. Bosshard and P. Günter, in *Nonlinear Optics of Organic Molecules and Polymers*, H. S. Nalwa and S. Miyata, Eds., Boca Raton, FL, CRC Press (1997), p. 391.
[6] D. M. Burland, R. D. Miller, and C. A. Walsh, *Chem. Rev.*, **94**, 31 (1994).
[7] S. R. Marder, L. T. Cheng, B. G. Tiemann, *et al.*, *Science*, **263**, 511 (1994).
[8] C. Bosshard, M. Bösch, I. Liakatas, M. Jäger, and P. Günter, in *Nonlinear Optical Effects and Materials*, vol. 72, P. Günter, Ed., Berlin Heidelberg NewYork, Springer Series in Optical Science (2000), pp. 163–300.
[9] M. G. Kuzyk, *Phys. Rev. Lett.*, **85**, 1218 (2000).
[10] K. Clays and B. J. Coe, *Chem. Mater.*, **15**, 642 (2003).
[11] L. R. Dalton, *J. Phys.: Condens. Matter*, **15**, R897 (2003).
[12] H. S. Nalwa and S. Miyata, Eds., *Nonlinear Optics of Organic Molecules and Polymers*, Boca Raton, FL, CRC Press (1997).
[13] M. Jazbinsek and P. Günter, in *Introduction to Organic Electronic and Optoelectronic Materials and Devices*, S. Sun and L. Dalton, Eds., Boca Raton, FL, CRC Press (2008), pp. 421–466.
[14] A.I. Kitaigorodskii, *Molecular Crystals and Molecules*, New York, Academic Press (1973).
[15] T. Tsunekawa, T. Gotoh, and M. Iwamoto, *Chem. Phys. Lett.*, **166**, 353 (1990).
[16] L.T. Cheng, W. Tam, S.H. Stevenson, *et al.*, *J. Phys. Chem.*, **95**, 10631 (1991).
[17] O.P. Kwon, B. Ruiz, A. Choubey, *et al.*, *Chem. Mater.*, **18**, 4049 (2006).
[18] G.R. Meredith, in *Nonlinear Optical Properties of Organic and Polymeric Materials*, D.J. Williams, Ed., Washington, DC, ACS Symposium Series Vol. 233 (1983) p. 27.
[19] S. Okada, A. Masaki, H. Matsuda, *et al.*, *Jpn. J. Appl. Phys.* 1, **29**, 1112 (1990).
[20] S.R. Marder, J.W. Perry, and W.P. Schaefer, *Science*, **245**, 626 (1989).
[21] S.R. Marder, J.W. Perry, and C.P. Yakymyshyn, *Chem. Mater.*, **6**, 1137 (1994).
[22] S. Okada, K. Nogi, Anwar, *et al.*, *Jpn. J. Appl. Phys.*, **42**, 668 (2003).
[23] Z. Yang, M. Jazbinsek, B. Ruiz, *et al.*, *Chem. Mater.*, **19**, 3512 (2007).
[24] B.J. Coe, J.A. Harris, I. Asselberghs, *et al.*, *Adv. Funct. Mater.*, **13**, 347 (2003).
[25] H. Figi, L. Mutter, C. Hunziker, *et al.*, *J. Opt. Soc. Am. B*, **25**, 1786 (2008).
[26] M.S. Wong, U. Meier, F. Pan, *et al.*, *Adv. Mater.*, **8**, 416 (1996).
[27] I. Liakatas, M.S. Wong, V. Gramlich, C. Bosshard, and P. Gunter, *Adv. Mater.*, **10**, 777 (1998).
[28] O.P. Kwon, M. Jazbinsek, J.I. Seo, *et al.*, *J. Phys. Chem. C*, **113**, 15405 (2009).
[29] M.C. Etter and G.M. Frankenbach, *Chem. Mater.*, **1**, 10 (1989).
[30] F. Pan, M.S. Wong, V. Gramlich, C. Bosshard, and P. Gunter, *J. Am. Chem. Soc.*, **118**, 6315 (1996).
[31] M.S. Wong, F. Pan, V. Gramlich, C. Bosshard, and P. Gunter, *Adv. Mater.*, **9**, 554 (1997).
[32] M.S. Wong, F. Pan, M. Bosch, *et al.*, *J. Opt. Soc. Am. B*, **15**, 426 (1998).
[33] C. Cassidy, J.M. Halbout, W. Donaldson, and C.L. Tang, *Opt. Commun.*, **29**, 243 (1979).
[34] J.A. Morrell, A.C. Albrecht, K.H. Levin, and C.L. Tang, *J. Chem. Phys.*, **71**, 5063 (1979).
[35] J.M. Halbout, S. Blit, W. Donaldson, and C.L. Tang, *IEEE J. Quantum Electron.*, **15**, 1176 (1979).
[36] K. Betzler, H. Hesse, and P. Loose, *J. Mol. Struct.*, **47**, 393 (1978).
[37] I. Ledoux and J. Zyss, *Chem. Phys.*, **73**, 203 (1982).

[38] I. Ledoux, J. Zyss, A. Jutand, and C. Amatore, *Chem. Phys.*, **150**, 117 (1991).
[39] J. Zyss, I. Ledoux, M. Bertault, and E. Toupet, *Chem. Phys.*, **150**, 125 (1991).
[40] C. Bosshard, G. Knopfle, P. Pretre, and P. Gunter, *J. Appl. Phys.*, **71**, 1594 (1992).
[41] B.F. Levine, C.G. Bethea, C.D. Thurmond, R.T. Lynch, and J.L. Bernstein, *J. Appl. Phys.*, **50**, 2523 (1979).
[42] G.F. Lipscomb, A.F. Garito, and R.S. Narang, *J. Chem. Phys.*, **75**, 1509 (1981).
[43] R. Morita, N. Ogasawara, S. Umegaki, and R. Ito, *Jpn. J. Appl. Phys., 2*, **26**, L1711 (1987).
[44] E.S.S. Ho, K. Iizuko, A.P. Freundorfer, and C.K.L. Wah, *J. Lightwave Technol.*, **9**, 101 (1991).
[45] J.O. White, D. Hulin, M. Joffre, et al., *Appl. Phys. Lett.*, **64**, 264 (1994).
[46] S.M. Rao, K.X. He, R.B. Lal, et al., *J. Mater. Sci.*, **30**, 179 (1995).
[47] P. Kerkoc, M. Zgonik, K. Sutter, C. Bosshard, and P. Gunter, *Appl. Phys. Lett.*, **54**, 2062 (1989).
[48] P. Kerkoc, M. Zgonik, K. Sutter, C. Bosshard, and P. Gunter, *J. Opt. Soc. Am. B*, **7**, 313 (1990).
[49] A. Carenco, J. Jerphagnon, and A. Perigaud, *J. Chem. Phys.*, **66**, 3806 (1977).
[50] J.G. Bergman and G.R. Crane, *J. Chem. Phys.*, **66**, 3803 (1977).
[51] Stevenso Jl, *J. Phys. D*, **6**, L13 (1973).
[52] J. Zyss, D.S. Chemla, and J.F. Nicoud, *J. Chem. Phys.*, **74**, 4800 (1981).
[53] M. Sigelle and R. Hierle, *J. Appl. Phys.*, **52**, 4199 (1981).
[54] R. Hierle, J. Badan, and J. Zyss, *J. Cryst. Growth*, **69**, 545 (1984).
[55] I. Ledoux, C. Lepers, A. Perigaud, J. Badan, and J. Zyss, *Opt. Commun.*, **80**, 149 (1990).
[56] C. Bosshard, K. Sutter, R. Schlesser, and P. Gunter, *J. Opt. Soc. Am. B*, **10**, 867 (1993).
[57] K. Sutter, C. Bosshard, W.S. Wang, G. Surmely, and P. Gunter, *Appl. Phys. Lett.*, **53**, 1779 (1988).
[58] M. Fujiwara, M. Maruyama, M. Sugisaki, et al., *Jpn. J. Appl. Phys. Part 1*, **46** (4A), 1528 (2007).
[59] M. Fujiwara, K. Yanagi, M. Maruyama, et al., *Jpn. J. Appl. Phys. Part 1*, **45**, 8676 (2006).
[60] H. Hashimoto, Y. Okada, H. Fujimura, et al., *Jpn. J. Appl. Phys. Part 1*, **36**, 6754 (1997).
[61] C. Bosshard, K. Sutter, P. Gunter, and G. Chapuis, *J. Opt. Soc. Am. B*, **6**, 721 (1989).
[62] P. Gunter, C. Bosshard, K. Sutter, et al., *Appl. Phys. Lett.*, **50**, 486 (1987).
[63] R.T. Bailey, F.R. Cruickshank, S.M.G. Guthrie, et al., *Opt. Commun.*, **65**, 229 (1988).
[64] R.T. Bailey, G.H. Bourhill, F.R. Cruickshank, et al., *J. Appl. Phys.*, **75**, 489 (1994).
[65] T. Kondo, R. Morita, N. Ogasawara, S. Umegaki, and R. Ito, *Jpn. J. Appl. Phys. 1*, **28**, 1622 (1989).
[66] T. Kondo, F. Akase, M. Kimagai, and R. Ito, *Opt. Rev.*, **2**, 128 (1995).
[67] S. Tomaru, S. Matsumoto, T. Kurihara, et al., *Appl. Phys. Lett.*, **58**, 2583 (1991).
[68] A. Yokoo, S. Tomaru, I. Yokohama, H. Itoh, and T. Kaino, *J. Cryst. Growth*, **156**, 279 (1995).
[69] J.D. Bierlein, L.K. Cheng, Y. Wang, and W. Tam, *Appl. Phys. Lett.*, **56**, 423 (1990).
[70] G. Knopfle, C. Bosshard, R. Schlesser, and P. Gunter, *IEEE J. Quantum Electron.*, **30**, 1303 (1994).
[71] S. Okada, A. Masaki, H. Matsuda, et al., *Proc. SPIE*, **1337**, 178 (1990).
[72] X.M. Duan, S. Okada, H. Nakanishi, et al., *Proc. SPIE*, **2143**, 41 (1994).
[73] G. Knöpfle, R. Schlesser, R. Ducret, and P. Günter, *Nonlinear Opt.*, **9**, 143–149 (1995).
[74] F. Pan, G. Knopfle, C. Bosshard, et al., *Appl. Phys. Lett.*, **69**, 13 (1996).
[75] U. Meier, M. Bosch, C. Bosshard, F. Pan, and P. Gunter, *J. Appl. Phys.*, **83**, 3486 (1998).

[76] Z. Yang, L. Mutter, M. Stillhart, *et al.*, *Adv. Funct. Mater.*, **17**, 2018 (2007).
[77] L. Mutter, F.D.J. Brunner, Z. Yang, M. Jazbinsek, and P. Gunter, *J. Opt. Soc. Am. B*, **24**, 2556 (2007).
[78] B.J. Coe, J.A. Harris, I. Asselberghs, *et al.*, *Adv. Funct. Mater.*, **12**, 110 (2002).
[79] S. Follonier, C. Bosshard, U. Meier, *et al.*, *J. Opt. Soc. Am. B*, **14**, 593 (1997).
[80] H. Figi, M. Jazbinsek, C. Hunziker, M. Koechlin, and P. Gunter, *Opt. Express*, **16**, 11310 (2008).
[81] H. Figi, M. Jazbinsek, C. Hunziker, M. Koechlin, and P. Gunter, *Proc. SPIE*, **7599**, 75991N-1 (2010).
[82] O.P. Kwon, S.J. Kwon, M. Jazbinsek, *et al.*, *Adv. Funct. Mater.*, **18**, 3242 (2008).
[83] C. Hunziker, S.J. Kwon, H. Figi, *et al.*, *J. Opt. Soc. Am. B*, **25**, 1678 (2008).
[84] H. Nakanishi, H. Matsuda, S. Okada, and M. Kato, in *Materials Research Society International Meeting on Advanced Materials*, M. Doyama, S. Somiya, and R.P.H. Chang, Eds., Pittsburgh, PA, Materials Research Society (1989), pp. 97–104.
[85] F. Pan, M.S. Wong, C. Bosshard, and P. Gunter, *Adv. Mater.*, **8**, 592 (1996).
[86] B. Ruiz, M. Jazbinsek, and P. Gunter, *Cryst. Growth Des.*, **8**, 4173 (2008).
[87] M. Jazbinsek, L. Mutter, and P. Gunter, *IEEE J. Sel. Top. Quantum Electron.*, **14**, 1298 (2008).
[88] C. Bosshard, R. Spreiter, L. Degiorgi, and P. Gunter, *Phys. Rev. B*, **66**, 205107 (2002).
[89] M. Jazbinsek and M. Zgonik, *Appl. Phys. B*, **74**, 407 (2002).
[90] M. Zgonik, R. Schlesser, I. Biaggio, *et al.*, *J. Appl. Phys.*, **74**, 1287 (1993).
[91] R. Spreiter, C. Bosshard, F. Pan, and P. Gunter, *Opt. Lett.*, **22**, 564 (1997).
[92] J.L. Oudar and D.S. Chemla, *J. Chem. Phys.*, **66**, 2664 (1977).
[93] J. Zyss and J.L. Oudar, *Phys. Rev. A*, **26**, 2028 (1982).
[94] K.D. Singer, M.G. Kuzyk, and J.E. Sohn, *J. Opt. Soc. Am. B*, **4**, 968 (1987).
[95] M. Thakur, J.J. Xu, A. Bhowmik, and L.G. Zhou, *Appl. Phys. Lett.*, **74**, 635 (1999).
[96] M. Thakur, A. Mishra, J. Titus, and A.C. Ahyi, *Appl. Phys. Lett.*, **81**, 3738 (2002).
[97] M. Thakur, J. Titus, and A. Mishra, *Opt. Eng.*, **42**, 456 (2003).
[98] R.K. Swamy, S.P. Kutty, J. Titus, S. Khatavkar, and M. Thakur, *Appl. Phys. Lett.*, **85**, 4025 (2004).
[99] F. Pan, K. McCallion, and M. Chiappetta, *Appl. Phys. Lett.*, **74**, 492 (1999).
[100] S. Sohma, H. Takahashi, T. Taniuchi, and H. Ito, *Chem. Phys.*, **245**, 359 (1999).
[101] X. Zheng, S. Wu, R. Sobolewski, *et al.*, *Appl. Phys. Lett.*, **82**, 2383 (2003).
[102] Z. Yang, S. Aravazhi, A. Schneider, *et al.*, *Adv. Funct. Mater.*, **15**, 1072 (2005).
[103] B. Ruiz, Z. Yang, V. Gramlich, M. Jazbinsek, and P. Gunter, *J. Mater. Chem.*, **16**, 2839 (2006).
[104] K. Akiyama, S. Okada, Y. Goto, and H. Nakanishi, *J. Cryst. Growth*, **311**, 953 (2009).
[105] J. Ogawa, S. Okada, Z. Glavcheva, and H. Nakanishi, *J. Cryst. Growth*, **310**, 836 (2008).
[106] Z. Glavcheva, H. Umezawa, Y. Mineno, *et al.*, *Jpn. J. Appl. Phys. Part 1*, **44**, 5231 (2005).
[107] T. Taniuchi, S. Ikeda, S. Okada, and H. Nakanishi, *Jpn. J. Appl. Phys. 2*, **44**, L652 (2005).
[108] C. Bosshard, R. Spreiter, and P. Gunter, *J. Opt. Soc. Am. B*, **18**, 1620 (2001).
[109] S.K. Kurtz and T.T. Perry, *J. Appl. Phys.*, **39**, 3798 (1968).
[110] M. Stillhart, A. Schneider, and P. Gunter, *J. Opt. Soc. Am. B*, **25**, 1914 (2008).
[111] B.J. Coe, D. Beljonne, H. Vogel, J. Garin, and J. Orduna, *J. Phys. Chem. A*, **109**, 10052 (2005).
[112] B. Ruiz, B.J. Coe, R. Gianotti, *et al.*, *Cryst. Eng. Comm.*, **9**, 772 (2007).
[113] S.R. Marder, D.N. Beratan, and L.T. Cheng, *Science*, **252**, 103 (1991).

[114] C.F. Shu, W.J. Tsai, and A.K.Y. Jen, *Tetrahedron Lett.*, **37**, 7055 (1996).
[115] S. Ermer, S.M. Lovejoy, D.S. Leung, et al., *Chem. Mater.*, **9**, 1437 (1997).
[116] C.F. Shu, Y.C. Shu, Z.H. Gong, et al., *Chem. Mater.*, **10**, 3284 (1998).
[117] Y.C. Shu, Z.H. Gong, C.F. Shu, et al., *Chem. Mater.*, **11**, 1628 (1999).
[118] K. Staub, G.A. Levina, S. Barlow, et al., *J. Mater. Chem.*, **13**, 825 (2003).
[119] U. Lawrentz, W. Grahn, K. Lukasznk, et al., *Chem. Eur. J.*, **8**, 1573 (2002).
[120] O.P. Kwon, S.J. Kwon, M. Jazbinsek, et al., *Adv. Funct. Mater.*, **17**, 1750 (2007).
[121] O.P. Kwon, S.J. Kwon, M. Stillhart, et al., *Cryst. Growth Des.*, **7**, 2517 (2007).
[122] S.J. Kwon, O.P. Kwon, J.I. Seo, et al., *J. Phys. Chem. C*, **112**, 7846 (2008).
[123] A. Choubey, O.P. Kwon, M. Jazbinsek, and P. Gunter, *Cryst. Growth Des.*, **7**, 402 (2007).
[124] O.P. Kwon, S.J. Kwon, H. Figi, M. Jazbinsek, and P. Gunter, *Adv. Mater.*, **20**, 543 (2008)
[125] R. Lemke, *Chem. Ber. Recl.* **103**, 1894 (1970).
[126] T. Kolev, Z. Glavcheva, D. Yancheva, et al., *Acta Crystallogr. E: Struct. Rep. Online*, **57**, o561 (2001).
[127] S.J. Kwon, M. Jazbinsek, O.P. Kwon, and P. Gunter, *Cryst. Growth Des.*, **10**, 1552 (2010).
[128] M. Jazbinsek, O.P. Kwon, C. Bosshard, and P. Gunter, in *Handbook of Organic Electronics and Photonics*, H.S. Nalwa, Ed., American Scientific Publishers (2008).
[129] K. Takayama, K. Komatsu, and T. Kaino, *Jpn. J. Appl. Phys.*, *1*, **40**, 5149 (2001).
[130] S. Manetta, M. Ehrensperger, C. Bosshard, and P. Gunter, *C. R. Physique*, **3**, 449 (2002).
[131] S.J. Kwon, C. Hunziker, O.P. Kwon, M. Jazbinsek, and P. Gunter, *Cryst. Growth Des.*, **9**, 2512 (2009).
[132] T. Yamashiki and K. Tsuda, *Opt. Lett.*, **28**, 316 (2003).
[133] T. Yamashiki, S. Fukuda, K. Tsuda, and T. Gotoh, *Appl. Phys. Lett.*, **83**, 605 (2003).
[134] L.G. Zhou and M. Thakur, *J. Mater. Res.*, **13**, 131 (1998).
[135] Y. Kubota and T. Yoshimura, *Appl. Phys. Lett.*, **53**, 2579 (1988).
[136] S. Gauvin and J. Zyss, *J. Cryst. Growth*, **166**, 507 (1996).
[137] P.M. Ushasree, K. Komatsu, T. Kaino, and T. Taima, *Mol. Cryst. Liq. Cryst.*, **406**, 313 (2003).
[138] A. Leyderman and Y.L. Cui, *Opt. Lett.*, **23**, 909 (1998).
[139] Z.F. Liu, S.S. Sarkisov, M.J. Curley, A. Leyderman, and C. Lee, *Opt. Eng.*, **42**, 803 (2003).
[140] M. Baldo, M. Deutsch, P. Burrows, et al., *Adv. Mater.*, **10**, 1505 (1998).
[141] H. Figi, M. Jazbinsek, C. Hunziker, M. Koechlin, and P. Gunter, *J. Opt. Soc. Am. B*, **26**, 1103 (2009).

7 Electrically poled organic materials and thermo-optic materials

Most prototype devices based on organic electro-optic (OEO) materials have been fabricated from materials prepared by electric field poling of dipolar molecules embedded in or covalently attached to polymeric or dendritic (macromolecular) material lattices. Such device-relevant electro-optic activity requires macroscopic as well as molecular non-centrosymmetric (acentric) symmetry and can be represented by the following equation (for the principal element, r_{zzz}, r_{333}, or r_{33}, of the electro-optic tensor of a poled material),

$$r_{33} = N\beta(\varepsilon, \omega) <\cos^3\theta> g(\varepsilon, n) \tag{7.1}$$

where N is the chromophore number density (molecules/cc), $\beta(\varepsilon,\omega)$ is the molecular first hyperpolarizability, $<\cos^3\theta>$ is the acentric (non-centrosymmetric) order parameter, and $g(\varepsilon,n)$ is the modified Lorentz–Onsager factor. ε is the dielectric permittivity of the material and ω is the optical frequency at which the electro-optic activity is evaluated (in other words, the operational optical frequency). The r_{33}, or more correctly the r_{333} (or r_{zzz}), tensor element corresponds to the principal elements of the molecular optical nonlinearity (β_{zzz}), the electrical field, and the optical field, all lying in the same (z) direction. For poled polymers, there are two non-zero elements of the electro-optic tensor (r_{33} and r_{13}). An attempt is usually made to utilize the larger element, r_{33}, in device applications, but some devices depend on a combination of these two elements, and other device configurations require use of the r_{13} element. For materials with low acentric order, $r_{33} \approx 3(r_{13})$.

To understand electro-optic activity in poled materials, it is necessary to understand the role played by intermolecular electrostatic interactions in determining poling-induced acentric order. It is also interesting to note that poled organic electro-optic materials represent excellent model systems for studying intermolecular electrostatic interactions and for testing various theoretical (statistical mechanical) methods to analyze such interactions. Poled materials exhibit non-zero values of both centric (even) and acentric (odd) order parameters, and lattice dimensionality can be defined from the ratio of these order parameters. Intermolecular electrostatic interactions also define dynamical properties such as viscoelasticity. A variety of experimental techniques can be employed to measure order parameters and viscoelastic properties, and the calculation of these static and dynamic properties by Monte Carlo (MC) and molecular dynamics (MD) theoretical methods provides an excellent test of such computational approaches and provides confidence for utilizing these theoretical methods for guiding the design of new materials.

A diverse array of poled organic electro-optic materials has been investigated, and each type of material requires different levels of theoretical treatment. Taking as the starting point for our consideration the treatment of the frequency (ω) and dielectric permittivity (ε) dependences of $\beta(\varepsilon,\omega)$ by quantum mechanical methods discussed in Chapter 4, we analyze electro-optic activity for (1) chromophore/polymer composites (chromophores physically dissolved in host polymer materials); (2) chromophores covalently incorporated into polymers, dendrimers, and dendronized polymers including heavily crosslinked materials; and (3) materials where specific spatially anisotropic intermolecular electrostatic interactions are introduced to enhance poling efficiency (called matrix-assisted-poling (MAP) materials), which include binary chromophore (BC) materials (chromophores physically incorporated as guests into chromophore-containing host materials) and materials prepared by laser-assisted electric field poling (LAEFP or LAP). For such materials, intermolecular electrostatic interactions may act to reduce freedom of chromophore motion, which can also be thought of as influencing the dimensionality of the lattice that surrounds the electro-optic chromophore. This restriction of chromophore motion or reduction in lattice dimensionality acts to increase poling-induced acentric order and thus electro-optic activity. Strong intermolecular electrostatic interactions are critical for the formation of crystalline electro-optic materials and materials formed by sequential-synthesis/self-assembly. Thus, the utilization of such interactions with electrically poled materials is a natural evolution and permits the properties of poled materials to approach some of the characteristics of materials discussed in Chapters 5 and 6. We will also discuss the interdependence of electro-optic activity on chromophore number density and dipole moment, explicitly showing that two maxima occur in a three-dimensional plot of r_{33} vs. number density vs. dipole moment. That is, for high-dipole-moment chromophores, the largest electro-optic activity can be observed for high number densities while the converse is true for intermediate-dipole-moment chromophores. Of course, it is important to consider properties other than electro-optic activity (such as optical loss, thermal stability, and photostability) to understand the practical utility of various types of materials. We defer discussion of these auxiliary properties to later in this chapter and turn our attention immediately to optimization of electro-optic activity.

7.1 Chromophore/polymer composites

One of the fastest ways to prepare organic electro-optic materials for prototype device fabrication and evaluation is to dissolve commercially available or newly synthesized chromophores into commercially available polymer materials such as polymethylmethacrylates (PMMAs), polycarbonates (PCs) including amorphous polycarbonate (APC), polyimides (PIs), or polyquinones (PQs). Electro-optic thin films of such materials are prepared by dissolving both chromophore guest and polymer host in a spin-casting solvent, depositing a thin film by spin casting on an appropriate substrate

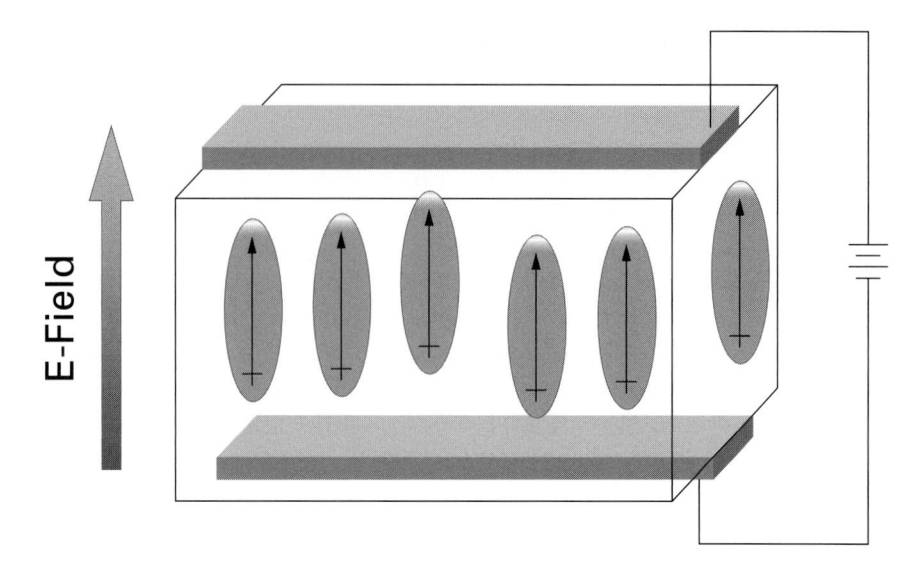

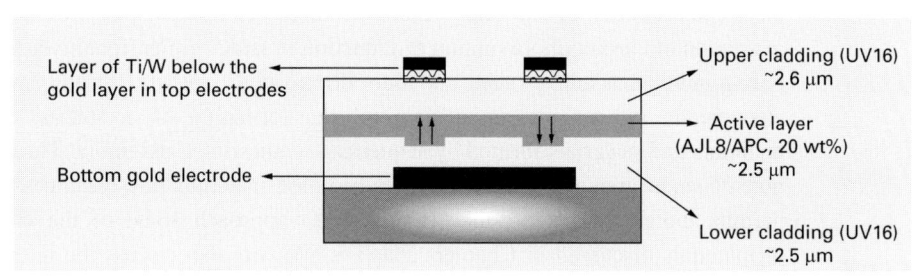

Figure 7.1 Top: A crude schematic of the poling of an electro-optic chromophore. Bottom: Cross-sectional view of a typical electro-optic device, showing poling electrodes, cladding layers, and organic electro-optic material. The very bottom layer is silicon. The black vertical arrows indicate the poling field directions (ideally the chromophore directions) in the two arms of the Mach–Zehnder modulator. This configuration is referred to as a push–pull modulator and requires half the drive voltage of a Mach–Zehnder device where modulation is applied only to one arm. UV16, UV-curable polymer; AJL8, chromophore dissolved in amorphous polycarbonate.

(such as electrically conductive indium tin oxide, ITO), and applying a poling field as the material is heated to near its glass transition temperature. Application of a poling field is typically accomplished by bringing a corona needle near the film (ITO acting as the ground electrode) or by depositing a layer of gold (or other appropriate electrode material) on top of the chromophore/polymer composite layer and then executing parallel plate poling (ITO again serving as the ground electrode). This is illustrated very crudely in Fig. 7.1 (top). In devices, top and bottom electrodes are frequently gold, as illustrated in Fig. 7.1 (bottom).

A fundamental consideration in creating chromophore/polymer composites is to choose host polymers with sufficiently high glass transition temperatures (e.g., $T_g > 130\ °C$) to permit adequate thermal stability to be achieved for prototype devices operating at temperatures below 85 °C (Telcordia standards) (see Section 7.6). Adding chromophores to a polymer host will cause plasticization (reducing T_g) of the composite materials relative to that of the host polymer, so starting with a host polymer that has a high T_g is important. Another important consideration is that there is sufficient compatibility of the dopant chromophore with the polymer host, and of both guest and host materials with the spin-casting solvent, so that good optical-quality glasses are formed. For devices to be of practical interest, total insertion loss must be less than 6–7 dB, which in turn implies that for active device lengths of 2–3 centimeters, electro-optic waveguide loss must be less than 2 dB/cm. Of course, parameters are inter-related, so it can be noted that large electro-optic coefficients permit shorter device lengths, and thus somewhat higher optical loss can be tolerated. As we shall see later in this chapter, organic electro-optic materials incorporated into silicon slot waveguides and into plasmonic waveguides involve shorter (e.g., millimeter or micrometer) device lengths, and thus optical loss of the organic material is less of a concern. For example, plasmonic waveguide losses are of the order of 0.5 dB/μm, which is much larger than the typical <0.02 dB/μm loss of organic electro-optic materials. That is, for a 10-μm-length plasmonic device, the contribution from an electro-optic material would be 0.2 dB, which is an insignificant contribution to the observed insertion loss of 12.5 dB.

In the early 1990s, it was generally assumed that chromophores, doped into conventional polymers to form composite materials, acted as independent particles influenced only by the applied poling field and thermal motions. This "independent particle approximation" (also called the gas model) assumes that the poling-induced order is not influenced by chromophore–chromophore interactions or by interactions between the chromophore guest and the polymer host. The result is that the order parameter is given by

$$<\cos^3\theta> = L_3(\mu f(\omega) E_{pol}/kT) \sim \mu f(\omega) E_{pol}/5kT \qquad (7.2)$$

where L is the Langevin function, E_{pol} is the applied electric poling field, μ is the chromophore dipole moment, $f(\omega)$ is a local field factor taking into account the effect of the dielectric environment defining the effective poling field felt by a chromophore, and kT is the thermal energy at the Kelvin poling temperature, T. The final approximate equality requires the additional assumption that the thermal energy is large compared with the energy of the chromophore dipole moment in the electric poling field (μE_{pol}). Such behavior is also referred to as the "Langevin" limit. For chromophores with dipole moments of 7 Debye or less and for modest chromophore loading (less than 20% guest/host by weight), the independent particle approximation is a useful and reliable assumption. An example of a system that essentially satisfies the independent particle approximation is the Disperse Red 1 (DR1) chromophore (see Fig. 7.2) dissolved in conventional, commercially available polymers such as PMMA, APC, and PI. However, when chromophores with dipole moments greater than 10 Debye are loaded into polymers at greater than 20wt% loading, significant deviation from the independent particle (Langevin limit) prediction is observed.

In 1997, Dalton and co-workers [1] suggested a simple analytical approach to the treatment of the effect of chromophore–chromophore intermolecular electrostatic interactions on the acentric order parameter $<\cos^3\theta>$ and thus the observed electro-optic

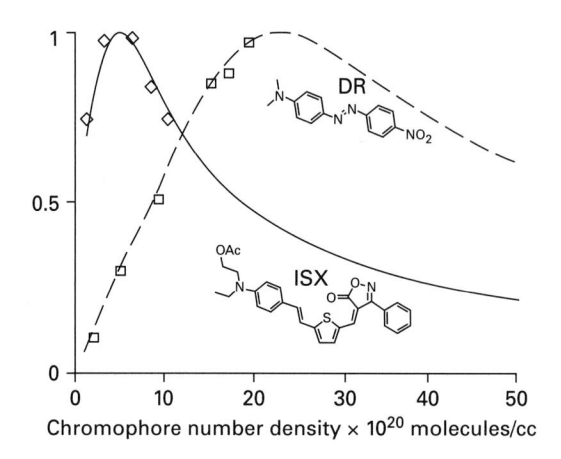

Figure 7.2 Overlay of normalized theoretical and experimental electro-optic coefficients for DR ($\mu\beta = 586 \times 10^{-48}$ esu, dashed line, squares) and the electro-optic chromphore ISX ($\mu\beta = 1960 \times 10^{-48}$ esu, solid line, diamonds) chromophores [1].

activity. They assumed a point dipole approximation for the treatment of intermolecular chromophore electronic electrostatic interactions, which leads to the following modification of the order parameter expression.

$$<\cos^3\theta> = L_3(\mu f(\omega)E_{pol}/kT)[1 - L^2(W/kT)] \quad (7.3)$$

where W is the chromophore dipole–dipole interaction energy. In the point dipole interaction approximation, the order parameter is predicted to decrease with increasing chromophore number density (note dipole–dipole interactions increase with a N^2 dependence). This, in turn, results in the prediction that electro-optic activity will transition from a linear dependence on N at low chromophore concentrations to an approximate inverse (N^{-1}) dependence at high concentrations. The maximum in the plot of r_{33} versus N is predicted to occur at $N_{max} \sim kT/\mu^2$.

A chromophore with improved $\beta(\varepsilon,\omega)$ is likely also to have an enhanced dipole moment, μ, so the practical consequence of the development of improved chromophores is often simply that the maximum in electro-optic activity (or equivalently, $N<\cos^3\theta>$) vs. N is shifted to lower chromophore concentration so that no net increase in material electro-optic activity is observed (see Fig. 7.2). Fortunately, chromophores do not behave as point dipoles, and the next approximation made by Dalton and co-workers [2,3] revealed an algorithm for improving electro-optic activity for chromophore/polymer composites. In addition to electronic intermolecular electrostatic interactions, nuclear electrostatic (repulsive or steric) interactions can also influence acentric order. These interactions play the critical role in determining how chromophores can pack in the material lattice, which is a particularly important consideration at high chromophore concentrations (number densities). Dalton and co-workers treated such interactions in the hard geometric object approximation. This requires modification of the limits of integration over orientation variables to explicitly take into account that chromophores in the hard object approximation are not permitted to interpenetrate (overlap). In the

7.1 Chromophore/polymer composites

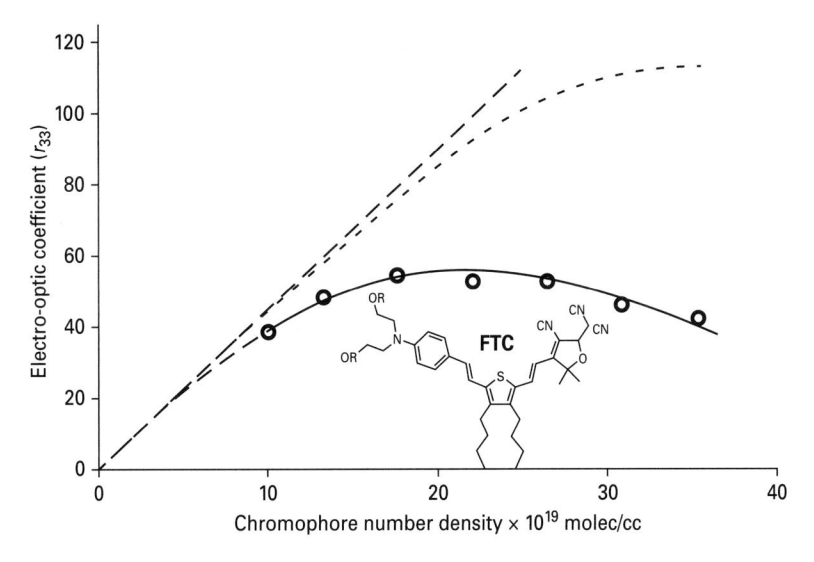

Figure 7.3 Overlay of theoretical and experimental results for butyl chain functionalized chromophore FTC dispersed in amorphous polycarbonate (circles). The solid line illustrates the theoretical results for a prolate ellipsoid in the hard object approximation. Results corresponding to spherical particles (dotted line) as well as the non-interacting model (Langevin limit, dashed line) are also shown [3].

accompanying Fig. 7.3, we illustrate the results of computation of the electro-optic activity for a typical high $\beta(\varepsilon,\omega)$ chromophore treated as spherical and as prolate ellipsoidal objects.

The prolate ellipsoid approximation to chromophore shape provides reasonable reproduction of experimentally observed data for the chromophore dissolved in amorphous polycarbonate, and the spheroid approximation illustrates that electro-optic activity can be greatly improved by changing the shape of the chromophore. It can be noted that dipole moment and molecular first hyperpolarizability are determined by the π-electron component of the chromophore while both π- and σ-components define chromophore shape and steric interactions. Thus, by adding "electro-optically inert σ-bonded" segments to the π-electron backbone of a chromophore, the shape of the chromophore can be modified with little effect on the molecular first hyperpolarizability.

Using the point dipole approximation, Dalton and co-workers defined a spatially anisotropic electronic intermolecular electrostatic potential function to be inserted into the partition function for statistical mechanical analysis of acentric order. There are two components of this potential – one favoring centric (centrosymmetric) order and one favoring acentric (non-centrosymmetric) order. Robinson and Dalton [3] subsequently treated the intermolecular electrostatic interactions of chromophore/polymer composite materials via a Monte Carlo method where chromophores are again treated in a hard object ("united atom") approach. This approach also permitted definition of the intermolecular electrostatic potential, describing the potential felt by a reference chromophore from surrounding chromophores. The potential functions from the analytical and Monte Carlo approaches agree quite well, as do the theoretical predictions of the

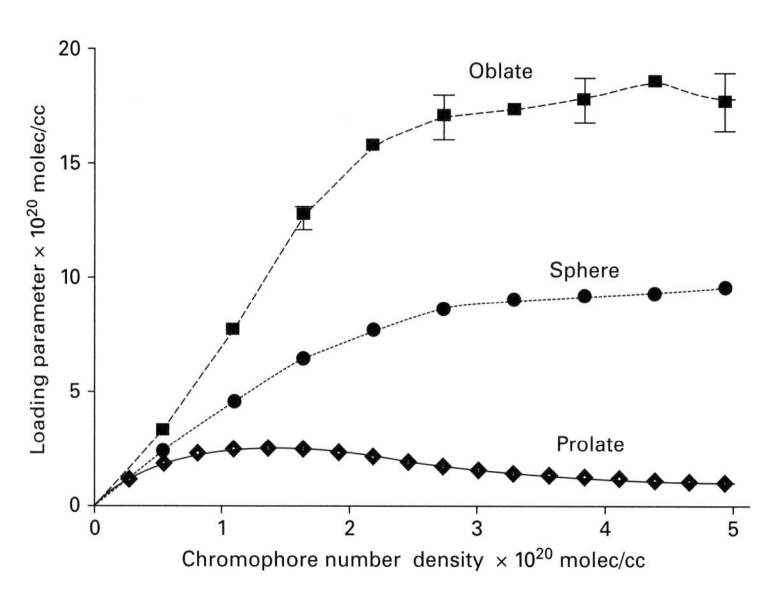

Figure 7.4 On-lattice (no chromophore aggregation) rigid body Monte Carlo (RBMC) simulations of the loading parameter ($N<\cos^3 \theta>$) as chromophore shape is modified from a prolate aspect ratio (2:1 length:width), to spherical (1:1), to oblate (1:2). Reprinted with permission from Ref. [4]. Copyright 2007, American Chemical Society.

variation of electro-optic activity with chromophore number density. Central to these early theoretical calculations was the assumption that chromophore aggregation did not occur and that a uniform spatial distribution of chromophores was maintained. With this approximation, electro-optic activity is predicted to improve as chromophore shape is modified from prolate ellipsoid to spherical to oblate ellipsoid, as can be seen in Fig. 7.4.

Subsequently, Robinson and co-workers [4] removed the restriction of a uniform distribution of chromophores, yielding the results shown in Fig. 7.5; namely, electro-optic activity is predicted to increase on going from prolate ellipsoidal chromophores to spheroidal chromophores but is predicted to decrease on going from spheroidal to oblate chromophores (assuming that the π-electron portion of the chromophore is the same in each of these cases).

It is interesting to examine equilibrium distributions of chromophores in the oblate ellipsoid case for typical trajectory runs. From Fig. 7.6, it is clear that although the acentric order parameter decreases with increasing chromophore number density in the oblate ellipsoid case, oligomeric aggregates of individual oblate chromophores contain chromophores organized in an acentric arrangement. This suggests the intriguing possibility of obtaining greatly improved electro-optic activity by combining electric field poling processing with surface functionalization discussed earlier in considering sequential synthesis methods (Chapter 5). Basically, this concept involves adding another spatially anisotropic force (potential function) to the spatially anisotropic functions associated with electronic and nuclear intermolecular electrostatic interactions and to the spatially anisotropic poling field-dipole moment interaction.

7.1 Chromophore/polymer composites

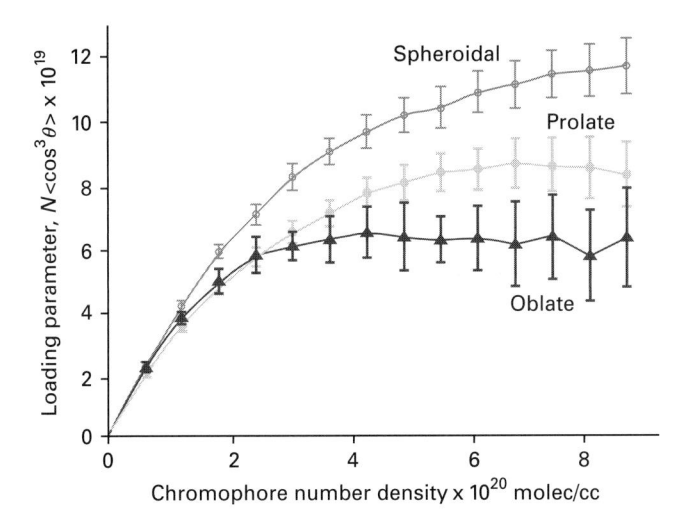

Figure 7.5 Off-lattice (chromophores allowed to translate/aggregate) RBMC simulations of the loading parameter ($N<\cos^3 \theta>$) as a function of chromophore number density for oblate, prolate, and spheroidal shaped molecules. Reprinted with permission from Ref. [4]. Copyright 2007, American Chemical Society.

Figure 7.6 "Snapshot" taken from RBMC simulations of dipolar oblate ellipsoids at high number density. Individual microdomains exhibit acentric order, while net acentric order is low in the bulk material. This illustrates the problem of build-up of dipolar forces in microdomain structures. Reprinted with permission from Ref. [4]. Copyright 2007, American Chemical Society.

Of course, having the right "footprint" is necessary in this approach to prevent chromophore dipolar interactions from tipping chromophores away from the normal to the deposition surface. This basically corresponds to eliminating the host polymer and duplicating the role played by the host polymer through modifying the chromophore into a macromolecular material of desired shape and controlled intermolecular electrostatic interactions. We shall see later in this chapter that it is possible to modify chromophores so that reasonable poling-induced acentric order can be achieved even for neat chromophore materials.

However, in the absence of spatially anisotropic forces other than those associated with untethered chromophores, optimization of electro-optic activity for chromophore/polymer composite materials is a matter of altering chromophore shape through chemical modification [5]. This approach has permitted the electro-optic activity of organic materials to be improved somewhat beyond values (32 pm/V) associated with the most commonly utilized inorganic electro-optic material, lithium niobate.

Several problems exist with the use of chromophore/polymer composite electro-optic materials in device applications. Sublimation of chromophores can occur at the temperatures required for electric field poling of high glass transition polymer hosts. Moreover, phase separation and electrophoretic migration of chromophores can occur during processing, leading to reduction of electro-optic activity and increase in optical loss. To address these problems and to achieve greater thermal and photochemical stability, covalent incorporation of chromophores into polymer and dendrimer materials has been pursued. Chromophores have been covalently attached as side-chains to polymer backbones, have been incorporated directly into polymer backbones (main chains), and have been functionalized at both ends to achieve highly crosslinked, three-dimensional lattices analogous to urethane polymers and sol-gel materials. Interpenetrating polymer networks have also been synthesized incorporating electro-optic chromophores.

7.2 Covalently incorporated chromophore materials

Theoretical treatment of electric field poling of polymers and dendrimers with covalently incorporated chromophores requires fully atomistic [6] or pseudo-atomistic Monte Carlo/molecular dynamic [3,4,7,8] methods. Fully atomistic methods provide greater accuracy but are sufficiently time-consuming that they prevent a significant range of systems and conditions from being explored. Pseudo-atomistic Monte Carlo/molecular dynamical calculations rely on π-electron conjugation restricting rotation about bonds and thus providing legitimacy to the use of the united-atom approximation to treat rigid π-electron segments with electron density distributions derived from quantum mechanics or treated within the point dipole approximation. In the pseudo-atomistic approach, σ-bonded segments are treated via a fully atomistic protocol with bond potential functions defined by quantum mechanical calculations. An example of the effectiveness of the pseudo-atomistic approach is illustrated in the quantitative simulation of electro-optic activity of multi-chromophore-containing dendrimers shown in Figs. 7.7 and 7.8.

7.2 Covalently incorporated chromophores

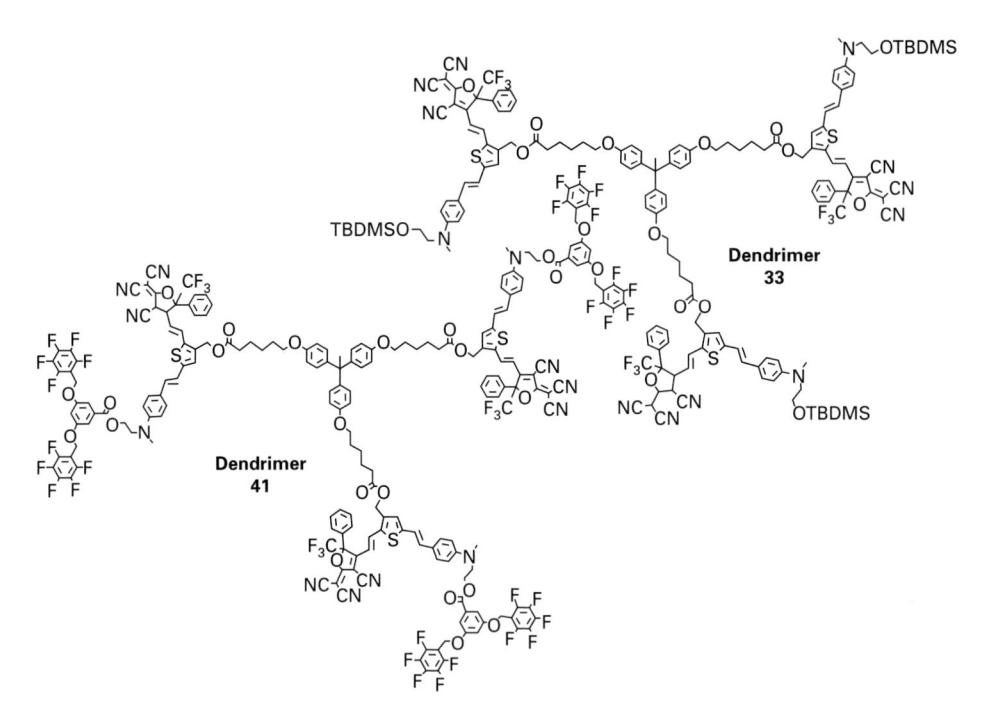

Figure 7.7 Multi-chromophore dendrimers 33 (PSLD33, also referred to as P3) and 41 (PSLD41). Both dendrimers contain three CF$_3$-FTC-type chromophores as their basic π-conjugated nonlinear optically active units. The two dendrimers differ mainly in the functionality chosen as the outer periphery. Reprinted with permission from Ref. [7]. Copyright 2007, American Chemical Society.

Both theory and experiment confirm that chromophores in these multi-chromophore-containing dendrimers behave as independent particles, i.e., a linear relationship is observed for the variation of electro-optic activity versus chromophore number density. Plotting r_{33}/E_{pol} (also denoted as r_{33}/E_p), rather than maximum observed r_{33}, leads to improved reliability since each ratio value is determined by least squares fitting of r_{33} values obtained for a number of poling fields and independent samples. In all cases, both theory and experiment yield a linear relationship between r_{33} and E_{pol} in the poling experiments, justifying linear least squares analysis.

Covalent incorporation of chromophores into polymer, dendrimer, and dendronized-polymer materials permits significant improvement in chromophore loading with less phase-separation or attenuation of electro-optic activity through centrosymmetric ordering. Higher chromophore loading also leads to higher dielectric permittivity and larger index of refraction values. As noted previously, increased dielectric permittivity can lead to increased $\beta(\varepsilon,\omega)$ values. The combination of increases in N, $\beta(\varepsilon,\omega)$, and $<\cos^3 \theta>$ can lead to significantly improved electro-optic activity. Of course, it should be noted that measured electro-optic activity will also depend strongly on the nature of the polymer and dendrimer structures into which chromophores are incorporated. For example, if the flexible spacers connecting chromophores to polymer and dendrimer

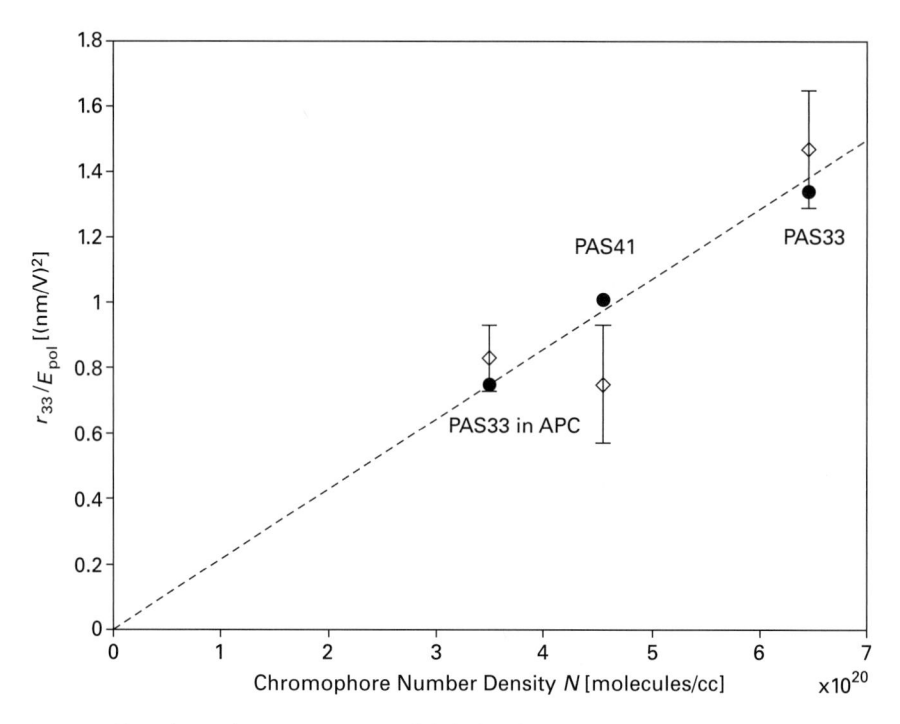

Figure 7.8 Experimental measurements (solid circles) for dendrimer 33 (PAS33 or PSLD33), dendrimer 33 in APC, and dendrimer 41 (PAS41 or PSLD41) are compared with pseudo-atomistic (rigid body) Monte Carlo simulation data (diamonds) assuming a poling field of E_{pol} (or E_p) = 150 V/μm, and hyperpolarizability of $\beta_{zzz}(\varepsilon,\omega)$ = 3200 × 10^{-30} esu. The error bars for the theoretical data reflect different computational trajectories and the variation of energy minima. Reprinted with permission from Ref. [7]. Copyright 2007, American Chemical Society.

core structures are too long, then the observed behavior will approach that exhibited by chromophore/polymer composite materials. The over-arching observation is that quantum and statistical mechanical modeling can provide useful prediction of observed behavior but each putative material must be analyzed in detail. The shape of chromophores, covalent bond potentials, and a variety of electronic inter- and intramolecular electrostatic interactions will play critical roles in defining observed results.

7.3 Matrix-assisted-poling (MAP) materials

Several interactions have been observed to enhance poling-induced order (and thus electro-optic activity) by impacting the rotational freedom of the electro-optic chromophore. Another way of stating this is that these interactions act to reduce the symmetry of the environment surrounding the electro-optic chromophore through the effect of a partially ordered host matrix on the chromophore. As can be seen from Fig. 7.9, any reduction in effective symmetry will result in an increase in the poling field induced

7.3 Matrix-assisted poling materials

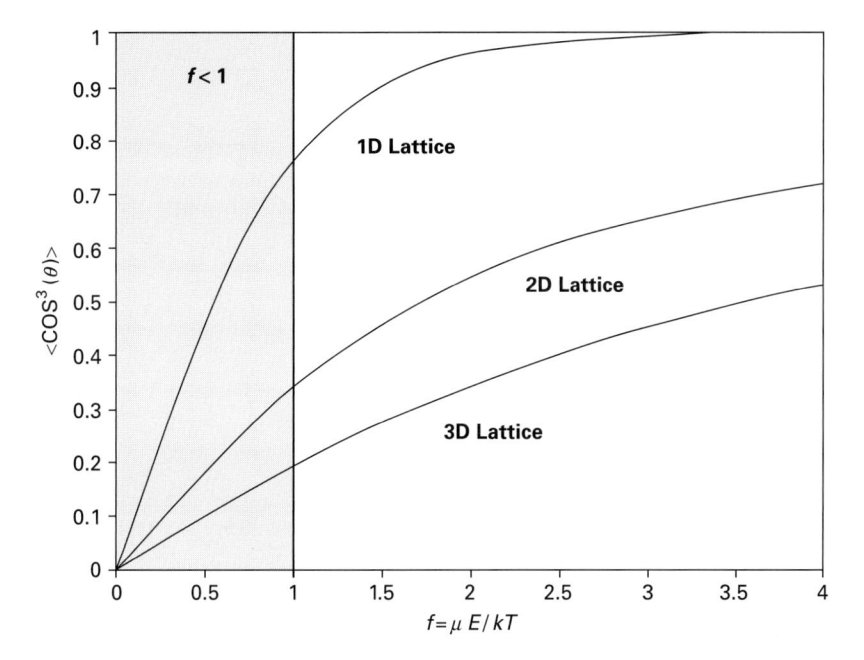

Figure 7.9 Rigid body Monte Carlo (RBMC) simulation generated plot of the expected ordering of dipolar united-atom ellipsoids as a function of the normalized poling energy, $f = \mu E/kT$. Ordering behavior is shown for restricted reorientation (reduced dimensionality). In the 3D case, chromophores can access any orientation, while in the 2D case chromophore reorientation is restricted to within the poling plane. Finally, in the case of a 1D lattice, chromophores may on be oriented either in the positive or negative direction (up or down) along the poling field axis. Reprinted with permission from Ref. [10]. Copyright 2008, American Chemical Society.

order parameter $<\cos^3 \theta>$ or equivalently $<\cos^3 \theta>/E_{pol}$ for a given chromophore number density, N.

Such increases in $<\cos^3\theta>/E_{pol}$ at high loading densities result in material loading parameter, $N<\cos^3\theta>$, and poling efficiency, r_{33}/E_{pol}, values that are enhanced relative to optimized guest–host or dendrimer/polymer-based EO materials behaving in the Langevin (3D independent particle) limit.

Combined with r_{33}/E_{pol} analysis and hyper-Rayleigh scattering (HRS)-derived measurements of $\beta(\varepsilon,\omega)$, VAPRAS or VASE (see below) measurements of the centrosymmetric order parameter $<P_2> = 0.5(3<\cos^2 \theta> - 1)$ can be used to quantitatively evaluate the dimensionality of a material using the relation

$$\langle \cos^3\theta \rangle \approx \sqrt{\left(\frac{9-2M}{2+M}\right)\left(\langle P_2 \rangle - \frac{3-M}{2M}\right)} \quad (7.4)$$

where M denotes the lattice dimensionality [9 to 11]. It has been shown that for MAP materials, the relationship between planar-centrosymmetric order and acentric dipolar order changes in a characteristic manner that is well described by the above approximation. As dimensionality is reduced, $<P_2>/<\cos^3 \theta>$ increases (Fig. 7.10).

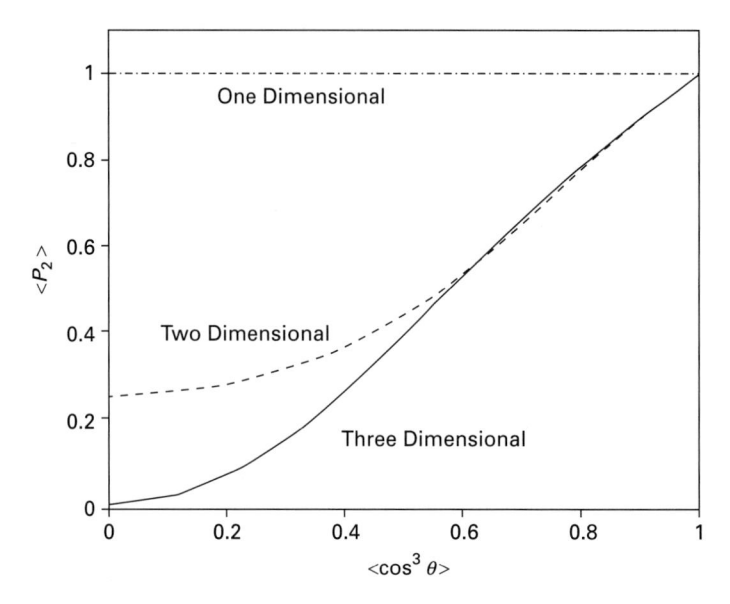

Figure 7.10 The relationship between the centrosymmetric order parameter, $<P_2> = 0.5$ $(3<\cos^2\theta>-1)$, and the acentric order parameter, $<\cos^3\theta>$, as lattice dimensionality changes is shown. Reprinted with permission from Ref. [9]. Copyright 2010, American Chemical Society.

Figure 7.11 Coumarin pendant functionalized chromophore C1, which is constructed from the same π-conjugated chromophore core as the analogous chromophore YLD156 (F2). Reprinted with permission from Ref. [9]. Copyright 2010, American Chemical Society.

Variable angle spectroscopic ellipsometry (VASE) and variable angle polarization-referenced absorption spectroscopy (VAPRAS) measurements on the SBLD-1 (C1) (Fig. 7.11) pendant-functionalized chromophore illustrate the impact of intermolecular interactions among (coumarin) pendant groups on poling-induced chromophore order (see Fig. 7.12) [9].

7.3 Matrix-assisted poling materials

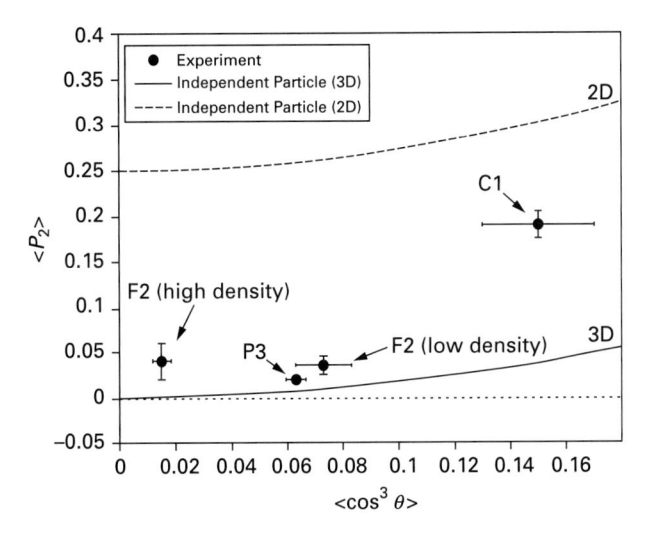

Figure 7.12 The centrosymmetric order parameter, $<P_2> = \phi = (3<\cos^2\theta> - 1)/2$, measured using VAPRAS, plotted with respect to the acentric order parameter $<\cos^3\theta>$, deduced from r_{33} and β, experimental measurements for pendant functionalized chromophore C1, two loading densities of F2 (YLD156 dissolved in PMMA), and pure dendrimer 33 (P3 – see Fig. 7.7). For reference, the expected relationships between $<P_2>$ and $<\cos^3\theta>$ (from the independent particle approximation) for perfectly 2D and 3D systems are also shown. Reprinted with permission from Ref. [9]. Copyright 2010, American Chemical Society.

Coumarin–coumarin interactions (as well as phenyl–perfluorophenyl (Ph–Ph^F) or more generally aryl–perfluoroaryl (Ar–Ar^F)) have been shown to result in a marked change in the dimensionality M, resulting in factors of >2 increase in acentric order, and thus electro-optic activity, above that expected for simple dipolar prolate chromophores such as YLD156 dispersed into an isotropic host (F2 chromophore/polymer composite). The effective lattice symmetry for C1 is approximately two-dimensional (Bessel lattice) – see Fig. 7.12. For F2 and P3, the effective lattice symmetry is approximately three-dimensional (Langevin lattice).

Further evidence of intermolecular cooperativity among coumarin dendrons (C1) and HD-FD dendrons (where FD is a dendron with fluorinated aromatic groups; HD, one with hydrogens instead of fluorine on the aromatic groups) is provided by shear modulation force microscopy (SM-FM) and intrinsic friction analysis (IFA) [12 to 18]. SM-FM provides critical temperature values that are indicative of subtle mobility (viscoelasticity) changes; such critical temperatures pinpoint non-covalent interactions and nanoscopic molecular cooperativity (order–disorder transitions). In polymers, critical temperatures are typically indicative of side-chain relaxations or backbone relaxations (glass transition temperature). Thermally activated transitions are determined from "kinks" in the SM-FM response curve; SM-FM data for the C1 and HD-FD materials are shown in Fig. 7.13.

Because of its small probing volume and the inherent sensitivity of the technique, SM-FM analysis is able to reveal more subtle thermal transitions than can be observed

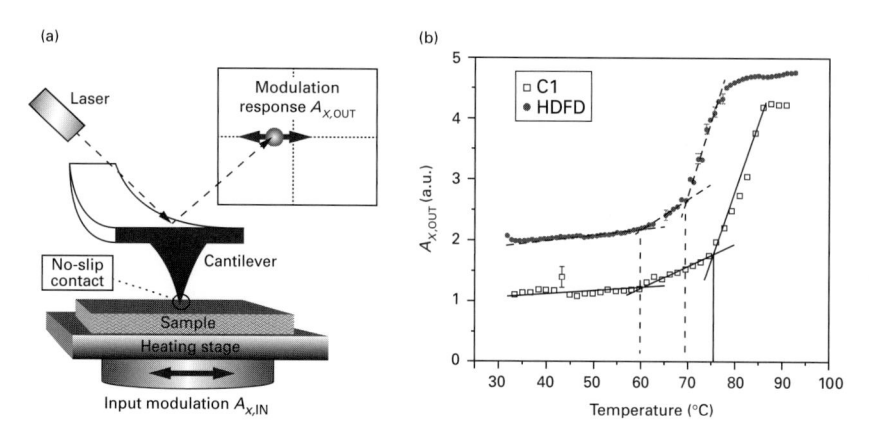

Figure 7.13 (a) Schematic diagram of the shear modulation force microscopy (SM-FM) apparatus. (b) SM-FM data for coumarin pendant functionalized chromophore C1 (open squares) and alternatively for fluorinated-/non-fluorinated-phenyl dendron-functionalized chromophore (HD-FD) (filled circles).

using differential scanning calorimetry (DSC). As is evident from Fig. 7.13, two distinct thermal transitions are observed for both C1 and HD-FD.

Intrinsic friction analysis is based on the well-established method of lateral force microscopy and can be used to quantify energies and molecular cooperativity associated with such transitions. While non-cooperative processes are described in terms of the activation enthalpy ΔH^* alone, cooperative processes also include the activation entropy, ΔS^*, as given by the activation Gibbs free energy $\Delta G^* = \Delta H^* - T\Delta S^*$. The two thermally activated transitions evident in the case of C1 occur at $T_1 = 61\,°C$ and $T_2 = 76\,°C$. The $T_2 = 76\,°C$ transition corresponds quite closely with the DSC derived $T_g = 80\,°C$. Using IFA analysis, three apparent activation energies of 16, 43, and 72 kcal/mol can be deduced. From these activation energy values, the relative enthalpic and entropic (cooperative) contributions can be estimated. For HD-FD, $T\Delta S^*$ is found to be 52–56 kcal/mol, while for C1, $T\Delta S^*$ is found to be 54–58 kcal/mol. Optimum poling efficiency is realized in the temperature regime where cooperative interactions are active. Indeed, optimum poling efficiency occurs at the temperature corresponding to maximum ΔS^* (or increased order), providing evidence of the direct effect of coumarin–coumarin interactions on poling efficiency and material order.

Many other spatially anisotropic interactions (e.g., hydrogen bonding, ionic interactions) could be used to influence lattice dimensionality. The important criterion is that the interactions are not so strong as to elevate the material's glass transition temperature above its decomposition temperature. If interactions are too strong, then processability will be lost. Positioning of moieties that define intermolecular cooperativity can be used to influence the effective dimensionality of the lattice, which in turn impacts chromophore order. An example of various potential attachment points to the high β chromophores FTC (heteroaromatic vinylene bridge) and CLD (polyene bridge) is given in Fig. 7.14.

7.4 Binary chromophore materials

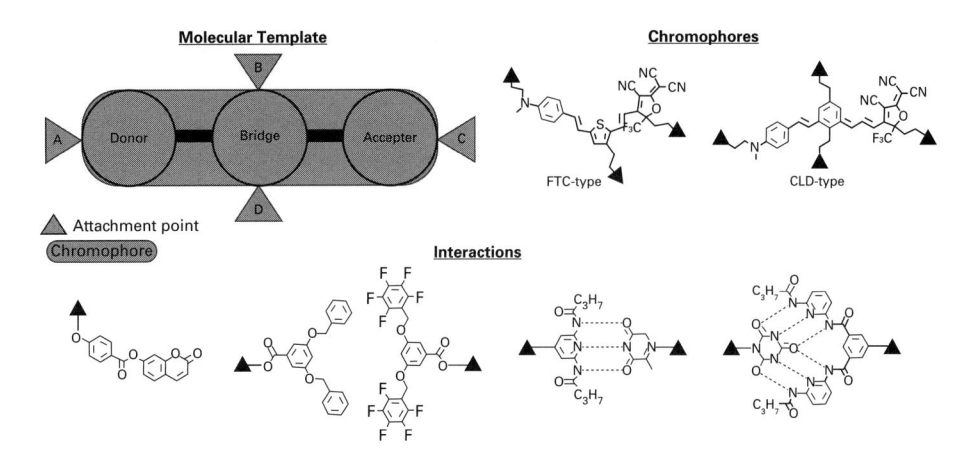

Figure 7.14 Various potential attachment points and chromophore modifications.

Figure 7.15 The structure of the DLD164 chromophore. This chromophore exhibits an electro-optic activity of 180 pm/V in devices when used as a neat chromophore material.

The stronger chromophore–chromophore dipolar interactions in CLD-type chromophores attenuate the effect of coumarin and HD-FD interactions. For example, the CLD analog of C1 yields a dimensionality, M, of 2.5 compared with 2.2 for C1. The larger molecular first hyperpolarizability of CLD-type chromophores leads to r_{33}/E_{pol} values of the order of 2.5 $(nm/V)^2$.

The chromophore shown in Fig. 7.15 appears to give very good results, although many variations remain to be evaluated (see Table 7.1), and indeed appears to exhibit a lattice dimensionality of slightly less than 2. Utilization of this chromophore in devices will be discussed later in this chapter.

7.4 Binary chromophore materials

A binary chromophore composite [19 to 21] consists of a chromophore guest doped into a chromophore-containing host. Guest chromophores can also be covalently coupled to the chromophore-containing host but since there are relatively few examples

Table 7.1 Values of poling efficiency, acentric order, and lattice dimensionality for several chromophore systems.

Material	E_p	$r_{33}/E_p{}^a$	$<P_2>$	$\beta_{zzz}(-\omega;0,\omega)^g$	$<\cos^3\theta>$	M^j
DLD164	35	3.1	0.31^c	$7000^{h,k}$	0.17	1.93
C1	50	1.92	0.19^d	3000^i	0.15	2.2
CLD-C1	50	2.52	0.12^d	7000^h	0.18	2.58
CLD-C1	80	2.52	0.28^e	7000^h	0.29	2.12
YLD156/ PMMA (LD)^m	50	0.45	0.035^d	3500^i	0.073	2.8
YLD156/ PMMA (HD)^n	60	0.15	0.04^f	3500^i	0.015	2.8
P3	50	1.42	0.019^d	$4033^{h,k}$	0.063	2.9

[a] Units = (nm/V)^2; [c] $<P_2>$ at E_p = 35 V/μm; [d] $<P_2>$ at E_p = 50 V/μm; [e] $<P_2>$ at E_p = 80 V/μm; [f] $<P_2>$ at E_p = 60 V/μm; [g] Units = 10^{-30} esu; [h] Estimated value; [i] Experimental value from HRS; [j] Dimensionality; [k] Value of β_{zzz} from DFT simulations (B3LYP); [m] Low density (20 wt%); [n] High density (40 wt%).

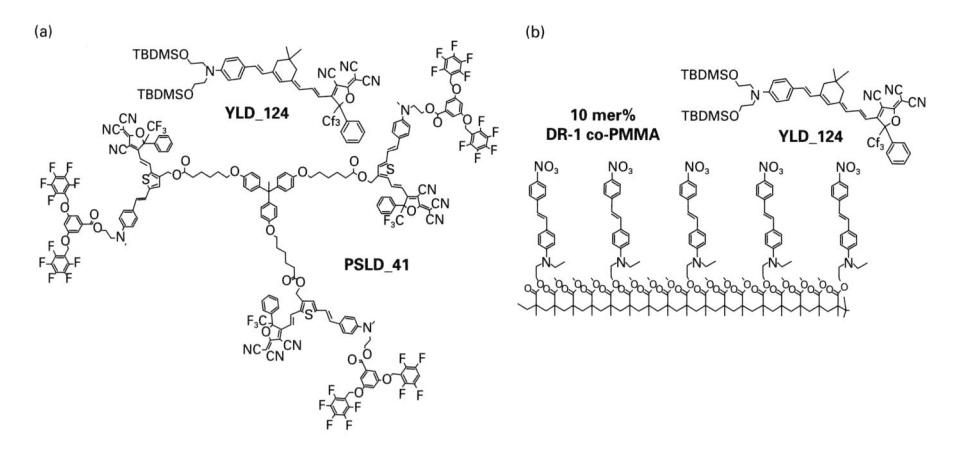

Figure 7.16 (a) Binary chromophore material consisting of host dendrimer PSLD41 (dendrimer 41) and guest chromophore YLD124. (b) Binary chromophore material consisting of host side-chain polymer DR-1-co-PMMA and guest chromophore YLD124. Reprinted with permission from Ref. [21].

of such materials, the present discussion is restricted to composite materials. Binary chromophore materials are also examples of matrix-assisted-poling (MAP) materials, i.e., cooperative interactions among host chromophores can influence the poling-induced order of guest chromophores and vice versa. As can be seen from a consideration of Figs. 7.16, 7.17, and 7.18, binary chromophore composites demonstrate several striking properties: (1) The slope of the graph of electro-optic activity versus chromophore number density depends on the nature of the matrix surrounding the dopant chromophore (see Figs. 7.16 and 7.17); (2) very high chromophore

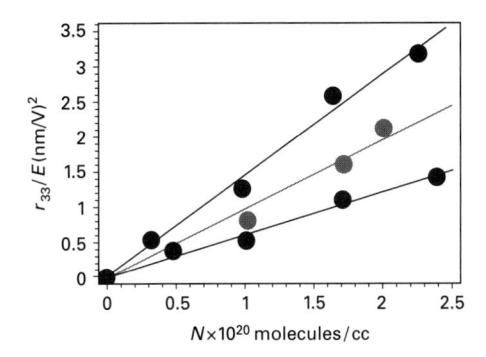

Figure 7.17 Experimental data showing the poling efficiency (r_{33}/E_{pol} or r_{33}/E_p) as a function of guest chromophore loading density corresponding to the YLD124 chromophore doped into (1) amorphous polycarbonate (APC) host (lower trace), (2) DR-1-co-PMMA host (middle trace), and (3) PSLD41 host (upper trace). Reprinted with permission from Ref. [21].

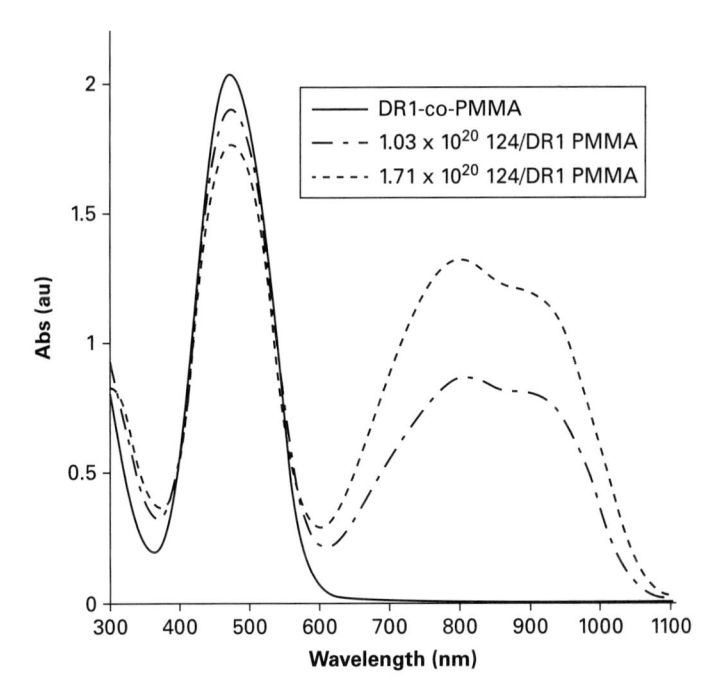

Figure 7.18 Optical absorption spectra of DR1-co-PMMA host polymer, and YLD124 guest chromophore doped into DR1-co-PMMA at increasing N. Reprinted with permission from Ref. [21].

loading can be achieved without attenuation of the linear dependence of electro-optic activity on number density and without phase separation [19]; and (3) an absence of solvatochromic shifts and spectral line broadening is observed in the linear optical (absorption) spectra (see Fig. 7.16) of the guest and host chromophores as

N increases. The combination of these properties leads to very favorable electro-optic activity and optical loss values relative to traditional chromophore/polymer composite materials.

How can the properties of binary chromophore materials be conceptually understood? The high loading without phase separation argues for favorable free energies (enthalpies and entropies) of mixing of the guest and host chromophore materials. This is not surprising given that these materials are formed by mixing a polar guest with a polar host unlike the case of a traditional composite materials where a highly polar guest (the chromophore) is mixed with a relatively non-polar (non-chromophore-containing) host. This also accounts for the disparity observed in the solvatochromic and spectral line broadening characteristics between binary chromophore and simple guest/host composites, i.e., the dielectric environment is relatively constant in the former and changes significantly in the latter with changing chromophore concentration. Explanation of the increase in slope of the r_{33} vs. E_{pol} plots is more challenging and requires at least a qualitative understanding of the effects of spatially anisotropic intermolecular electrostatic interactions. Indeed, the increased slope observed for binary chromophore composites relative to traditional composites suggests that the chromophores are experiencing electrostatic interactions that are more favorable to poling-induced acentric order (i.e., MAP). Theoretical calculations [20] suggest that two types of spatially anisotropic intermolecular electrostatic interactions may be important for enhancing acentric order. The first are specific guest–host chromophore interactions [20]. The second stems from the fact that both guest and host chromophore interact with the poling field and this interaction reduces the dimensionality of the lattice surrounding an individual chromophore. A quite general theoretical observation is that the acentric order parameter, $<\cos^3 \theta>$, of an electrically poled chromophore will increase as the dimensionality of the lattice that confines that chromophore decreases from 3D to 2D to 1D (see Figs. 7.9 and 7.10).

Recently, Olbricht and co-workers [21 to 23] have provided a convincing demonstration of the role of guest–host chromophore interactions in enhancing electric field poling-induced acentric order. They investigated the binary chromophore composite material shown in Figs. 7.16(b) and 7.18. The host in this case is DR1-co-PMMA (the Disperse Red 1 chromophore covalently attached to polymethylmethacrylate). It is well known that the order of the host chromophore can be improved by optically assisted poling using polarized light [22,24]. Linearly polarized light at the wavelength of the interband (charge transfer) transition of trans-DR1 drives the azobenzene chromophore into its cis-conformation. The cis-conformation rapidly relaxes back to the trans-conformation with the net effect being a change of orientation for the DR1 chromophore. Because both the electric field vector of the linearly polarized light and the transition dipole moment of the DR1 chromophore are axially symmetric, the chromophore will continue reorienting until the transition moment (matrix) is zero. This relationship between optical polarization and transition dipole results in the chromophores being driven into a plane in orientation space (i.e., a two-dimensional or Bessel lattice). When this optically induced ordering occurs in a symmetry-breaking electric poling field, the acentric order of the DR1 chromophore is enhanced relative to

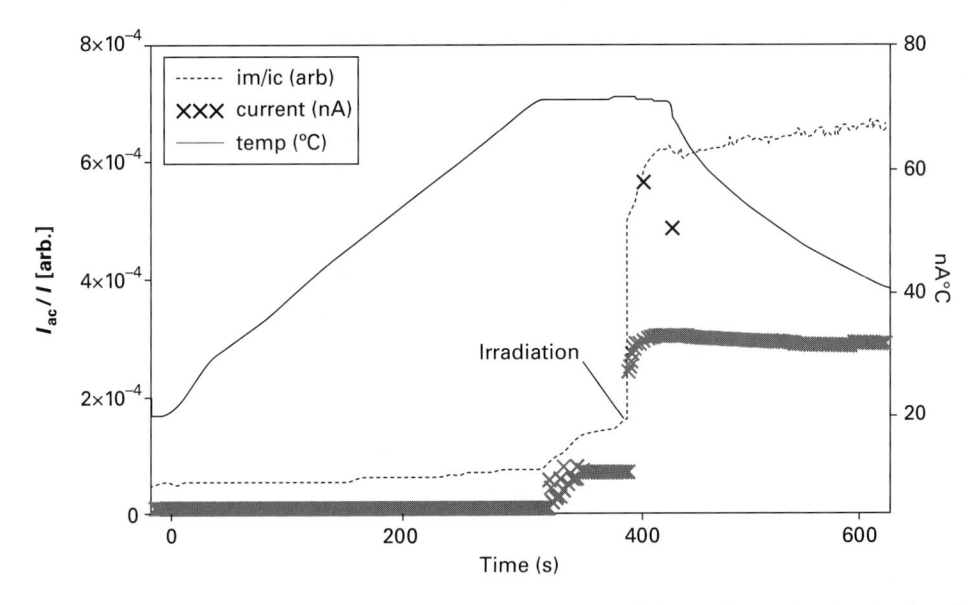

Figure 7.19 Real-time monitoring of electro-optic activity ($I_{ac}/I \propto r_{33}$), using fixed angle ellipsometry, during laser-assisted poling of the YLD124 guest / DR-1-co-PMMA host binary chromophore material. The point labeled "irradiation" denotes the introduction of the polarized 532-nm pump laser. Reprinted with permission from Ref. [22]. Copyright 2008, American Chemical Society.

the order observed without optically assisted poling (also referred to as laser-assisted poling). If there is a defined interaction between guest and host chromophores, then the acentric order of the non-photochromic guest (no photochromic conformation transitions possible) may be enhanced as the result of optically assisted poling of the DR1-containing host chromophore material. In Fig. 7.19, we demonstrate the enhancement of the poling-induced order of a guest chromophore by applying optically assisted electric field poling to the host chromophore.

A 2.5-fold enhancement in the electro-optic activity of the guest chromophore is observed employing LAP. The number density has not changed in the experiment and the absence of significant solvatochromic shifts and line broadening (Fig. 7.18) suggests that the dielectric environment of the guest chromophore has not changed significantly; thus, it is most likely that the increase in electro-optic activity is driven by an increase in acentric order of the guest chromophore caused by a guest–host order interaction. In other words, when the host order is increased (host dimensionality is reduced) owing to its interaction with the polarized laser light, the order of the guest is also enhanced. Of course, the extent of enhancement depends on the nature and strength of electrostatic interactions between guest and host chromophores but the experiments of Olbricht et al. [22,23] leave little doubt as to the existence of such interactions. In passing, it can be noted that Olbricht and co-workers also investigated the detailed dependence of electro-optic activity upon optical power driving the photochromic trans–cis–trans isomerization of the azobenzene chromophores and upon poling temperature [22].

In other binary chromophore materials, a variety of additional spatially anisotropic interactions between guest and host materials have been suggested as contributing to enhanced acentric order, including quadrupolar interactions between fluorinated and protonated aromatic dendrons. Spatially anisotropic ionic and hydrogen bonding interactions may also contribute. Optically assisted electric field poling has also been demonstrated in a Mach–Zehnder modulator [23].

Binary chromophore materials represent an important step toward engineering organic electro-optic materials with crystalline-like order. For the present, these solution-processed binary chromophore composite materials yield electro-optic activities in the range of 200–500 pm/V (at telecommunication wavelengths) and r_{33}/E_{pol} values of the order of 3–4 with optical loss of the order of 2 dB/cm. These materials currently define the state-of-the-art of organic electro-optic materials, although values for some coumarin and arene–perfluoroarene modified materials are approaching those of BC materials.

The understanding of intermolecular electrostatic interactions discussed in this chapter is also relevant to understanding the performance of materials prepared by sequential assembly by Merrifield or Langmuir–Blodgett methods and the non-centrosymmetric (acentric) crystallization of materials such as DAST. In sequential assembly, chromophores tend to tilt away from the normal to the assembly surface as the result of chromophore–chromophore dipolar interactions. This tilt reduces electro-optic activity and increases optical loss associated with increased disorder. Note that disorder increases progressing from the assembly surface, since disorder associated with defects will propagate and dipolar interactions will build up as more chromophores assemble in acentric columns. Order can be improved by employing chromophores with the appropriate "footprint" to disfavor side-by-side dipolar interactions, which favor a centrosymmetric (or tilted) arrangement of chromophores. Note that dipole–dipole interactions go to zero when chromophores are oriented at the magic angle. Order will also be improved by increased dielectric screening of chromophores.

Stronger intermolecular electrostatic interactions (e.g., ionic) are frequently operative in crystalline organic materials although hydrogen bonding and strong π–π interactions are involved in some materials. If melt processability is lost, then microdomain formation and growth anisotropy can be problematic with utilization of crystalline materials in devices. A few crystalline organic electro-optic materials do exhibit melting temperatures below decomposition temperatures and for such materials optically assisted electric field poling, as demonstrated in Fig. 7.20, can be used to achieve desired orientation of individual EO-active molecules in a thin film format [25]. In this case, optically assisted electric field poling involves orientation-selective melting of molecules in the solid state. Because the intermolecular interaction is strong and drives acentric organization of the component molecules, an acentric order parameter of unity is achieved. This result serves to indicate the common principles involved in preparing organic electro-optic materials by crystal engineering, poling of polymers and dendrimers, and sequential-synthesis/self-assembly fabrication. The interplay of electronic and nuclear (steric) interactions, chromophore mass, and thermal energy defines the material structure and dynamics in all cases.

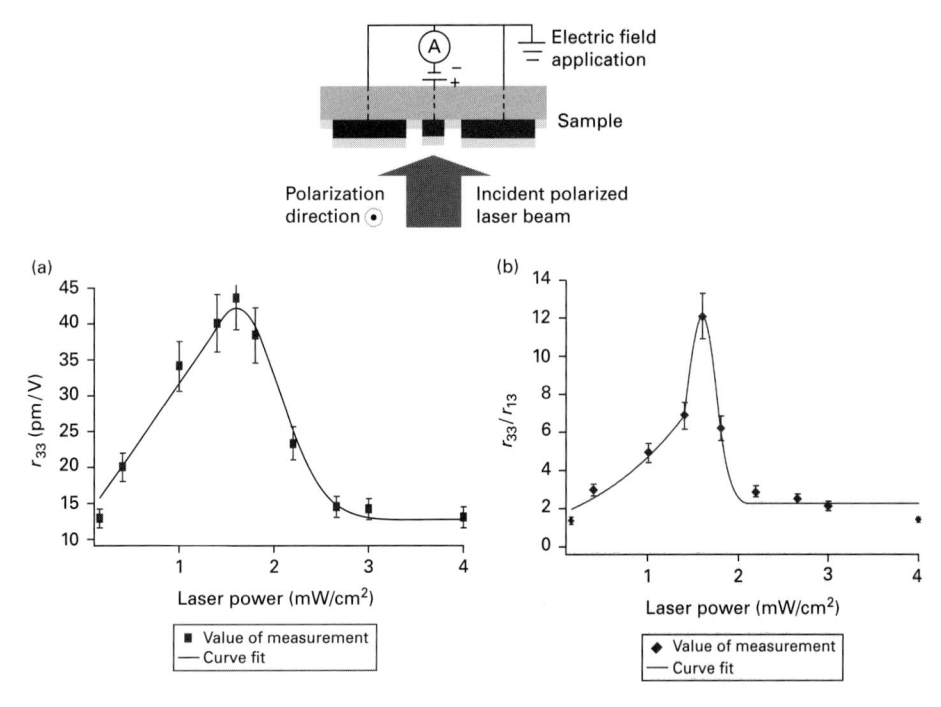

Figure 7.20 Laser-assisted poling of N-benzyl-2-methyl-4-nitroaniline (BNA) using a coplanar electrode configuration and two-slit interferometric measurement of r_{33} and r_{13} is shown. Reprinted with permission from Ref. [25]. Copyright 2008, Optical Society of America.

As a caveat, it should be noted that when metal electrodes and optical fields interact, plasmonic effects may influence poling including via plasmonic heating effects.

7.5 Complexity

In Chapters 5, 6, and 7, we have seen how an improved understanding of intermolecular electrostatic interactions can be translated to preparation of improved organic electro-optic materials. We have also seen how the properties of such materials can be tuned over a wide range and how crystalline, sequential-synthesis/self-assembly, and electrically poled materials are approaching each other in physical properties and design principles. With electrically poled polymer and dendrimer materials, theory has played an increasingly important role in designing improved materials, particularly in terms of using (non-covalent bonding) dipolar and quadrupolar interactions to influence assembly. However, the interplay of steric (nuclear repulsive), dipolar, quadrupolar, ionic, and general van der Waals interactions can be complex, and it may be difficult to predict the exact properties of many new organic electro-optic materials. For example, steric interactions can play a role in determining how closely chromophores can approach each other and thus whether excitonic features will be observed in optical spectra.

Figure 7.21 The structures of the triptycene-modified chromophore (AJPL172) and reference chromophore (AJPL170).

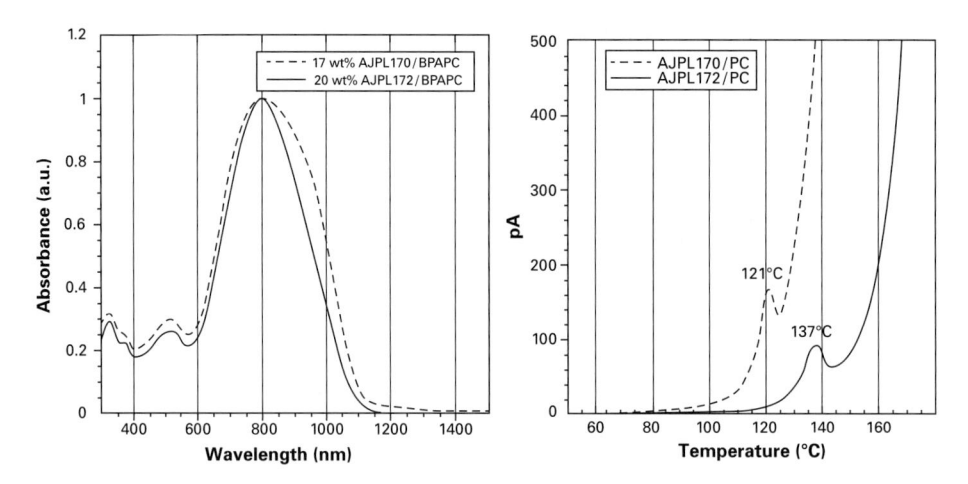

Figure 7.22 The optical spectra (left) and conductivity (right) of the chromophores of Fig. 7.21. Adapted from [26] with permission of the American Chemical Society.

Recent studies [26] of triptycene-modified chromophores (see Figs. 7.21 and 7.22) illustrate this point. The optical spectra (Fig. 7.22 left) of triptycene-modified chromophores show less aggregation; indeed, they approach the solution spectra of these chromophores. Less conductivity (Fig. 7.22 right) is observed at poling temperatures for the triptycene-modified chromophore. Moreover, triptycene-modified chromophores also exhibit improved stability due to enhanced interaction of the triptycene-modified chromophore with polymer hosts such as poly-(bisphenol A carbonate) [26]. While statistical mechanical simulations have not been executed for these materials, and centric and acentric order parameters have not been measured, it is likely that the steric bulk of the triptycene moieties inhibits "J-type" aggregate formation and facile charge hopping from chromophore to chromophore.

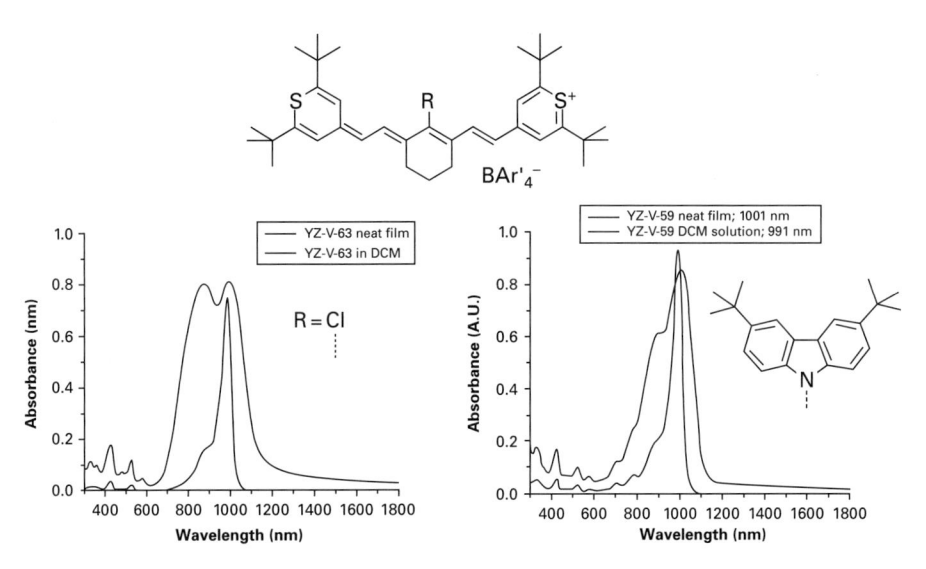

Figure 7.23 (Top): The structure of a third-order chromophore modified (by replacement of a chlorine with a carbazole group) to inhibit excitonic interactions in the solid state. (Bottom): The reduction of excitonic interactions in the solid state is seen by comparing the optical spectra of neat thin films of chromophores YZ-V-63 (R = Cl) and YZ-V-59 (R = carbazole). Figure prepared by Professor Seth Marder and co-workers.

These results illustrate the continued importance of steric interactions and more complex chromophore–host interactions in the control of properties of organic electro-optic materials.

Use of steric interactions to control chromophore aggregation has recently become an important research activity for third-order organic nonlinear optical materials (e.g., charged third-order chromophores) as shown by recent results from Marder and co-workers (see Fig. 7.23) [27]. Functionalization of polyene bridges introducing steric interactions is clearly shown to inhibit aggregation through the attenuation of excitonic contributions and line broadening in optical spectra. The material design template is essentially that of Fig. 7.14.

7.6 Thermal stability issues

Understanding relaxation of poling-induced acentric order in nonlinear optical polymeric and dendritic (macromolecular) materials is of fundamental importance in order to evaluate the long-term stability of poled polymers in devices. The actual usefulness of organic electro-optic materials relies on a sufficient stability of the poling-induced order of the nonlinear optical chromophores within these polymers.

An essential requirement for stabilizing electrically poled polymeric nonlinear optical materials is the formation of a glassy state at relatively high temperatures. Amorphous polymers, among other glasses, typically show evidence of a phase transition from a

Table 7.2 Azo dye substitution patterns and glass transition temperatures T_g of selected modified polyimide polymers.

	n	R_1	R_2	T_g (°C)
A-095.11	3	CH$_3$	H	137
A-097.07	3	CH$_3$	Cl	149
A-148.02	2	H	H	172

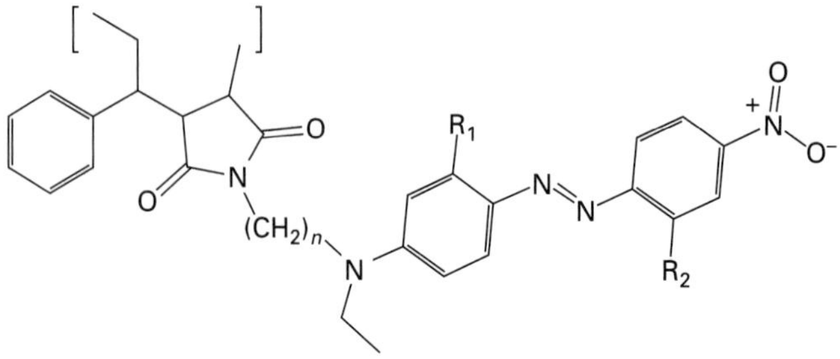

Figure 7.24 The molecular structure of the electro-optic side-chain polyimides based on Disperse Red.

liquid-like to a glassy state when cooled from high temperatures. The temperature at which this transition occurs, known as the glass transition (T_g), is recognized to be primarily a kinetic phenomenon, whether or not it is valid to classify it as a true phase transition. The physical origin of the glass transition is primarily associated with the cooperative motions of large-scale molecular segments of the polymer. The actual experimental observation of a glass transition is most easily and commonly probed by measuring enthalpic changes in the polymer as a function of temperature via differential scanning calorimetry (DSC). As discussed earlier in this chapter, techniques such as shear modulation force microscopy can also be used to measure glass transition temperatures (and pre-transitions). Several phenomenological theories describe the primary aspects of the glass transition, at least as they are experimentally observed.

Relaxation processes in nonlinear optical polymers have been studied by several authors [28 to 32]. One of the most thorough investigations was carried out for polyimide polymers with side-chain azo chromophores listed in Table 7.2 and Fig. 7.24, having glass transition temperatures in the range of 140 °C < T_g < 170 °C [30 to 33]. The relaxation and other model parameters have been studied by differential scanning calorimetry, dielectric relaxation, and second-harmonic generation experiments.

Figures 7.25 and 7.26 show the relaxation of the nonlinear optical susceptibility d_{31} for the polymer polyimide side-chain polymer A-148.02 (at different temperatures from 80 °C to the glass transition temperature) and the same chromophore as guest molecule in the polymer. The T_g for A-148–02 is 172 °C. The lower thermal stability of the guest–host system is clearly observed. In all these experiments it was possible to model the results of the relaxation behavior of nonlinear optical chromophores both above and

7.6 Thermal stability issues 143

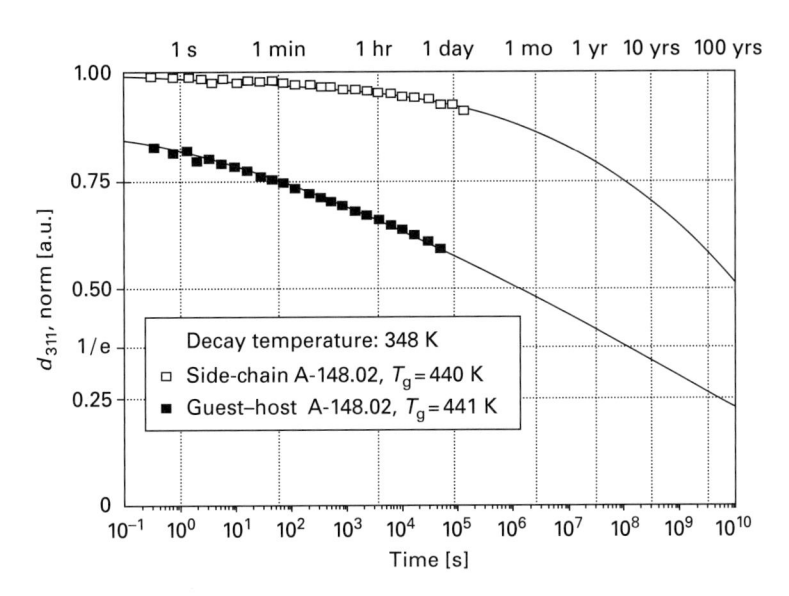

Figure 7.25 Comparison of the second-harmonic decay at 75 °C for the side-chain and guest–host system A-148-02.

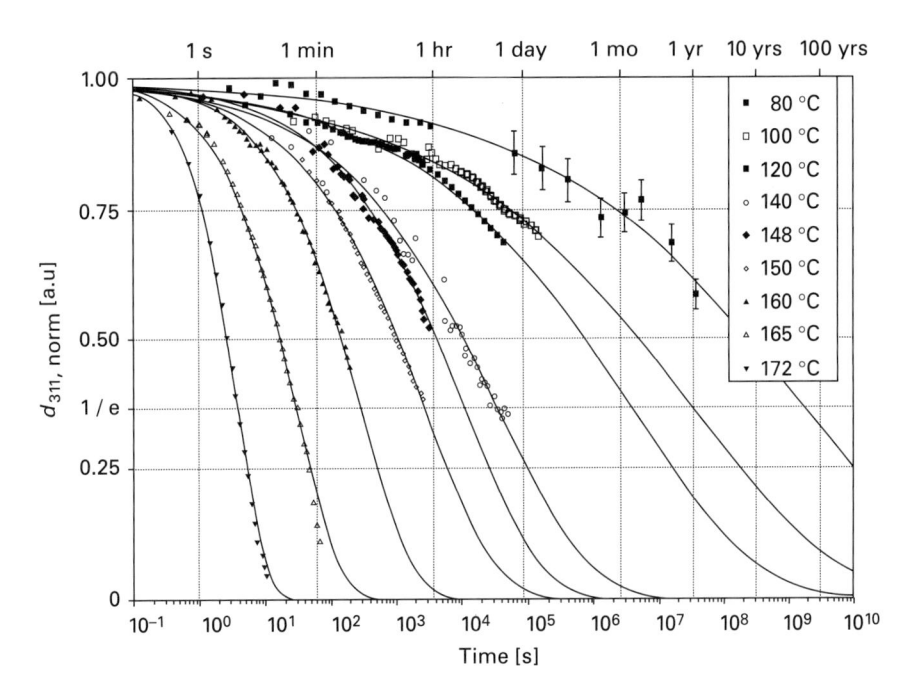

Figure 7.26 Second-harmonic relaxation for the polyimide side-chain polymer A-148-02 as a function of the decay temperature. Solid lines are fits using the model of references [33 to 35].

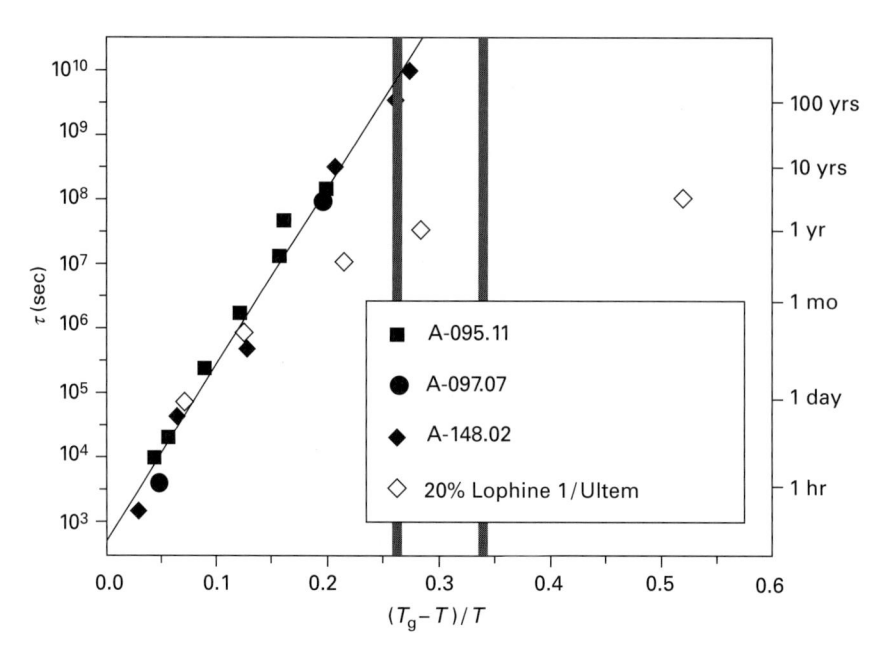

Figure 7.27 Temperature scaling of normalized dielectric loss and relaxation times of second-harmonic generation are shown with respect to the scaling variable $(T_g - T)/T$ for the three side-chain polymers of Table 7.2 and one guest–host polymer (20% Lophine 1/Ultem). It is clearly seen that the nonlinear optical moiety has to be coupled to the polymer main chain for increased stability. The two thick vertical lines are related to examples discussed in the text.

below the glass transition temperature over more than 15 orders of magnitude in time using the Tool–Narayanaswamy procedure incorporating the appropriate Williams–Landel–Ferry (WLF) parameters for the nonlinear optical polymers [33 to 35]. The model description allows the scaling of relaxation times in the glassy state with the simple scaling parameter $(T_g - T)/T$ as seen in Fig. 7.25. The relaxation time at any operating temperature T below T_g depends only on one parameter: T_g. Figure 7.25 and the underlying theory permit definition of the T_g required to achieve a certain level of stability. As an example, for a polymer with $T_g = 172\ °C$, the relaxation time, $(1/e)$ at an operating temperature of 80 °C, is about 100 years. With a T_g of 200 °C, this relaxation time greatly increases. For better stability, polymers and dendrimers are frequently crosslinked to elevate the material's T_g (to harden the material lattice).

Obviously, thermal stability will not, in general, be an issue for crystalline organic electro-optic materials or for materials prepared by sequential-synthesis/self-assembly exploiting strong covalent bond formation or ionic interactions. The strong intermolecular electrostatic interactions existing in such systems elevate material glass transition temperatures (or melting temperatures) well above material decomposition temperatures (i.e., >300 °C). With electrically poled materials, realization of adequate thermal stability typically requires choosing a high glass transition host for chromophore/polymer composite materials or effecting lattice hardening by crosslinking subsequent to introducing acentric order by electric field poling. Care must be exercised in tuning poling and crosslinking processes so that they do not interfere with each other. When both are thermally activated processes, poling and lattice hardening (crosslinking) temperatures must be coordinated. One method circumventing the interdependence of poling and crosslinking processes is utilization of photo-induced crosslinking. To the present, photo-induced crosslinking has not been very successful, largely owing to the competition of the electro-optic chromophores and the photo-initiator for light (used for crosslinking). Coumarin-containing dendrimers discussed earlier in this chapter may be an exception to this statement, but further study is required. While methods such as two-photon-initiated crosslinking and development of photo-initiators with critical absorptions that do not overlap those of the electro-optic chromophores (such as the case with coumarins) may ultimately be ways around the current impasse, they have yet to be demonstrated.

Over the years, a number of crosslinking chemistries have been applied to the hardening of organic electro-optic materials including condensation [29, 36 to 38] reactions as well as free radical [39,40] and Diels–Alder/retro-Diels–Alder [41,42] cycloaddition reactions. Condensation reactions, as illustrated by sol-gel [37] and urethane [38] chemistries, involve elimination products such as water or hydrochloric acid, which can cause lattice disruptions as the result of the expulsion [38]. The reaction stoichiometry of such reactions can be sensitive to atmospheric moisture [38] and the "effective" molecular weight of final products achieved by condensation chemistry is not as high as can be achieved by free radical reactions. By and large, lattice hardening based on condensation reactions has largely given way to addition reactions based on "soft" free radical chemistry or Diels–Alder/retro-Diels–Alder cycloaddition reactions. The most popular soft free radical reaction involves the cycloaddition reaction of fluorovinyl ether moieties to form cyclobutyl crosslinks. This chemistry was initially pioneered by researchers at Dow Chemical and more recently championed by Dennis Smith at Clemson University (now at University of Texas-Dallas) [39]. This approach has been successfully employed to achieve organic electro-optic materials with final glass transition temperatures of the order of 200 °C [39,40].

An even more commonly utilized and versatile reaction is the Diels–Alder/retro-Diels–Alder reaction, where the reversibility of the reaction and the final glass transition temperature can be tuned by the choice of diene and dienophile reactants [41 to 44]. Again, a number of organic electro-optic materials have been prepared with glass transition temperatures of the order of 200 °C. Both fluorovinyl ether and Diels–Alder chemistries can be implemented without the introduction of significant scattering loss. Diels–Alder chemistry has also been applied to the fabrication of organic electro-optic waveguide circuitry by nano-imprint lithography, where the material glass transition temperature is adjusted to be compatible with the stamping temperature [45].

Of course, in lattice hardening, care must be exercised to avoid chemical decomposition of the electro-optic chromophores. "Hard" free radical chemistries that involve attack on the chromophore or processing temperatures that induce thermal decomposition are not acceptable. A significant concern for electro-optic materials is photochemical decomposition associated with the photo-production of singlet oxygen.

Studies of such decomposition clearly demonstrate that degradation is inhibited by dense, hard lattices; thus, crosslinking typically leads to improved photostability as well as improved thermal stability.

It is critical to note that material glass transition temperature is determined not only by the number and positioning of crosslinks but also by the segmental flexibility of components of the crosslinked material. Thus, design of materials to be crosslinked as well as choice of crosslinking chemistries is important to the performance of the hardened material.

Lattice hardening is also important to optical loss and can have either positive or negative influence on final waveguide propagation loss. Crosslinking must not introduce material density inhomogeneity, or scattering loss will increase. Crosslinking frequently results in improvement in optical loss following processing of waveguides by reactive ion etching as waveguide wall smoothness is enhanced, i.e., smoother walls are observed with hard materials than for soft materials (uncrosslinked polymers).

While the above remarks provide general guidance for the utilization of lattice hardening to increase operational thermal stability, the remarks must be adapted to each new electro-optic material system considered.

7.7   Optical loss issues

The two categories of material optical loss that are particularly important for poled organic electro-optic materials are intrinsic material loss (also called absorption loss) and processing-associated loss (also called scattering loss). Intrinsic absorption loss is associated with two contributions at telecommunication operational wavelengths: (1) loss arising from contributions of electronic transitions and (2) loss arising from contributions associated with overtone vibrational transitions of hydrogen from either chromophore or non-chromophore (matrix or lattice) components of the material. The former is typically dominated by the lowest energy or interband electronic transition. In this regard, it is important to avoid contributions associated with excitonic bands arising from chromophore–chromophore (orbital–orbital rather than dipolar) interactions. For telecommunication applications (operation at wavelengths centered around λ = 1.3 or 1.55 µm), avoidance of significant absorption loss from electronic transitions means keeping chromophore charge-transfer electronic transition wavelength maxima below 850 nm and avoiding low-energy excitonic contributions.

Although solvatochromic shifts and line broadening effects are virtually absent in binary chromophore organic glass materials (see Fig. 7.18), the high chromophore concentrations of these materials can be problematic for optical loss unless the position of interband electronic transition is carefully controlled. Because of the higher proton densities encountered with traditional polymer hosts, contributions from vibrational overtone absorptions are typically most problematic for conventional chromophore/polymer composites. Indeed, it is difficult to realize optical loss values of less than 1.2 dB/cm for such materials unless some of the protons are replaced with fluorines (or otherwise eliminated, e.g., through use of dendritic or unsaturated structures).

If contributions from interband electronic absorption are avoided and proton concentrations are minimized, optical loss values as low as 0.2 dB/cm can be achieved. However, realization of such low optical loss in waveguide devices also requires heroic efforts to minimize processing losses, as well as absorption losses. Such extreme minimization of optical loss is, in general, not practical, and a more realistic target is a waveguide propagation loss value of 2 dB/cm or less. This is particularly the case with materials with very high electro-optic activity, which permit short active waveguide lengths (which are also necessary for realization of high bandwidth devices).

Processing losses arise from spin casting, electric field poling, deposition of cladding layers, fabrication of buried channel waveguides, and deposition of electrodes. Spin casting and electric field poling are most commonly problematic because spatial variation in chromophore concentration can occur. Deposition of cladding layers can cause pitting (surface roughness) due to attack on the electro-optic material by the spin-casting solvent used in the deposition of the cladding material. For UV-curable cladding materials, the radicals produced by UV irradiation can sometimes attack the electro-optic core material. Reactive ion etching fabrication of waveguides can result in surface roughness leading to noticeable optical loss. This roughness can be caused by mask resolution or control of the kinetic energy of the reactive ions. With exercise of proper control of processing conditions, these types of losses can be very low (e.g., of the order of a few tenths of a dB); however, losses can be very high particularly when working with new materials where processing protocols have not been optimized.

Optical loss can also arise from other materials, e.g., electrode and cladding materials. Indeed, the reason for using cladding materials is to prevent evanescent optical fields from interacting with the very lossy metal electrodes.

A final source of optical loss contributing to total device insertion loss is coupling loss. There are two components of coupling loss: (1) loss arising from mode size and shape mismatch associated with transitioning of light from a laser source or transmission line (e.g., traditional silica optical fiber) into the electro-optic waveguide, and (2) loss associated with index of refraction mismatch between passive and active waveguides. For organic electro-optic waveguides coupled to silica fiber optics, the former is much more problematic than the latter and typically requires a mode size transformer [46 to 48]. With use of mode transformers, coupling loss can be kept to a few tenths of a dB per facet. As will be discussed shortly, coupling loss is an even more severe problem when dealing with the small dimensions of silicon photonic or plasmonic circuitry and the coupling of such circuits to silica fiber optics or a laser light source.

7.8   Photochemical stability

Photochemical stability has been an on-going concern with organic electro-optic material as with other electroactive organic materials (e.g., organic light-emitting and photovoltaic materials). Most photochemical stability measurements have been carried out via pump-probe methods [49 to 54]. Initially, these measurements focused on visible

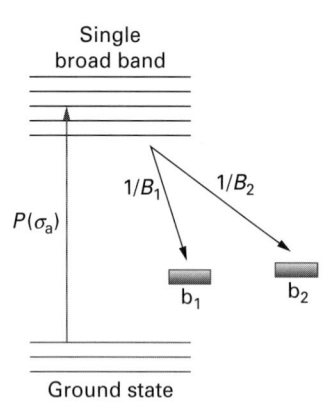

Figure 7.28 Schematic representation of a dual pathway photochemical degradation model. Optical irradiation excites the chromophores into one of several excited states. These excited states can then react either to form one bleached product, b$_1$ (probability 1/B_1), or react in a different manner to produce a different photodegradation product, b$_2$ (probability 1/B_2) [54].

optical wavelengths but more recently measurements have been extended to telecommunication wavelengths [53,54]. These measurements have been analyzed by very simple models representing photochemical decay from the lowest-lying unoccupied molecular orbital (LUMO) by one or two pathways. This type of analysis permits a simple photochemical figure-of-merit, B/σ, where B^{-1} is the probability of decay from the LUMO state of the interband (charge transfer) transition and σ is the absorption coefficient of the chromophore at the pump wavelength. For two pathways, two figure-of-merit values can be defined (Fig. 7.28).

For most materials, generation of singlet oxygen and subsequent attack on reactive chromophore sites is the most prevalent mechanism of photochemical degradation. When photo-decay associated with singlet oxygen is inhibited, the photochemical figures-of-merit can improve by up to five orders of magnitude. Indeed, accelerated testing at high optical powers suggests that, with exclusion of oxygen, greater than 10 years' stability can be obtained for current telecommunication power levels. Methods of retarding singlet oxygen kinetics include steric protection of reactive chromophore sites, lattice hardening (crosslinking) to inhibit oxygen diffusion, introduction of singlet oxygen quenchers, and packaging to exclude oxygen. Organic single crystals, such as DAST, and materials prepared by sequential-synthesis/self-assembly do not exhibit significant photo-instability, reflecting the dense lattices associated with these materials.

B/σ values of greater than 250×10^{32} m^{-2} are required to satisfy the requirement of 10 years' stability at current telecommunication power levels [53 to 55]. Table 7.3 provides values for typical air-saturated samples. The chromophore is the same in these two cases; in one case it is physically incorporated into amorphous polycarbonate (APC) to form a composite material while in the second case it is covalently incorporated into the SBLD-1 (C1) material (see Fig. 7.11 for the structure of SBLD-1 (C1)).

Table 7.3 Photostability of SBLD-1 (C1) compared with its guest/host counterpart, FTC loaded into amorphous polycarbonate [54 to 56].

Material	$B/\sigma \times 10^{32}$ (m^{-2})	$B/\sigma_1 \times 10^{32}$ (m^{-2}); contribution	$B/\sigma_2 \times 10^{32}$ (m^{-2}); contribution
SBLD-1	61 ± 14	15 ± 4 (28%)	210 ± 59 (72%)
FTC in APC	7.2 ± 0.7	1.0 ± 0.1 (40%)	6 ± 1 (60%)

In Table 7.3 data are shown for both single and multiple decay pathways. Note that even the somewhat denser structure of the SBLD-1 (C1) material leads to a significant improvement in photostability. It can be noted that commercial electro-optic devices, based on organic electro-optic materials, that are marketed by Gigoptix Corporation satisfy Telcordia standards.

7.9 Experimental methods for evaluating r_{33} in poled EO materials

A number of experimental methods have been commonly employed to evaluate the electro-optic coefficient, r_{33}, resulting from electric field poling of EO materials. These methods assess the magnitude either of second-harmonic generation (SHG) or of the linear electro-optic effect.

Using various SHG techniques, first-order nonlinear susceptibility $\chi^{(2)}_{333}$ may be calculated from the second-harmonic intensity, $I_{2\omega}$ [57]. However, the frequency dispersion relationship of the measured value differs with respect to that of r_{33}. In order to compare this value (d_{33}) with those from direct r_{33} measurements, the frequency dependence may be accounted for using either a two-state approximation [58,59] or by using time-dependent quantum mechanical computer modeling (i.e., time-dependent density functional theory, TDDFT; or real-time, time-dependent density functional theory, RT-TDDFT) [8,10,55,60,61]. Assuming a two-state dispersion model, the electro-optic coefficient is related to the first nonlinear susceptibility by

$$r_{33} = \frac{2}{n^4} \frac{(3\omega_0^2 - \omega^2)\left(\omega_0^2 - \omega'^2\right)\left(\omega_0^2 - 4\omega'^2\right)}{3\omega_0^2 (\omega_0^2 - \omega^2)^2} \chi^{(2)}_{333} \qquad (7.5)$$

where the operating (fundamental) frequencies for r_{33} and SHG are given by ω and ω', respectively, the material resonance frequency (related to the reciprocal of λ_{max}) is denoted as ω_0, and n is the refractive index at ω [58]. A problem associated with the use of second-harmonic generation is the absorption of the second-harmonic light by the chromophore.

Direct r_{33} measurement methods typically detect a change in output intensity caused by electric-field-induced modulation of the refractive index. Such experiments include non-contact methods such as single-beam reflection ellipsometry (Teng–Man technique) [62,63] and two-slit interference [25,64], as well as contact-prism coupling methods such as attenuated total reflection (ATR) [65 to 67]. Additionally, r_{33} can be calculated from the operating voltage, $V\pi$, of a working (e.g., Mach–Zehnder) optical modulator [36,68] or by Fabry–Perot interferometry.

The two most practical and common routine screening methods are the TMT and ATR methods. In the TMT method, a time-varying electric field induces an anisotropic change in refractive index, altering the phase angle, ψ, between surface and plane polarizations of the incident light. The incident phase angle is set by a Soleil–Babinet compensator and is therefore known. As ψ changes owing to the linear electro-optic effect, a change in net polarization results. The magnitude of change in ψ is governed by r_{13} and r_{33}, respectively. For small order parameters such as those typically encountered in simple guest/host poled EO polymers, it can be assumed that $r_{33} \sim 3r_{13}$. When the sample is placed between two crossed polarizers, the resulting time-varying intensity change (measured by a lock-in amplifier) is proportional to r_{33}.

To perform ATR measurements, a high refractive index prism is pressed against a contact electrode poled sample, typically using a pneumatic plunger [66,67]. The sample/prism assembly is rotated to vary the incidence angle of the fundamental beam. At an angle determined by the refractive index of the sample, the fundamental is no longer reflected from the prism/sample interface, but coupled into the sample. This phenomenon causes an attenuation of the reflected fundamental intensity as the incidence angle is scanned past the coupling angle. Simultaneously, a time-varying electric field is applied using the poling electrodes, modulating the refractive index. This index modulation produces a time-varying shift in the prism-coupling angle and a change in reflection intensity. This intensity change, measured using a lock-in amplifier, is proportional to r_{33}. Unlike the TMT method, ATR can also be used to measure r_{13} directly.

The simple TMT method assumes that there is no dichroism (anisotropic absorption) and that $n_o \approx n_e$. The accuracy of the TMT may suffer from complications related to multiple reflections at sample layer interfaces [69]. However, similar to SHG, the TMT is easily modified to monitor relative r_{33} during the poling process [70 to 76]. Additionally, *in-situ* TMT has been used to monitor laser-assisted poling experiments [22] and thermal stability of r_{33} [7,72]. In comparison, ATR instrumentation is more difficult to construct and operate than TMT. ATR *in-situ* measurements are difficult, and sample damage is possible owing to prism contact. However, ATR can be used to measure both r_{33} and r_{13}, does not suffer from multiple reflection difficulties, and can be easily adapted to perform measurements at multiple wavelengths. ATR data also yield additional information such as sample thickness and refractive index from a single measurement. Unlike SHG, both TMT and ATR experiments require only relatively inexpensive optics and low-power laser sources. Two advantages afforded by SHG are ease of *in-situ* monitoring, and the ability to pole without a top contact electrode (corona poling). SHG is also a non-contact technique that can be used to evaluate very thin material layers (monolayers), and materials that are difficult to process (crystals).

7.10 Optical measurement of poling-induced order: VAPRAS and VASE

No direct, independent measurement of the order parameter $<\cos^3\theta>$ is currently available. Recent work has demonstrated that by using relatively simple experimental

7.10 Optical measurement of poling-induced order

techniques such variable angle polarization referenced absorption spectroscopy (VAPRAS) [77] or variable angle spectroscopic ellipsometry (VASE), it is possible to evaluate the centrosymmetric order parameter, $\langle P_2 \rangle = \phi = (3\langle\cos^2\theta\rangle - 1)/2$ [9,10,74,77,78]. In simple guest/host materials (simple three-dimensional systems at low density), $\langle P_2 \rangle$ is related to the acentric order parameter $\langle\cos^3\theta\rangle$ quite simply by $\langle\cos^3\theta\rangle = [3/5\langle P_2 \rangle]^{1/2}$ [4,8]. However, in reduced-dimensionality materials (MAP, binary chromophore materials, etc.) this relationship becomes more complicated (see discussion above). Because, in each individual material, both centrosymmetric and acentric order parameters are measures of the same underlying molecular orientation distribution, the relationship between the two can always be assessed using rigid body Monte Carlo computer simulations. In fact, when chromophores are loaded into an ordered host (matrix-assisted poling, binary chromophore materials, etc.), this relationship may be used in concert with nonlinear optical measurements and computer simulation to assess the effects of the host order on poling-induced guest order [9].

Ordered materials composed of ensembles of dipolar dyes (chromophores) exhibit both anisotropic nonlinear and linear optical susceptibilities. The magnitude of anisotropy in the imaginary (dichroism) and real (birefringence) components of the refractive index is related to the nature and extent of molecular order. Assuming a dipolar dye in which the transition dipole tensor element, coincident with the permanent dipole moment, is much greater than any other component, the extent to which polarized light is absorbed is related to the square of the cosine of the angle between the chromophore dipolar axis and the E-field vector of the incident light. The quantitative relationship between absorption of polarized light of a given wavelength, λ, can be written in terms of the absorbance, $\langle a_\lambda \rangle$, which is useful for absorption spectroscopy. Equivalently, this relationship may be expressed in terms of the imaginary component of the refractive index, k_λ, where the complex refractive index is $n_c = n_\lambda + ik_\lambda$. This form is more compatible with ellipsometric measurements. For light polarized parallel to the z-axis (film normal)

$$\langle a_\lambda \rangle_P = \bar{a}_\lambda \{(1 - \langle P_2 \rangle) + 3\langle P_2 \rangle \sin^2\psi\} \tag{7.6}$$

where ψ is the angle at which light propagates through the film (with respect to the z-axis), and $\bar{a}_\lambda$ is the absorbance of an isotropic (unpoled) sample. In a sample possessing poling-induced order, $\langle a_\lambda \rangle_P$ increases as ψ approaches 90°. In contrast, the absorbance of light polarized perpendicular to the z-axis does not depend on ψ (for uniaxial order) and

$$\langle a_\lambda \rangle_\perp = \bar{a}_\lambda \{1 - \langle P_2 \rangle\}. \tag{7.7}$$

In the simplest absorption spectroscopy method, the absorption of unpolarized light passed through a sample at normal incidence is measured before and after poling. In this normal incidence method (NIM), as the chromophores are ordered along the z-axis, light absorption decreases. Rearranging the equations above, $\langle P_2 \rangle$ can be calculated from the NIM data using

$$\langle P_2 \rangle = \phi = 1 - \frac{\langle a_\lambda \rangle_\perp}{\bar{a}_\lambda} \tag{7.8}$$

In this case, $\bar{a}_\lambda$ and $\langle a_\lambda \rangle_\perp$ are simply the absorbances of the sample measured before and after poling [75 to 77]. It has been suggested that the NIM provides only a crude estimation of $<P_2>$. Although precautions (de-poling experiments) [75 to 77] may be taken, the NIM potentially has difficulty accounting for poling-induced decomposition, and can be subject to error associated with sample alignment and referencing. In order to address these potential pitfalls, the VAPRAS method was introduced [77], extending the original dichroism-based method proposed by Graf *et al.* [78]. In VAPRAS, $\langle a_\lambda \rangle_P$ is measured at increasing angles of incidence, θ. Rigorous analysis of the absorption angle dependence exhibited by multilayer (i.e., glass, ITO, EO material) poled samples can be accomplished using a Jones Matrix analysis method [10,79]. However, such rigorous analysis requires detailed information about the optical constants (real and imaginary) and thickness of each sample layer. Although such rigorous analysis sets the standard for accuracy in measurements of the small values of $<P_2>$ typically encountered in poled films, detailed refractive index and thickness data must be known for each of the multiple sample layers involved prior to analysis. In order to reduce this complication and increase the accessibility of $<P_2>$ measurements, VAPRAS has the clear advantage of being able to remove the layer thickness dependence and permit determination of an effective sample refractive index over the course of a single experiment [77].

In VAPRAS, both $\langle a_\lambda \rangle_P$ and $\langle a_\lambda \rangle_\perp$ are measured as a function of angle. From these data, $<P_2>$ can be calculated using the ratio

$$\frac{\langle a_f(\lambda_{\exp})\rangle_P}{\langle a_f(\lambda_{\exp})\rangle_\perp} = 1 + \sin^2\theta\left(\frac{3r\langle P_2\rangle}{1 - r\langle P_2\rangle}\right) \qquad (7.9)$$

In the VAPRAS analysis $\langle a_\lambda \rangle_\perp$ is essentially a reference. Effective n_c (sufficiently accurate for sample analysis) can be deduced by identifying the refractive index where $\langle a_\lambda \rangle_\perp$ does not change with respect to θ (Fig. 7.29).

Another complementary method for assessing $<P_2>$ is through variable angle spectroscopic ellipsometry (VASE) [55,80 to 83]. VASE is an established technique by which thin-film thickness as well as real and imaginary components of refractive index may be determined for both isotropic and anisotropic materials [82,83]. By fitting the data using a Kramers–Kronig consistent oscillator layer model, real and imaginary components of the complex refractive index are determined across a broad wavelength range. In the case of poled EO materials, a uniaxial anisotropic model can be used to assess optical constant anisotropy. In the uniaxial anisotropic model, the dielectric properties only differ between $E_P(k_P)$ and $E_\perp(k_\perp)$ electric field vector orientations. Such a model corresponds to differences only in the diagonal elements of the 2×2 Jones Matrix similar to the rigorous VAPRAS analysis. Using VASE data, $<P_2>$ values can be determined according to

$$\langle P_2 \rangle = \frac{k_P - k_\perp}{k_P + 2k_\perp} \qquad (7.10)$$

derived analogously from equations (7.6) and (7.7) above.

An advantage of VASE is that the orientations of multiple components of the organic electro-optic material can be simultaneously defined, e.g., both courmarin (pendant) and chromophore orientation can be simultaneously measured as illustrated in Fig. 7.30.

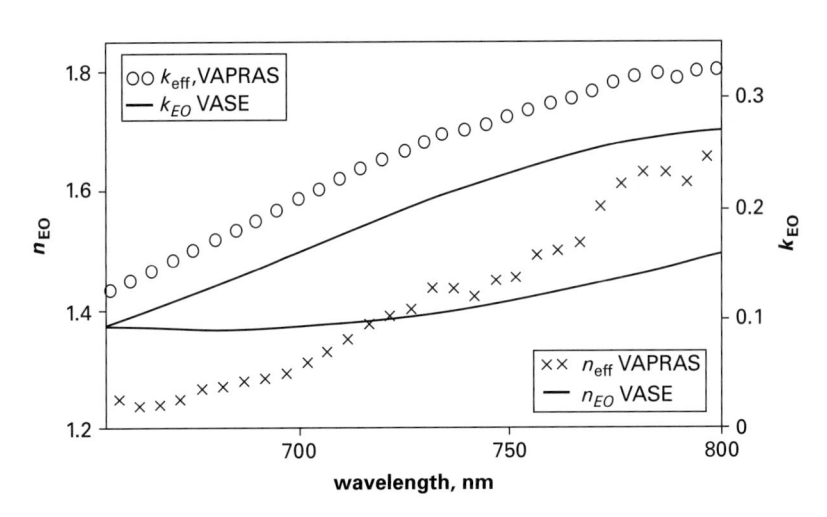

7.10 Optical measurement of poling-induced order

Figure 7.29 Comparison of the real, n, and imaginary, k, components of the refractive index of an EO material film plotted with respect to optical wavelength as determined by VASE (n_{EO}, k_{EO}) and VAPRAS (n_{eff}, k_{eff}). Reprinted with permission from Ref. [77]. Copyright 2011, American Chemical Society.

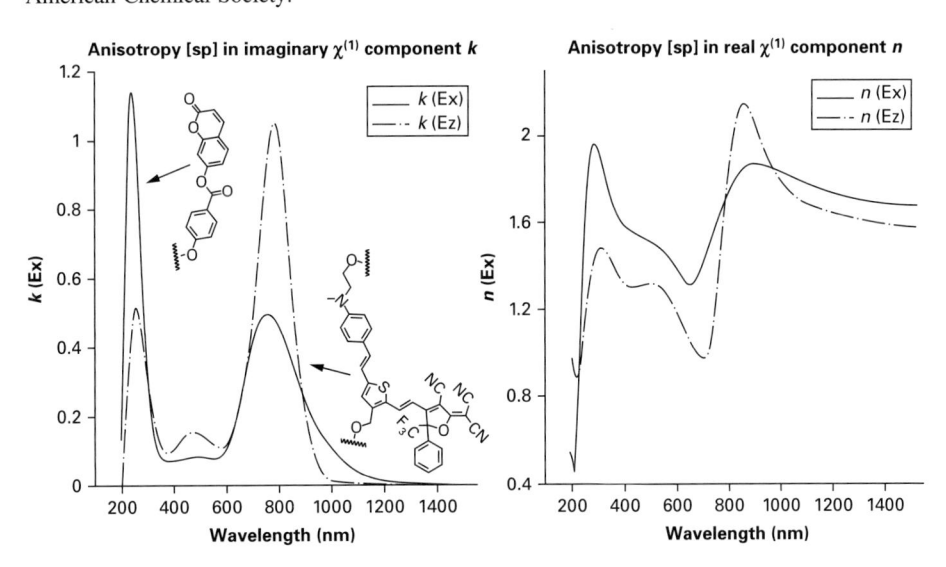

Figure 7.30 The VASE of C1, illustrating that coumarin and chromophore moieties are restricted to orthogonal planes. The $<P_2>$ order parameters are $<P_2> = 0.24$ (chromophore) and $<P_2> = -0.19$ (coumarin).

For example, as shown in Fig. 7.30, the chromophores and coumarin of C1 are restricted to orthogonal planes.

The NIM and VAPRAS methods for assessing $<P_2>$ have the advantages of simplicity and cost. Both experiments essentially employ modified UV–visible spectrophotometers. In contrast, VASE requires quite expensive analytical equipment and complex data analysis software. However, the NIM and VAPRAS methods are both transmission experiments. As such, these experiments are limited to relatively low optical densities or thin samples and low substrate absorption (e.g., ITO) properties. The VASE technique can be operated in reflection mode, alleviating some of these difficulties and allowing a very wide spectral window for sample analysis.

Recently, difference-frequency generation (DFG) studies of the CN stretch in the spectral region 2150–2350 cm^{-1} has been employed to estimate poling-induced order of electro-optic chromophores [84]. This, of course, assumed that chromophore order is the same as the order of the cyano-containing acceptor region of the chromophore.

7.11 Processing options

An advantage of organic electro-optic materials is that they are amenable to a variety of processing options including soft/nano-imprint lithography [45,85] and lift-off techniques [86] (see Figs. 7.31 and 7.32).

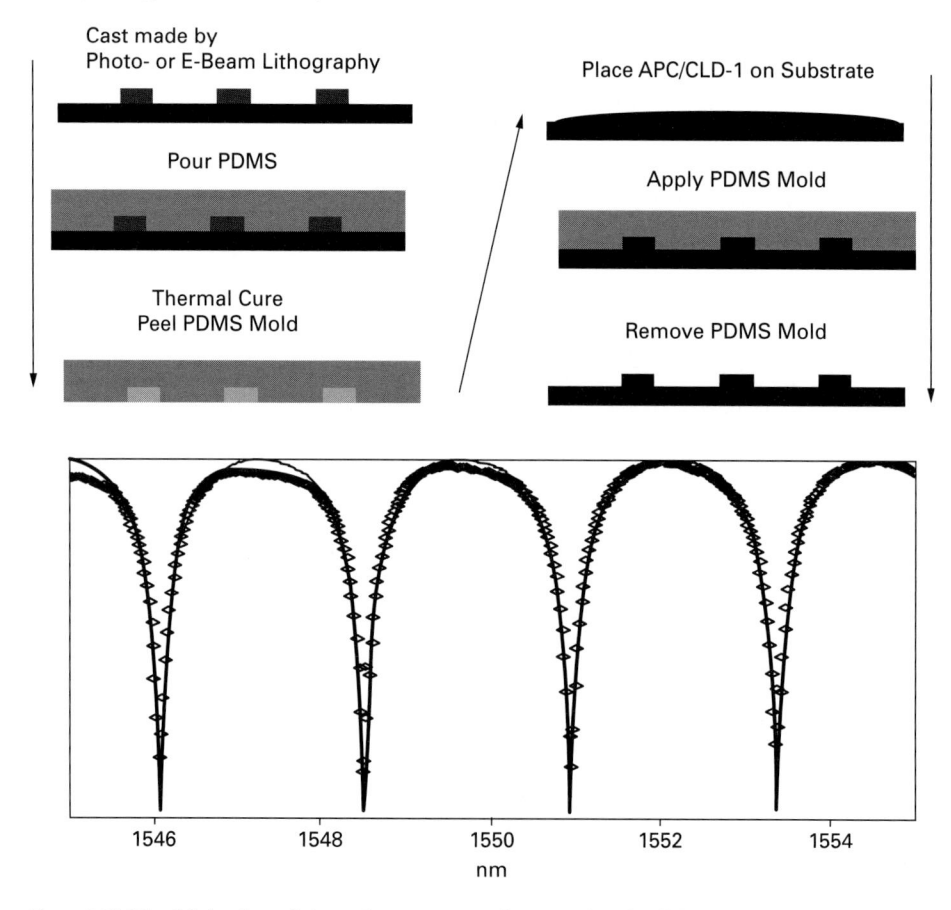

Figure 7.31 The fabrication of ring microresonators by nano-imprint lithography (top), together with demonstration of performance (the line is the theoretical simulation while the diamonds are experimental points). Adapted from Ref. [45] with permission of the American Chemical Society.

7.12 Fabrication of all-organic devices 155

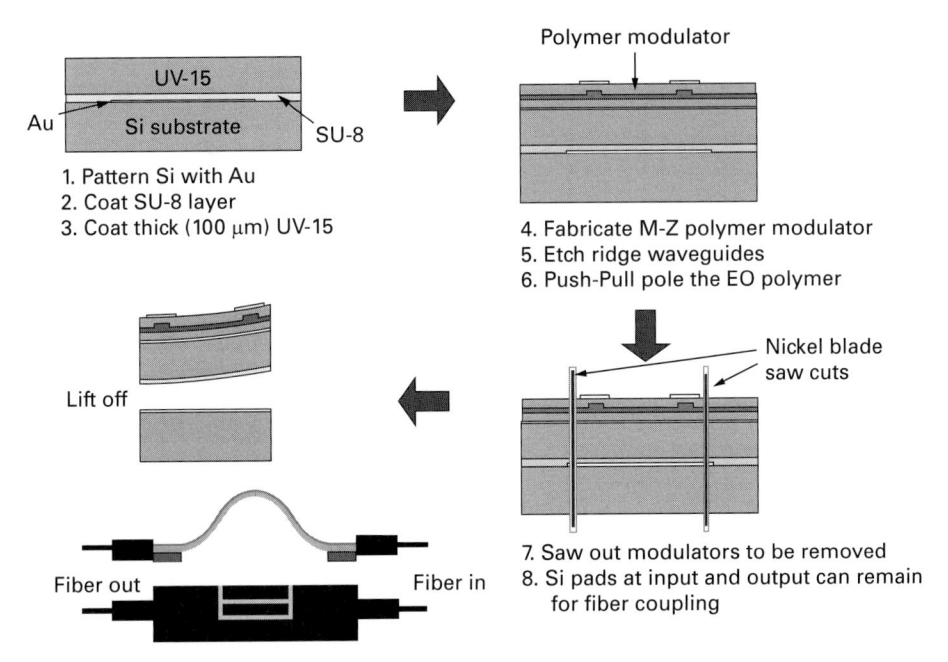

Figure 7.32 The fabrication of flexible modulators by a lift-off technique, and a test bed for measurement of the performance with bending, shown at the bottom left. Adapted from Ref. [86] with permission.

Soft and nano-imprint lithography can be used to "stamp out" complex circuits such as multiple ring microresonators [45] and Mach–Zehnder modulators [85]. The performance of such circuits is not noticeably different from those prepared by precision etching techniques such as electron-beam etching [45]. Lift-off techniques have been used to fabricate conformal and flexible devices [86]. Again, quite impressive performance can be realized for such devices. Properties such as drive voltage (see Fig. 7.33), insertion loss, and bias voltage are little changed even with extreme bending. Organic electro-optic devices join organic light-emitting devices in affording a very viable approach to conformal and flexible devices. As already noted, control of material glass transition temperature by Diels–Alder/retro-Diels–Alder crosslinking chemistry is very useful in facilitating soft and nano-imprint lithography techniques.

With use of planarizing polymers, organic electro-optic (photonic) circuitry can be fabricated on top of VLSI electronic circuitry [87]. Neither the performance of the electronic nor the photonic circuitry is compromised by this process.

7.12 Fabrication of all-organic devices

All-organic electro-optic devices are typically three-layer devices consisting of lower cladding layer, active electro-optic core layer, and upper cladding layer, which are, of course, surrounded by lower and upper electrodes (see Fig. 7.1 bottom). As noted

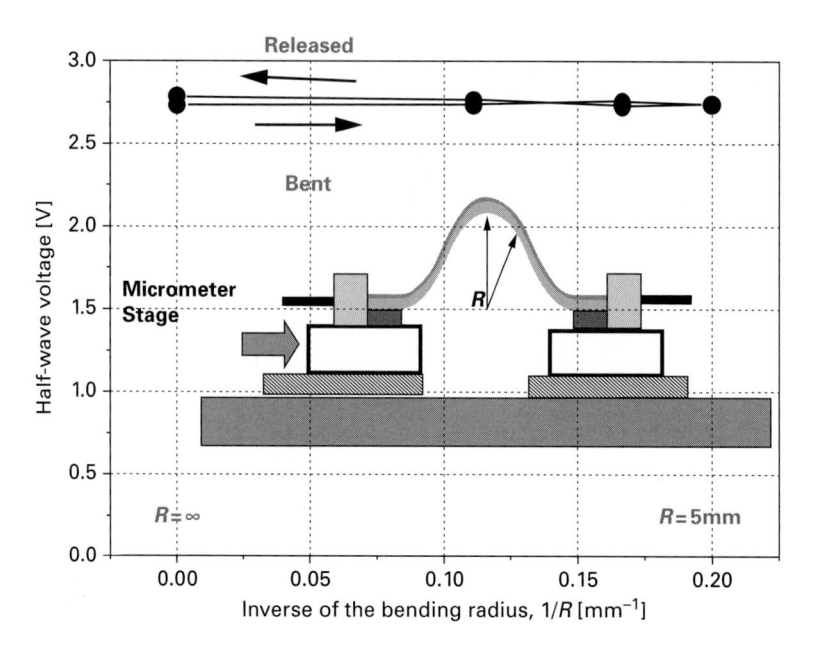

Figure 7.33 Variation of half-wave (V_π) voltage with bending radius.

previously, an advantage of all-organic materials is that they can be prepared by exploiting a variety of processing options to produce complicated device structures, including by soft and nano-imprint lithographic techniques, and conformal/flexible devices by lift-off techniques. There are some distinct challenges related to the production of all-organic devices, including problems associated with deposition of cladding layers. Cladding layers are required to prevent fringing fields of light propagating in the electro-optically active core waveguide from interacting with the metal electrodes. Cladding layers must be sufficiently thick that such optical loss is within acceptable limits (a few tenths of a dB/cm). Of course, thicker cladding layers lead to an unwanted increase in electrode separation and thus drive voltage, see Eq. (7.11), and the trade-off between optical loss and drive voltage must be balanced in selection of cladding layer thickness.

$$V_\pi = \lambda h / n^3 r_{33} L \Gamma \tag{7.11}$$

where λ is the operating wavelength, h is the electrode spacing, L is the electrode length, r_{33} is the electro-optic coefficient, n is the index of refraction, and Γ is the modal overlap parameter. The electrode spacing, h, is significantly influenced by cladding layer thickness. Cladding layer thickness can also impact poling efficiency when poling is effected through the cladding layers.

The optical loss of cladding layers is often ignored in device design, which as can be seen from Table 7.4 is potentially a very problematic assumption. Moreover, it can be a mistake to ignore the photostability of cladding materials.

This is particularly a problem for UV-curable epoxy cladding materials. Another problem that can arise with respect to cladding layers is that the spin-casting solvent

7.12 Fabrication of all-organic devices

Table 7.4 Commonly employed cladding polymers and their optical properties.

Materials	Index @ 1310 nm	Loss dB/cm @ 1310 nm	Index @ 1550 nm	Loss dB/cm @ 1550 nm
SU-8	1.567	0.5	1.565	4.0
NOA61	1.545	0.3	1.541	1.1
UFC170	1.490	0.5	1.488	3.0
UV15	1.510	0.9	1.504	4.2
Teflon AF 1601	1.300	–	1.297	–
ZPU12-R1		0.14	1.45–1.47	0.3
ZPU13-R1		0.13	1.41–1.45	0.24
CLD/APC EO polymer	1.614 TM	1.0	1.612 TM	1.7

TM, transverse magnetic mode

used in deposition of the cladding layer can sometimes attack the active core layer onto which the cladding layer is being deposited. This attack can lead to pitting and unacceptable optical loss due to scattering from the pitted interface. Crosslinking of the core layer prior to deposition of a cladding layer can help prevent such solvent attack. UV curing of cladding layers can also initiate unwanted free radical reactions in the electro-optic core layer, and the composition of both core and cladding must be considered in deciding upon processing options.

The electrical and optical properties of the cladding materials must be taken into account when analyzing the performance of all-organic electro-optic devices prepared by electric field poling through cladding layers. Ideally, the optical loss of the cladding material should be low and the electrical conductance greater than that of the active core (so that the poling and drive electric fields are dropped across the cladding layers to the active EO core layer). James Grote and co-workers [88 to 93] have discussed the role of cladding materials at length. High conductance through electro-optic materials is particularly a problem for binary chromophore organic glasses and other materials containing high chromophore number densities and has required the deposition of a thin charge injection blocking layer (e.g., 25–200 nm of titanium dioxide or benzocyclobutene) as illustrated in Fig. 7.34 [73,94]. Without the use of charge injection blocking layers, electrical conductance under electric field poling conditions can limit poling fields to 70 volts/micron or even lower. Thus, charge blocking layers can lead to 30–50% improvement in poling induced electro-optic activity. Such conductance is not typically a problem at lower temperatures (normal operational temperatures of devices). Conductance can be an even greater problem with materials prepared using laser-assisted electric field poling, owing to photo-induced charges.

A wide variety of all-organic device structures have been fabricated, including both non-resonant (stripline) and resonant (microring resonator and etalon) devices. For both resonant and non-resonant device structures, drive voltage (V_π) and bandwidth are interdependent (but in different ways). In Fig. 7.35, the interdependence of device

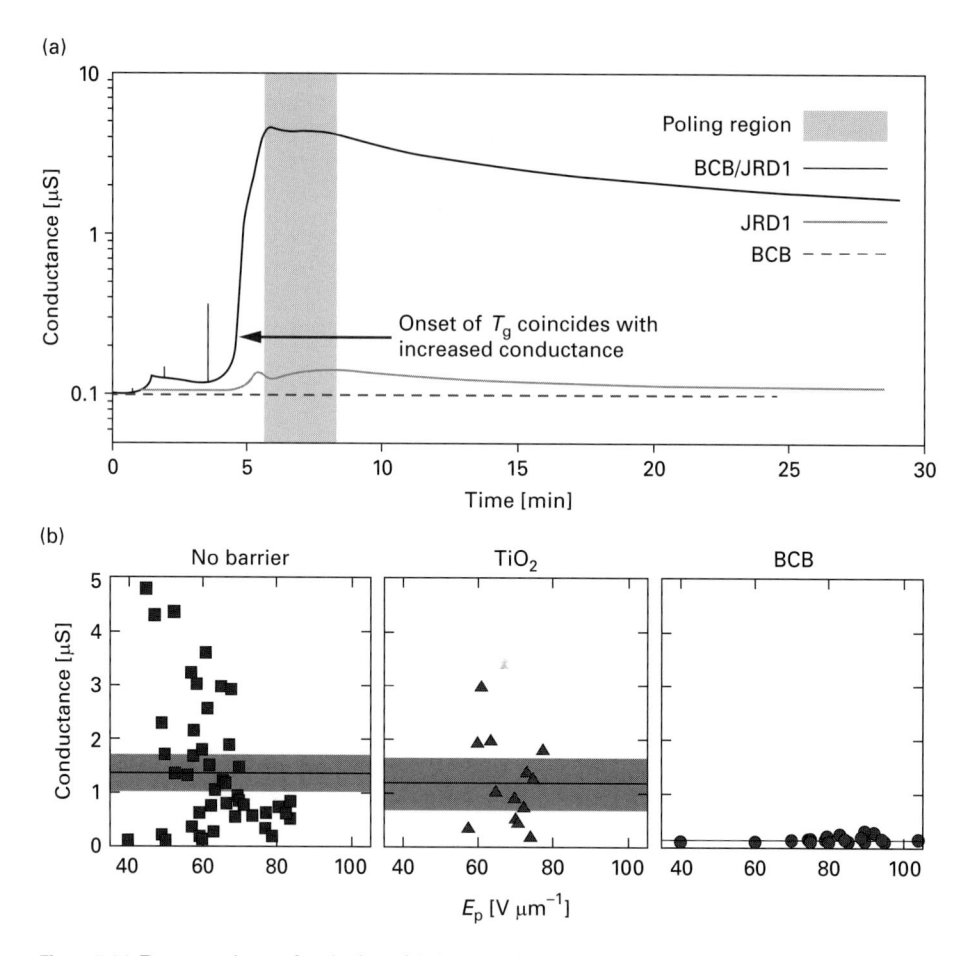

Figure 7.34 Data are shown for devices fabricated with neat JRD1 chromophore with and without various barrier layers. The JRD1 chromophore is essentially the YLD124 chromophore with two phenyl groups attached to the Si atoms on the donor end (replacing methyl groups). This additional bulk facilitates the examination of neat chromophore films. (a) Conductance (G) of a JRD1 device as a function of time during poling with applied poling fields of 95 V/μm. The upper solid line trace is without any buffer layer. The middle solid line trace is with a BCB barrier layer [73] and the lower dashed line trace is for a BCB layer. The shaded region is the duration of the poling process (the period of time that the sample is at the glass transition temperature). (b) The average conductance during the poling process of JRD1 devices with no barrier layer (squares, $G = 1.37 \pm 0.18$ μS), with a TiO_2 buffer layer (triangles, $G = 1.18 \pm 0.22$ μS), and with a BCB buffer layer (circles, $G = 0.142 \pm 0.008$ μS). The shaded bands represent the 95% confidence intervals around the mean conductance. Reproduced from [73] with permission of the American Chemical Society.

length, bandwidth (switching speed), and drive voltage is illustrated for stripline (Mach–Zehnder) devices employing gold metal drive electrodes (microwave loss of 0.75 dB(GHz)$^{1/2}$/cm).

For resonant devices, the quality factor (lifetime of the photons in the resonant device) defines both the drive voltage requirement and the bandwidth. Thus, the critical

7.12 Fabrication of all-organic devices 159

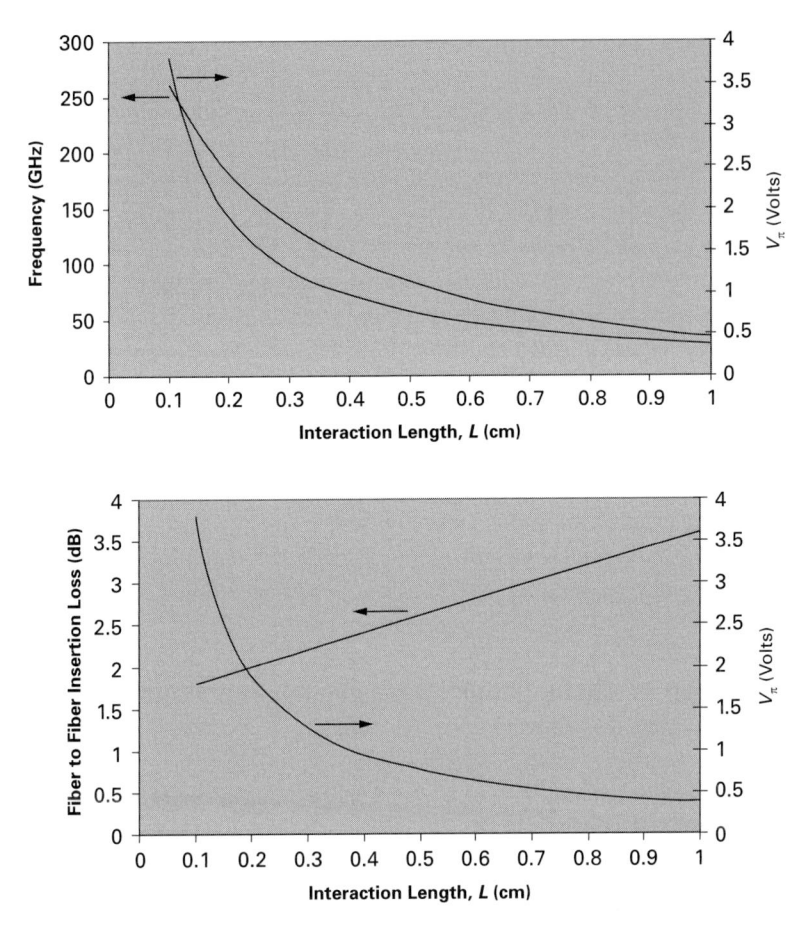

Figure 7.35 Upper figure: The relationships between half-wave voltage (V_π), device bandwidth (frequency), and electrode length (or more precisely, electrical–optical field interaction length). Lower figure: The relationships between half-wave voltage, fiber to fiber insertion loss, and electrode length (interaction length).

parameter is the bandwidth/voltage ratio. Resonant devices fabricated from electro-optic materials can be conveniently viewed as voltage-controlled bandpass filters. The bandwidth/voltage ratio defines how far the bandpass notch can be shifted with application of a given applied voltage. Effective modulation requires an approximate shift equal to the full width at half maximum of the notch bandpass. Typical Q values (of a few thousand) limit information processing bandwidths to a few tens of gigahertz. However, if active wavelength division multiplexing (color coding of information) is being employed, high information throughput can still be achieved. For example, if information is multiplexed into 10 wavelengths then a factor of 10 improvement of effective bandwidth is achieved. (See Chapter 9 for a more detailed discussion of waveguide devices.)

The bandwidth performance of stripline organic electro-optic devices is typically defined by the resistivity of metal electrodes. For 2-cm gold electrodes, the bandwidth is typically limited to less than 100 GHz (see Fig. 7.35). Higher bandwidth requires shorter device lengths, which results in increased drive voltage requirements. The situation for organic electro-optic stripline devices is quite different than for lithium niobate modulators where the bandwidth and drive voltages can be limited by the velocity mismatch between electrical and optical waves propagating in the material. The high dielectric constant of lithium niobate ($\varepsilon = 32$) and relatively low index of refraction ($n = 2.2$) results in substantial velocity mismatch. Although clever device design has addressed the velocity mismatch problem (leading to devices where bandwidth is essentially defined by the resistivity of metal electrodes as with organic electro-optic materials), it is still very difficult to fabricate Mach–Zehnder modulators from lithium niobate that exhibit operational bandwidths greater than 40 GHz.

Different types of devices have different drive voltage requirements. For radiofrequency (RF) photonic applications, it is desirable to achieve the electrical-to-optical-to-electrical signal transduction process with no loss of signal or even better with the realization of gain. Typically, achievement of gain requires a V_π voltage of 0.2 V or less, which in turn requires an electro-optic coefficient of the order of 300–1000 pm/V (depending on device length and operational bandwidth).

7.13 Fabrication of silicon photonic, plasmonic, and photonic crystal hybrid devices

One of the most exciting events in optics and nonlinear optics has been the advent of silicon photonics. This is because the implementation of photonics as well as electronics in a CMOS silicon platform has the potential for photonic circuit production to leverage the substantial investment in CMOS electronics. Also, the high index of refraction of silicon facilitates circuit miniaturization. For example, light at telecommunication wavelengths can be confined in waveguides of a few hundred nanometers lateral dimension. Moreover, light can be guided more effectively around sharp bends. Thus, photonic circuit dimensions can be greatly reduced relative to the dimensions of conventional silica fiber optics. Furthermore, light can be further compressed into sub-150-nm vertical slots cut into silicon waveguides [95].

An equally exciting observation is that lossless (or very low loss) transition of light from silicon waveguides can be made into active organic nonlinear optical materials by filling such materials into 25–150 nm vertical slots cut into silicon waveguides (silicon–organic hybrid, SOH, devices; see Fig. 7.36 [96 to 99, 101 to 112]. More recently, horizontal slot organic electro-optic/silicon photonic hybrid devices have been demonstrated (see Fig. 7.37) [99].

The concentration of light (and "low frequency" electric fields by reduction of electrode spacing) leads to marked enhancement in the "effective" optical nonlinearity of organic materials [95 to 122]. Indeed, millivolt drive voltages for electro-optic modulation [97 to 99, 101 to 112] and milliwatt optical control powers for all-optical modulation [100] are feasible by exploiting organic nonlinear optical materials incorporated into silicon photonic circuitry. Both silicon ring microresonator and stripline

7.13 Fabrication of hybrid devices

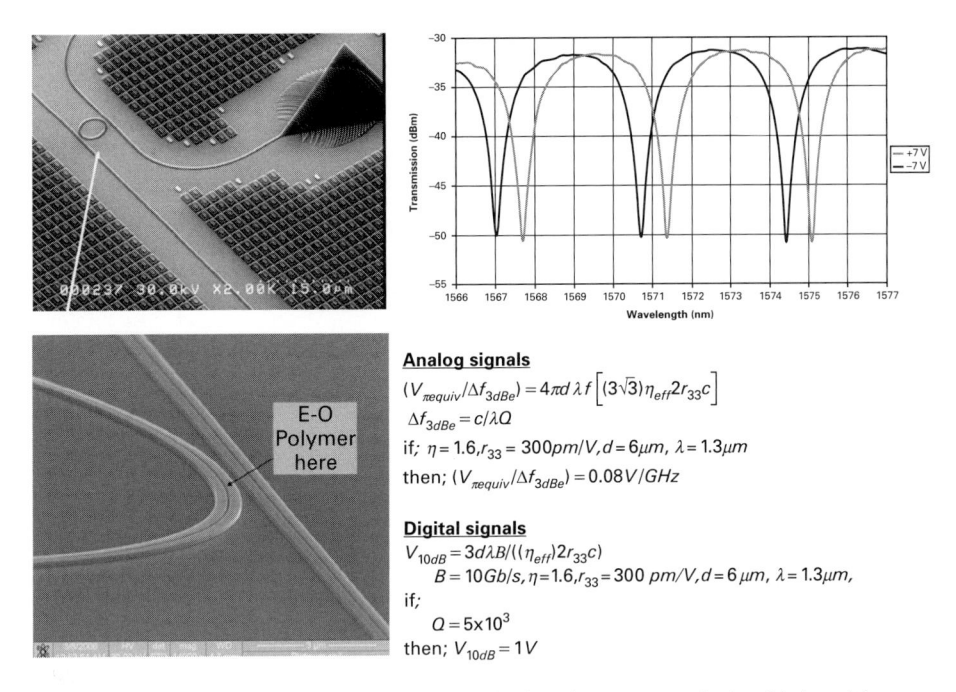

Analog signals

$(V_{\pi equiv}/\Delta f_{3dBe}) = 4\pi d\,\lambda f\left[(3\sqrt{3})\eta_{eff}2r_{33}c\right]$

$\Delta f_{3dBe} = c/\lambda Q$

if; $\eta = 1.6, r_{33} = 300\,pm/V, d = 6\mu m, \lambda = 1.3\mu m$

then; $(V_{\pi equiv}/\Delta f_{3dBe}) = 0.08\,V/GHz$

Digital signals

$V_{10dB} = 3d\lambda B/((\eta_{eff})2r_{33}c)$

$B = 10\,Gb/s, \eta = 1.6, r_{33} = 300\,pm/V, d = 6\mu m, \lambda = 1.3\mu m,$

if;

$Q = 5 \times 10^3$

then; $V_{10dB} = 1\,V$

Figure 7.36 Electron micrograph image of a slotted microring resonator device fabricated in silicon (left). Illustration of performance characteristics (right) [55, 96, 97]. Reprinted with permission from [55]. Copyright 2010, American Chemical Society.

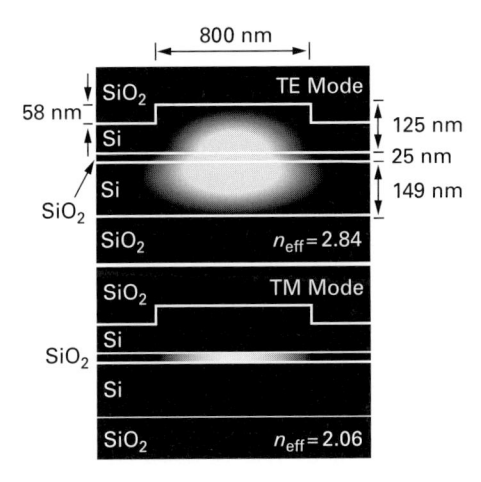

Figure 7.37 The confinement of TE and TM optical modes in a horizontal slot modulator with 25 nm thickness.

devices have been demonstrated. The major problem encountered to the present with silicon photonics is that of optical loss (both waveguide propagation loss and coupling loss). Propagation loss is typically defined by mask resolution, which determines wall roughness. Care must also be exercised to maintain bending loss to within acceptable limits. However, as experience in device fabrication has grown, propagation losses have

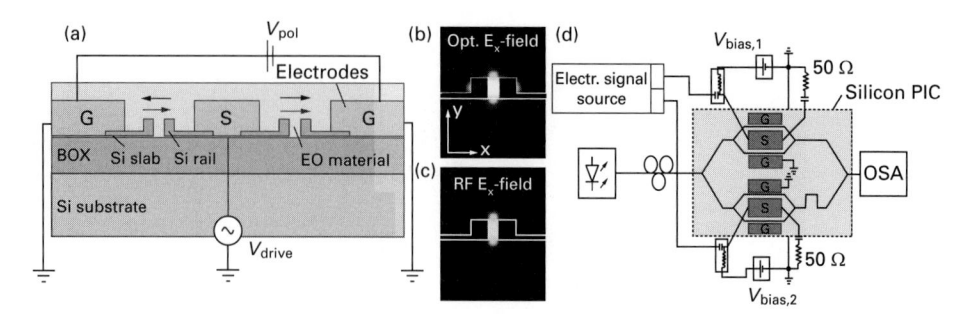

Figure 7.38 (a) A schematic representation of a slotted silicon waveguide modulator filled with an organic electro-optic material (SOH device). Adapted from Ref. [111] with permission. (b) and (c) Optical and RF field distributions. (d) A schematic representation of the IQ (in-phase quadrature) modulator configuration.

been reduced to acceptable values of a few dB/cm, which become insignificant for SOH devices 1 mm in length (or even shorter). Coupling losses have been greatly reduced by employment of AWG (array waveguide gratings) and various tapers including reverse tapers. Both metal electrodes and doped silicon electrodes have been employed with silicon–organic hybrid photonic devices. The latter afford greatly reduced electrode spacing, but the relatively low conductivity (relative to high-conductivity metals) of doped silicon limits bandwidth performance and makes electric field poling more difficult.

Researchers at Boeing, California Institute of Technology, and the University of Washington have demonstrated construction of a $1\times4\times1$ reconfigurable optical add/drop multiplexer (ROADM) based on using silicon microring resonators filled with an organic nonlinear optical material [101]. Researchers at the Universities of Texas and Washington [120 to 122] have recently demonstrated that vertical slow silicon photonic waveguides can be integrated with silicon photonic bandgap (slot wave) structures to achieve a voltage-length product of 0.29 V-mm, i.e., an effective drive voltage of 0.29 V for a 1 mm long device. Of course, this performance is achieved with accompanying bandwidth limitations, and such devices require great care to avoid unacceptable coupling loss. Recently, Chen and co-workers have extended this research to the integration of an organic electro-optic/silicon photonic sensor into a bowtie antenna structure [122]. Still more recently, researchers in Europe (Institutes IPQ and IMT of the Karlsruhe Institute of Technology and the Institute of Electromagnetic Fields of the ETH Zürich) and at the University of Washington, have demonstrated a voltage–length product of 0.5 V-mm for the SOH device structure shown in Fig. 7.38 [105 to 112]. This 1-mm device operating at 40 Gbit/s permits an energy efficiency of approximately 1 fJ/bit to be achieved (this is three orders of magnitude better than silicon injection/depletion devices). The poling-induced electro-optic activity realized for the DLD164 chromophore (see Fig. 7.38) in this device structure is approximately 180 pm/V. Higher in-device electro-optic activity for SOH devices has been demonstrated more recently [107, 110]. Advanced modulation formats (QPSK, 16QAM, etc.) have

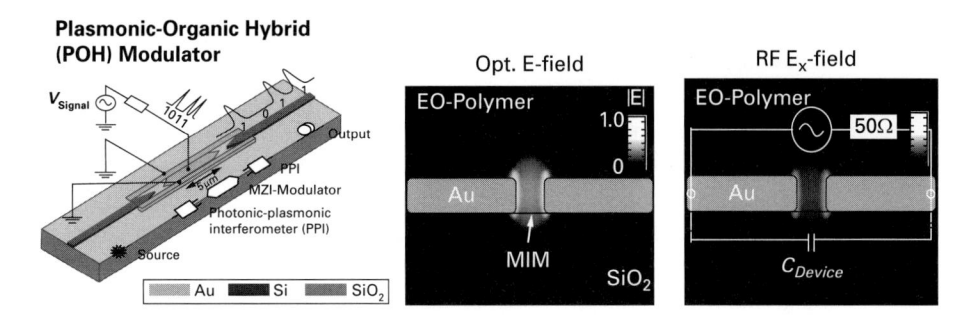

Figure 7.39 A schematic representation of a POH modulator is shown (left) together with optical and RF field distributions for a POH (right).

been demonstrated using these devices, and a variety of device architectures (for comb generation, bidirectional polarization-independent modulation, difference-frequency generation, etc.) have been implemented based on SOH integration.

Organic electro-optic materials have also been integrated into horizontal slot silicon photonic devices. Such devices are attractive for use with materials prepared by sequential synthesis/self-assembly and by crystal growth as a convenient deposition surface is available.

It is highly likely that innovations in silicon photonic device architectures will lead to further improved performance. Moreover, organic materials are beginning to be integrated with plasmonic and metamaterial device architectures, permitting the realization of active devices [113 to 119]. For plasmonic devices, including both insulator–metal–insulator (IMI) and metal–insulator–metal (MIM) devices, a reduction in optical loss has been demonstrated by exploiting nanostructured gold and silver. Reduced drive voltage–length parameters have been demonstrated for plasmonic–organic hybrid (POH) devices [115 to 117]. An example of plasmonic–organic hybrid (POH) integration is shown in Fig. 7.39. Devices with lengths of 10, 12.5, 25, and 50 μm have been fabricated. POH devices permit very high (>100 GHz analog and >100 Gbit/s digital) bandwidths to be achieved owing to both low resistivity and capacitance (low RC time constants) of such short devices. Voltage–length parameters of 50 V-microns have been observed for modulators exhibiting a plasmonic loss of 0.5 dB/micron (leading to an insertion loss of 12.5 dB for the shortest devices) and power efficiencies of 12.5 fJ/bit. Excellent extinction (24 dB) is achieved with POH devices. Bit error rates of 6×10^{-6} or lower are observed for operation of devices at rates of 72 Gbit/s or lower. High bandwidth operation has been demonstrated for advanced modulation formats (OOK, BPSK, and 4-APSK) leading to some of the best overall performance observed to date for Pockels effect modulation.

7.14 Synthetic strategies for covalently incorporated chromophore materials

The general class of covalently incorporated chromophore materials includes a wide array of chemical structures, as discussed in the preceding sections and in a number of recent reviews [55,71]. Various structural motifs have been demonstrated in which chromophores are bound to polymers, dendrimers, pendant groups, and to each other, in order to create nano-engineered bulk EO materials [55,71,123,124]. The intent of these varied attachment strategies is to increase optimum chromophore loading densities (improve solubility), to control intermolecular electrostatic interactions, and to direct acentric chromophore order (MAP). Here chemical synthesis methodologies relevant to the preparation of multi-chromophore dendrimers, dendronized polymers, and side-chain Diels–Alder "click" dendronized polymers are briefly introduced. Diels–Alder and fluorovinyl ether crosslinking chemistries are also discussed.

7.14.1 Multi-chromophore dendrimers

Arguably the most commonly employed method for the preparation of multi-chromophore dendrimers and dendronized chromophores alike is through the use of mild esterification and/or amidization chemistries [7,40,75,123,124].

Figure 7.40 illustrates the synthesis of multi-chromophore dendrimers 33 and 44 (Fig. 7.7). Core 1 was prepared via Williamson ether synthesis [72]. Chromophore 2 was synthesized as outlined in Chapter 4 of this book. Chromophore–core coupling was accomplished under DCC/DPTS esterification conditions at room temperature to afford tert-butyl dimethysilyl (TBDMS) protected dendrimer 33 in 50% yield. Subsequent acid deprotection followed by a second esterification reaction gave dendrimer 44 in 62% yield. Little or no chromophore decomposition was observed during the reaction sequence. The intermediates and target materials were purified by silica column chromatography.

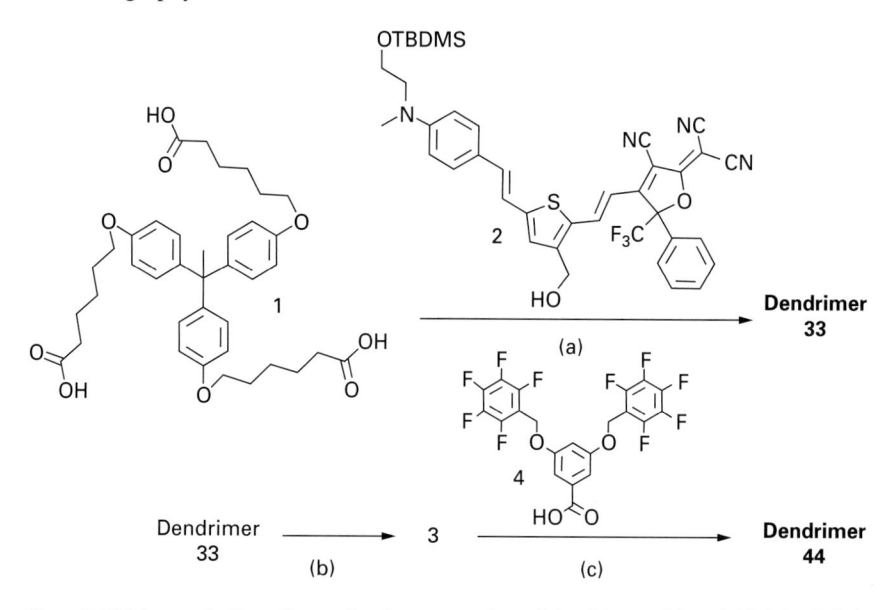

Figure 7.40 The synthetic pathway for the preparation of dendrimers 33 and 44 (Fig. 7.7) is shown. Reagents and conditions are as follows: (a) DCC, DPTS, DCM, RT 48 hr; (b) 1N HCl, MeOH, 1 hr; (c) DCC, DPTS, DCM, RT 24 hr.

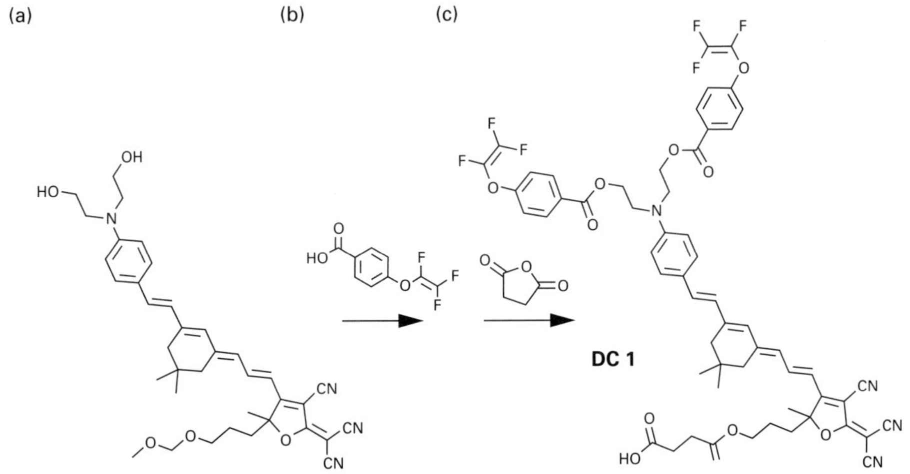

Figure 7.41 The synthesis of a dendritic CLD-type chromophore DC 1 is shown. Reagents and Conditions are as follows: (a) DCC/DPTS, DCM, 20 hr; (b) TMSBr, DCM, −30 °C, 4.5 hr; (c) pyridine, DCM.

7.14.2 Dendronized polymers

The synthesis of dendronized polymers has also been demonstrated through the use esterification chemistry similar to that discussed above. In order to avoid side-reactions, with some of the newest generations of high-β chromophores, post-polymerization chromophore/dendrimer attachment schemes have been employed [19, 40].

Figure 7.41 illustrates the synthesis of polyene (CLD) type dendronized chromophore 1 (DC 1) [40]. The donor terminus of the polyene bridge chromophore, bearing a methyl-O-methyl (MOM)-protected TCF acceptor, is derivatized with trifluorovinyl ether moieties at 67% yield. The MOM group is then removed using a Lewis acid (TMSBr) at 82%. Finally, an attachment point (carboxylic acid) is created by ring opening of succinic anhydride at the acceptor terminus in 81% yield.

Chromophore DC 1 may then be attached under esterification conditions to polyhydroxystyrene, followed by addition of a bulky crosslinkable dendritic unit to produce dendronized polymer P1 (Fig. 7.42).

This system not only provides site-isolation of the active chromophores, it also allows for thermal crosslinking during the poling process (Fig. 7.43). Upon heating, a thermally induced cyclodimerization reaction occurs between fluorovinyl ether units, hardening the lattice, and greatly enhancing thermal stability.

7.14.3 Diels–Alder "click" dendronized polymers

Diels–Alder "click" chemistry schemes have been introduced to improve control over chromophore attachment and crosslinking. Using these methods, polymer matrices bearing chromophores as side-chains may be prepared *in-situ* simply by mixing the diene- or dienophile-bearing components, and heating [40,123,124].

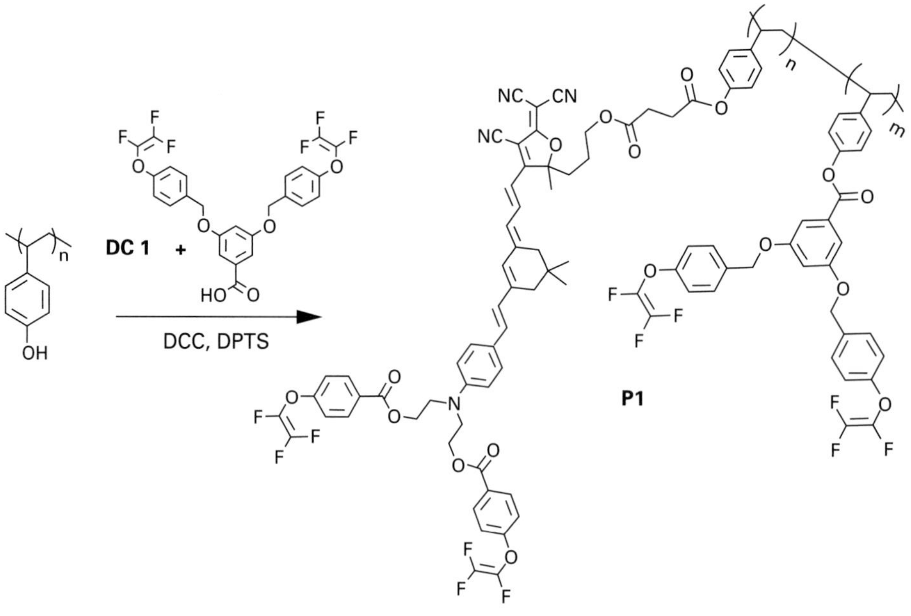

166   **Electrically poled and thermo-optic materials**

Figure 7.42 The synthesis of thermally crosslinkable dendronized polymer P1.

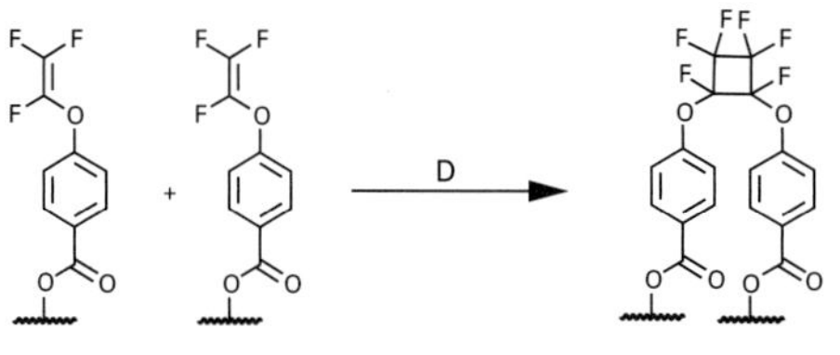

Figure 7.43 Fluorovinyl ether crosslinking through thermal cycloaddition.

Synthetic preparation of a side-chain dendronized polymer using Diels–Alder "click" chemistry is illustrated in Fig. 7.44. Diels–Alder Polymer 1 (DA P1) is prepared by free-radical co-polymerization of anthracene methacrylate and methyl methacrylate. Here, the anthracene moiety serves as the Diels–Alder diene. Preparation of maleimide (dienophile) functionalized Diels–Alder Chromophore 1 (DA C1) is quite similar to that employed in the case of dendrimers 33 and 44. Co-dissolution of DA P1 and DA C1, such that an excess of anthracene groups is present, followed by thin-film (e.g., spin) casting allows preparation of an initially guest–host-like EO material. When this film is then heated during the poling process, the Diels–Alder reaction occurs, covalently linking the chromophore/dendritic unit to the polymer host matrix. In this way, chromophore molecules are initially quite free to reorient in response to the poling field. However, as the Diels–Alder reaction nears completion, the chromophore is sequestered to form an *in-situ* generated side-chain polymer, and mobility is reduced. Thus, chromophore relaxation is slowed, EO effect thermal stability is improved, and the composite material is stabilized against phase separation. Further improvement in thermal stability may be achieved by the addition of a complementary crosslinking

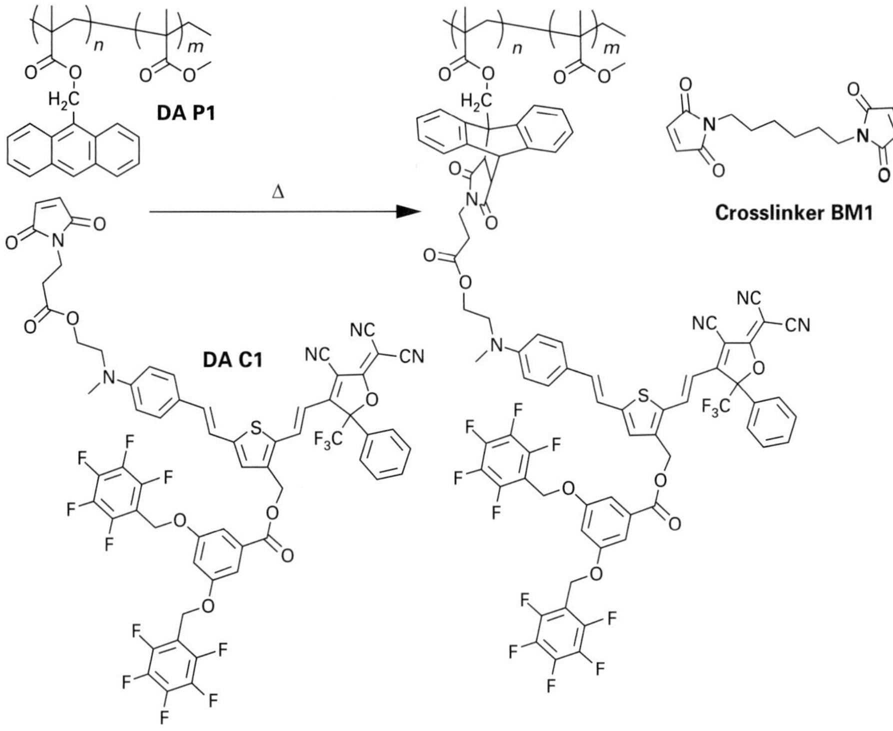

Figure 7.44 The preparation of Diels–Alder "click" dendronized polymers.

agent. As an example, bis-maleimide crosslinker 1 (BM1) is shown in Fig. 7.44. BM1 may be added to the Diels–Alder system in order to further harden (cure) the material during poling.

7.14.4 Donor and bridge functionalization of chromophore

Modification of donor, bridge, and acceptor regions of the chromophores to control intermolecular interactions is critical, as discussed at length in this chapter. Figure 7.45 illustrates one route to modification of donor and bridge segments of a representative (CLD-type) chromophore.

7.15 Summary of macromolecular electro-optic materials

Two decades ago, electrically poled, chromophore-containing polymers were characterized by low chromophore loading and poor poling efficiencies (small acentric order parameters). The indices of refraction of such materials were correspondingly low (~1.6), and the materials exhibited little birefringence. Optical loss was largely determined by the host material surrounding the electro-active chromophore, and stability (both thermal and photochemical) was modest (determined by the glass transition temperature

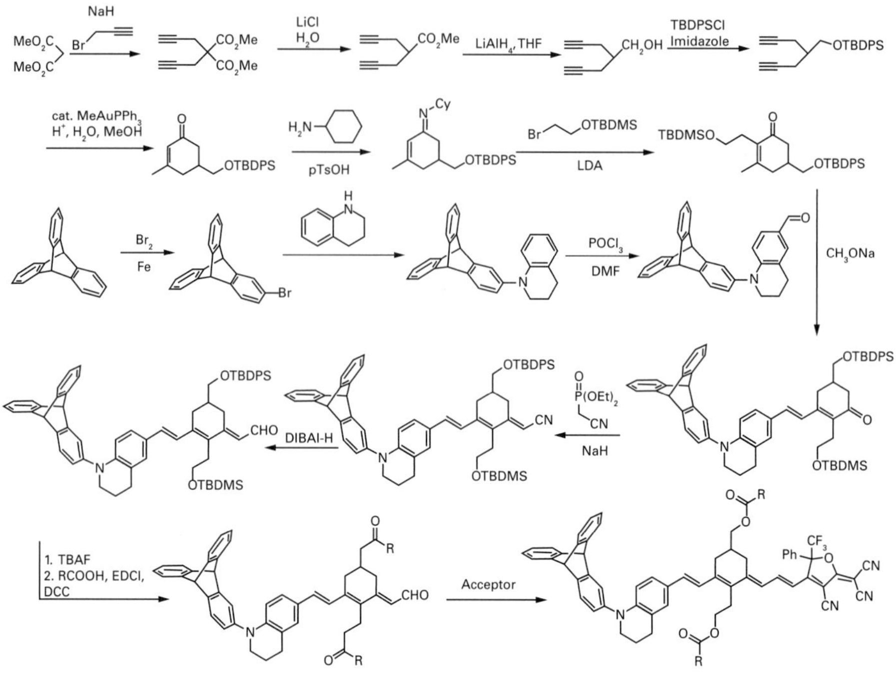

Figure 7.45 Synthetic modification of donor and bridge segments.

of the host polymer). Through control of intermolecular electrostatic interactions for the purpose of inhibiting centrosymmetric ordering and aggregation, chromophore loading has been increased to values as high as 75% and acentric order parameters have been increased to values ~0.2. Such high loading and improved acentric order has led to a corresponding increase in electro-optic activity, index of refraction, dielectric permittivity, and birefringence. Thus, electrically poled macromolecular (polymer and dendrimer) materials are beginning to approach the characteristics of organic electro-optic crystalline materials and materials prepared by sequential-synthesis/self-assembly, e.g., index of refraction values approaching 2.0. It is important for device engineers to recognize these changes; for example, an electro-optic material with an index of refraction of the order of 2.0 is not usable with silicon nitride waveguide devices.

The critical observation is that the properties of electrically poled macromolecular materials are rapidly evolving and in some cases are approaching the properties of more highly ordered materials such as crystalline organic electro-optic materials (although acentric order parameters are still relatively low). The continuing advantage of electrically poled materials is the great variety of processing options and material modifications that can be exploited. Materials exhibiting electro-optic activity as high as 1000 pm/V with optical loss values of 2 dB/cm or less are not out of the question. Index of refraction and dielectric permittivity can be tuned over substantial ranges, e.g., n can be tuned from 1.5 to 2.2 by control of chromophore concentration and the surrounding host environment. This flexibility in processing options and in tuning properties will likely be particularly important for electronic/photonic integration, including chipscale integration.

Substantial advances in theoretical methods have been achieved. Quantum mechanical methods have been improved to the point of yielding reliable prediction of linear and nonlinear optical properties, although a single TD-DFT functional has not been identified that yields simultaneous prediction of linear and nonlinear optical properties. Significant advances have been achieved in coarse-grained (pseudo-atomistic) statistical mechanical calculations. The importance of utilizing the appropriate level of detail (LOD) in describing chromophore structure in Monte Carlo simulations has been identified. A single prolate ellipsoid (a LOD of unity) has been shown to be adequate for the simulation of chromophore composite materials and for chromophores covalently incorporated into polymers and dendrimers. However, a level of detail of two is required for high-number-density neat chromophore materials. Methods have been developed for significantly accelerating statistical mechanical computations. The adiabatic volume adjustment (AVA) method introduced by Tillack and co-workers [125] has accelerated realization of steady-state chromophore distributions and simulates the condensation of chromophores that occurs in crystal growth.

In is interesting to consider how far the performance of Pockels effect modulators has been improved. In the year 2000, the voltage–length performance of a state-of-the-art organic electro-optic modulator was 24 V-mm and that of a lithium niobate modulator was 30–50 V-mm. Today, SOH modulators yield values of <0.5 V-mm and POH modulators yield values of the order of 0.05 V-mm. SOH and POH devices yield greatly improved digital power efficiencies, e.g., 1 fJ/bit (or less) and 12.5 fJ/bit, respectively. High 3 dB$_e$ bandwidths (e.g., 100 GHz and even greater) are realized for short SOH and POH devices. The excellent signal quality of Pockels effect modulation permits high extinction and excellent bit error rates. It can be anticipated that a factor of 2 improvement of in-device electro-optic activity for organic materials is very likely, and such an advance will lead to further improvement in the above device characteristics.

7.16 Thermo-optic materials

The thermo-optic effect (change of refractive index, n, of a material with temperature, T) is due to two effects: (1) a temperature-dependent change in density, ρ; and (2) a temperature-dependent change in index of refraction at constant density. Mathematically, the thermo-optic effect is expressed as

$$(dn/dT) = (\partial n/\partial \rho)_T (\partial \rho/\partial T) + (\partial n/\partial T)_\rho \qquad (7.12)$$

which can be rewritten in terms of the coefficient of volume (thermal) expansion, α, as

$$(dn/dT) = -\alpha(\rho \partial n/\partial \rho)_T + (\partial n/\partial T)_\rho \qquad (7.13)$$

The Lorentz–Lorenz equation permits $(\rho \partial n/\partial \rho)_T$ to be expressed in terms of the strain polarizability constant, Λ_0 (which takes into account the effect of density changes on atomic polarizability) as

Table 7.5 Thermo-optic and thermal properties of various polymers.

Polymer	dn/dT × 10^4 (C^{-1})	α × 10^4 (C^{-1})
Polymethylmethacrylate	−1.3	2.3
Polystyrene	−1.2	2.2
Polycarbonate	−0.9	1.7
Urethane acrylate elastomer	−4.2	7.2
Silicone	−3.1	5.9
Sol-gel glass	−2.1	3.5
Epoxy	−1.0	1.7

$$(\rho \partial n/\partial \rho)_T = \{(1 - \Lambda_0)(n^2 + 2)(n^2 - 1)\}/6n \qquad (7.14)$$

From the above it can be seen that the sign and relative magnitudes of the thermo-optic effect for different materials will be determined by the competition of the two terms of Eq. (7.13). We will shortly provide an explicit example.

Low optical loss, low cost, and processing simplicity are important requirements for materials to be used in waveguide applications including as optical switches (OSs), variable optical attenuators (VOAs), optical couplers/splitters, and arrayed waveguide gratings (AWGs). Thus, although many organic and inorganic materials exhibit large thermo-optic coefficients and are used for a variety of specialized optical applications, most commercial waveguide (e.g., telecommunications) applications have focused on two classes of materials: fused silica and organic polymers.

For fused silica, the coefficient of thermal expansion is low ($\alpha \sim 10^{-6}$/°C). Moreover, Λ_0 = 0.4 and n = 1.5; thus, the first term becomes −0.3 ×10^{-6}/°C. For fused silica, the second term is $\sim 10^{-5}$/°C and thus dominates, leading to a positive thermo-optic coefficient. For polymers, the coefficient of thermal expansion, α, is relatively large at $\sim 2 \times 10^{-4}$/°C. The value of Λ_0 for polymethylmethacrylate (PMMA) is 0.15 and for polycarbonate (PC) it is 0.18. For both of these polymers (and many others) the index of refraction is approximately 1.5. Thus, the first term of Eq. (7.13) is approximately -10^{-4}/°C. For polymers, the thermal change of refractive index at constant density is small ($\sim 7 \times 10^{-6}$/°C) and so the first term clearly dominates, leading to a negative thermo-optic coefficient. Commercial thermo-optic switches based on organic polymers (e.g., BeamBox™ by Akzo Nobel Photonics) have been available for about a decade.

While loss, cost, and processing simplicity are important material requirements, other important requirements of optical switches include polarization and wavelength independence, low cross talk, low drive power, small size, reliability, and acceptable switching speed. A severe limitation of thermo-optic switches is switching speed, which is limited to milliseconds by thermal diffusion. The magnitude of thermo-optic coefficients is also critical to limit power consumption (particularly in dense photonic circuits) to acceptable levels. Representative thermo-optic coefficients for common polymer materials are shown in Table 7.5. More extensive tabulations are available elsewhere [126]. Clearly, thermo-optic switches and VOAs are not promising for high bandwidth applications and for applications requiring high density integration.

References

[1] L. R. Dalton, A. W. Harper, and B. H. Robinson, *Proc. Natl. Acad. Sci. USA*, **94**, 4842 (1997).
[2] L. R. Dalton, B. H. Robinson, A. K.-Y. Jen, W. H. Steier, and R. Nielsen *Opt. Mater.*, **21**, 19 (2003).
[3] B. H. Robinson and L. R. Dalton, *J. Phys. Chem. A*, **104**, 4785 (2000).
[4] H. L. Rommel and B. H. Robinson, *J. Phys. Chem C*, **111**, 18765 (2007).
[5] Y. Shi, C. Zhang, H. Zhang, et al., *Science*, **288**, 119 (2000).
[6] M. R. Leahy-Hoppa, P. D. Cunningham, J. A. French, and L. M. Hayden, *J. Phys. Chem. A*, **110**, 5792 (2006).
[7] P. A. Sullivan, H. Rommel, Y. Liao, et al., *J. Am. Chem. Soc.*, **129**, 7523 (2007).
[8] P. A. Sullivan, H. L. Rommel, Y. Takimoto, et al., *J. Phys. Chem. B*, **113**, 15581 (2009).
[9] S. J. Benight, L. E. Johnson, R. Barnes, et al., *J. Phys. Chem. B.*, **114**, 11949 (2010).
[10] L. R. Dalton, S. J. Benight, L. E. Johnson, et al., *Chem. Mater.*, **23**, 430 (2011).
[11] P. A. Sullivan, L. R. Dalton, *Accts. Chem. Res.*, **43**, 10 (2010).
[12] R. M. Overney, C. Buenviaje, R. Luginbuhl, and F. Dinelli, *J. Therm. Anal. Calorim.*, **59**, 205 (2000).
[13] T. K. Gray, M. Haller, J. Luo, A. K.-Y. Jen, and R. M. Overney, *Appl. Phys. Lett.*, **86**, 211908 (2005).
[14] S. Ge, Y. Pu, W. Zhang, et al., *Phys. Rev. Lett.*, **85**, 2340 (2000).
[15] S. Stills, R. M. Overney, W. Chau, et al., *J. Chem. Phys.*, **120**, 5334 (2004).
[16] T. Gray, T. D. Kim, D. B. J. Knorr, et al., *Nano Lett.*, **8**, 754 (2008).
[17] D. B. J. Knorr, T. Gray, and R. M. Overney, *Ultramicroscopy*, **109**, 991 (2009).
[18] S. J. Benight, D. B. Knorr, Jr., L. E. Johnson, et al., *Adv. Mater.*, **24**, 3263 (2012).
[19] T.-D. Kim, J. Luo, Y.-J. Cheng, et al., *J. Phys. Chem. C*, **112**, 8091 (2008).
[20] Y. V. Pereverzev, K. N. Gunnerson, et al., *J. Phys. Chem. C*, **112**, 4355 (2008).
[21] L. R. Dalton, P. A. Sullivan, D. H. Bale, and B. C. Olbricht, *Solid State Electronics*, **51**, 1263 (2007).
[22] B. C. Olbricht, P. A. Sullivan, G.-A. Wen, et al., *J. Phys. Chem. C*, **112**, 8091 2008).
[23] S. Kim, Q. Pei, H. R. Fetterman, B. C. Olbricht, and L. R. Dalton, *IEEE Photon. Tech. Lett.*, **23**, 845 (2011).
[24] M. Dumont, S. Hosotte, G. Froc, and Z. Sekkat, *Proc. SPIE*, **2042**, 2 (1994).
[25] Z. Wang, W. Sun, A. Chen, et al., *Opt. Lett.*, **36**(15) 2853–2855 (2011).
[26] B. M. Polishak, S. Huang, J. Luo, et al., *Macromolecules*, **44**, 1261 (2011).
[27] S. R. Marder, Annual Review MURI Center for Organic Materials for All-Optical Switching, Atlanta, GA, October 27, 2011.
[28] M. Straehelin, C. A. Walsh, D. M. Burland, et al., *J. Appl. Phys.*, **73**, 8471 (1993).
[29] D. M. Burland, R. D. Miller, and C. A. Walsh, *Chem. Rev.*, **94**, 31 (1994).
[30] P. Prêtre, U. Meier, U. Stalder, et al., *Macromolecules*, **31**, 1947 (1998).
[31] P. Kaatz, P. Pretre, U. Meier, et al., *Macromolecules*, **29**, 1666 (1996).
[32] C. Weder, P. Neuenschwander, U. W. Suter, et al., *Macromolecules*, **27**, 2181 (1994).
[33] A. Q. Tool, *J. Am. Ceram. Soc.*, **29**, 240 (1946).
[34] O. S. Narayanaswamy, *J. Am. Ceram. Soc.*, **54**, 471 (1971).
[35] M. L. Williams, R. F. Landel, and J. D. Ferry, *J. Am. Chem. Soc.*, **77**, 3701 (1955).
[36] L. R. Dalton, A. W. Harper, R. Ghosn, et al., *Chem. Mater.*, **7**, 1060 (1995).
[37] H. W. Oviatt, K. J., Shea, S. Kalluri, et al., *Chem. Mater.*, **7**, 493 (1995).

[38] S. S. H. Mao, Y. Ra, L. Guo, et al., *Chem. Mater.* **10**, 146 (1998).
[39] S. Suresh, S. Chen, C. Topping, J. Ballato, and D. Smith Jr., *Proc. SPIE*, **4991**, 530 (2003).
[40] J. Luo, M. Haller, H. Ma, et al., *J. Phys. Chem. B*, **108**, 8523 (2004).
[41] M. Haller, J. Luo, H. Li, et al., *Macromolecules*, **37**, 688 (2004).
[42] T.-D. Kim, J. Luo, Y. Tian et al., *Macromolecules*, **39**, 1676 (2006).
[43] Z. Shi, S. Hau, J. Luo, et al., *Adv. Funct. Mater.*, **17**, 2557 (2007).
[44] T.-D. Kim, Z. Shi, J. Luo et al., *Proc. SPIE*, **6470**, 64700D1 (2007).
[45] Y. Huang, G. T. Paloczi, A. Yariv, C. Zhang, and L. R. Dalton, *J. Phys. Chem. B*, **108**, 8606 (2004).
[46] A. Chen, V. Chuyanov, F. I. Marti-Carrera, et al., *Proc. SPIE*, **3005**, 65 (1997).
[47] Y. Enami, D. Mathine, C. T. DeRosa, et al., *Appl. Phys. Lett.*, **92**, 193508 (2008).
[48] M.-C. Oh, C. Zhang, H.-J. Lee, W. H. Steier, and H. R. Fetterman, *IEEE Phot. Tech. Lett.*, **14**, 1121 (2002).
[49] Q. Zhang, M. Canva, and G. Stegeman, *Appl. Phys. Lett.*, **73**, 912 (1998).
[50] A. Galvan-Gonzalez, M. Canva, G. I. Stegeman, et al., *Opt. Lett.*, **24**, 1741 (1999).
[51] A. Galvan-Gonzalez, M. Canva, G. I. Stegeman, et al., *Opt. Lett.*, **25**, 332 (2000).
[52] A. Galvan-Gonzalez, M. Canva, G. I. Stegeman, et al., *J. Opt. Soc. Am. B*, **17**, 1992 (2000).
[53] M. E. DeRosa, M. He, J. S. Cites, S. M. Garner, and Y R. Tang, *J. Phys. Chem. B*, **108**, 8725 (2004).
[54] D. Rezzonico, M. Jazbinsek, P. Gunter, et al., *J. Opt. Soc. Amer. B*, **24**, 2199 (2007).
[55] L. R. Dalton, P. A. Sullivan, and D. H. Bale, *Chem. Rev.*, **110**, 25 (2010).
[56] S. J. Benight, D. H. Bale, B. C. Olbricht, and L. R. Dalton, *J. Mater. Chem.*, **19**, 7466 (2009).
[57] M. W. Becker, L. S. Sapochak, R. Ghosen, et al., *Chem. Mater.*, **6**, 104 (1994).
[58] K. D. Singer, M. G. Kuzyk, J. E. Sohn, *J. Opt. Soc. Am. B*, **4**, 968 (1987).
[59] H. Kang, P. Zhu, Y. Yang, A. Facchetti, T. J. Marks, *J. Am. Chem. Soc.*, **126**, 15974 (2004).
[60] Y. Takimoto, C. Isborn, B. E. Eichinger, J. J. Rehr, and B. H. Robinson, *J. Phys. Chem. C*, **112**, 8016 (2008).
[61] Y. Takimoto, F. D. Vila, and J. J. Rehr, *J. Chem. Phys.*, **127**, 154114 (2007).
[62] C. C. Teng and H. T. Man, *Appl. Phys. Lett.*, **56**, 1734 (1990).
[63] J. S. Schildkraut, *Appl. Opt.*, **29**, 2839 (1990).
[64] S. Kalluri, S. Garner, M. Ziari, et al., *Appl. Phys. Lett.*, **69**, 275 (1996).
[65] V. Dentan, Y. Levy, M. Dumont, and R. E. Chastaing, *Opt. Commun.* **69**, 379 (1989).
[66] A. Chen, V. Chuyanov, S. Garner, W. H. Steier, and L. R. Dalton, in *Organic Thin Films for Photonic Applications*, vol. 14, Washington, DC, Optical Society of America (1997), pp. 158–159.
[67] J. A. Davies, A. Elangovan, P. A. Sullivan, et al., *J. Am. Chem. Soc.*, **130**, 10565 (2008).
[68] L. R. Dalton, *Adv. Polym. Sci.*, **158**, 1 (2002).
[69] D. H. Park, C. H. Lee, and W. N. Herman, *Opt. Express*, **14**, 8866 (2006).
[70] F. Michelotti, E. Toussaere, R. Levenson, J. Liang, and J. Zyss, *Appl. Phys. Lett.*, **67**, 2765 (1995).
[71] S. K. Lee, M. J. Cho, J.-I. Jin, and D. H. Choi, *J. Polym. Sci. A*, **45**, 531 (2007).
[72] P. A. Sullivan, A. J. P. Akelaitis, S. K. Lee, et al., *Chem. Mater.*, **18**, 344 (2006).
[73] W. Jin, P. V. Johnston, D. L. Elder, et al., *Appl. Phys. Lett.*, **104**. 243304 (2014).
[74] M. A. Mortazavi, A. Knoesen, S. T. Kowel, B. G. Higgins, and A. Dienes, *J. Opt. Soc. Am. B*, **6**, 733 (1989).

[75] T.-D. Kim, J.-W. Kang, J. Luo, et al., *J. Am. Chem. Soc.*, **129**, 488 (2007).
[76] V. Rodriguez, F. Adamietz, L. Sanguinet, T. Buffeteau, and C. Sourisseau, *J. Phys Chem. B*, **107**, 9736 (2003).
[77] B. C. Olbricht, P. A. Sullivan, P. C. Dennis, et al., *J. Phys. Chem. B*,, **115**, 231 (2011).
[78] M. H. Graf, O. Zobel, A. J. East, and D. Haarer, *J. Appl. Phys.*, **75**, 3335 (1994).
[79] M. Mansuripur, *J. Appl. Phys.*, **67**, 6466 (1990).
[80] J. N. Hilfiker, C. M. Herzinger, T. Wanger, et al., *Thin Sol. Films*, **455–456**, 591 (2004).
[81] J. N. Hilfiker, B. Johs, C. M. Herzinger, et al., *Thin Sol. Films*, **455–456**, 596 (2004).
[82] H. Fujiwara, *Spectroscopic Ellipsometry Principles and Applications*, West Sussex, John Wiley & Sons (2007).
[83] J. A. Woollam, in *Wiley Encyclopedia of Electrical and Electronics Engineering*, New York, Wiley (2000), pp. 109–117.
[84] L. E. Johnson, M. T. Casford, D. L. Elder, P. B. Davies, and M. S. Johal, *Proc. SPIE*, **8817**, 8817P1 (2013).
[85] G. T. Paloczi, Y. Huang, A. Yariv, J. Luo, and A. Jen, *Appl. Phys. Lett.*, **85**, 1662 (2004).
[86] H. C. Song, M. C. Oh, S. W. Ahn, and W. H. Steier, *Appl. Phys. Lett.*, **82**, 4432 (2003).
[87] L. R. Dalton, A. W. Harper, A. S. Ren, et al., *Ind. Eng. Chem. Res.*, **38**, 8 (1999).
[88] J. S. Grote, J. S. Zetts, J. P. Drummond, et al., *Proc. SPIE*, **3950**, 108 (2000).
[89] J. S. Grote, J. S. Zetts, C. H. Zhang, et al., *Proc. SPIE*, **4114**, 101 (2000).
[90] J. G. Grote, J. S. Zetts, R. L. Nelson, et al., *Opt. Eng.*, **40**, 2464 (2001).
[91] J. G. Grote, J. S. Zetts, R. L. Nelson et al., *Proc. SPIE*, **4470**, 10 (2001).
[92] M. Leovich, P. P. Yaney, C. Zhang, et al., *Proc. SPIE Int. Soc. Opt. Eng.*, **4652**, 97 (2002).
[93] D. E. Diggs, J. G. Grote, A. A. Davis, et al.,, *Proc. SPIE Int. Soc. Opt. Eng.*, **4813**, 94 (2002).
[94] S. Huang, T.-D. Kim, J. D. Luo, et al., *Appl. Phys. Lett.*, **96**, 243311 (2010).
[95] V. R. Almeida, Q. Xu, R. R. Panepucci, C. A. Barrios, and M. Lipson, *Mat. Res. Soc. Symp. Proc.*, **797**, V6.10.3 (2004).
[96] T. Baehr-Jones, M. Hochberg, G. Wang et al., *Optics Express*, **13**, 5216 (2005).
[97] M. Hochberg, T. Baehr-Jones, G. Wang, et al., *Optics Express*, **15**, 8401 (2007).
[98] T. Baehr-Jones, B. Penkov, J. Huang, et al., *Appl. Phys. Lett.*, **92**, 163303-1 (2008).
[99] H. Figi, D. H. Bale, A. Szep, L. R. Dalton, and A. Chen, *J. Opt. Soc. Amer. B*, **28**, 2291 (2011).
[100] M. Hochberg, T. Baehr-Jones, G. Wang, et al., *Nature Materials*, **5**, 703 (2006).
[101] J. Takayesu, M. Hochberg, T. Baehr-Jones, et al., *IEEE J. Lightwave Technol.*, **27**, 440 (2008).
[102] T. Baehr-Jones and M. Hochberg, *J. Phys. Chem. C*, **112**, 8085 (2008).
[103] D. Korn, R. Palmer, H. Yu, et al., *Optics Express*, **21**, 13219 (2013).
[104] J. Leuthold, W. Freude, J.-M. Brosi, et al., *Proc. IEEE*, **97**, 1304 (2009).
[105] J. Leuthold, C. Koos, W. Freude, et al., *IEEE J. Sel. Top. Quantum Electron.*, **19**, 3401413 (2013).
[106] C. Weimann, P. C. Schindler, R. Palmer, et al., *Optics Express*, **22**, 2629 (2014).
[107] R. Palmer, S. Koeber, D. L. Elder, et al., *IEEE J. Lightwave Technol.*, **32**, 2726 (2014).
[108] M. Lauermann, S. Wolf, R. Palmer, et al., *Proc. SPIE*, **9516**, 951607 (2015).
[109] M. Lauermann, R. Palmer, S. Koeber, et al., *Opt. Express*, **22**, 29927 (2014).
[110] M. Lauermann, S. Wolf, P. C. Schindler, et al., *IEEE J. Lightwave Technol.*, **33**, 1210 (2015).
[111] S. Koeber, R. Palmer, M. Lauermann, et al., *Light: Sci. Applic.*, **4**, e255 (2014).

[112] C. Koos, J. Leuthold, W. Freude, *et al.*, "Femtojoule modulation and frequency comb generation in silicon–organic hybrid (SOH) devices," *16th International Conference on Transparent Optical Networks*, invited presentation (2014).

[113] S.-K. Kim, N. Sylvain, S. J. Benight, *et al.*, *Proc. SPIE*, **7754**, 7754-3-1 (2010).

[114] L. R. Dalton, A. K.-Y. Jen, B. H. Robinson, *et al.*, *Proc. SPIE*, **7935**, 793502-1 (2011).

[115] A. Melikyan, L. Alloatti, A. Muslija, *et al.*, *Nat. Photon.*, **8**, 229 (2014).

[116] J. Leuthold, A. Melikyan, L. Alloatti, *et al.*, "Smaller, faster, and more energy efficient modulators – from silicon organic hybrid to plasmonic modulation," *Proc. European Conference on Optical Communication 2014*, invited presentation (2014).

[117] A. Meilkyan, K. Koehnle, M. Laueermann, *et al.*, *Opt. Express*, **23**, 9924 (2015).

[118] C. Koos, W. Freude, J. Leuthold, *et al.*, "Silicon–organic hybrid (SOH) and plasmonic-organic hybrid (POH) integration," Invited presentation, *Proc. OFC'2015*, March 22–26, Los Angeles, 2015.

[119] W. Heni, C. Haffner, Y. Fedoryshyn, *et al.*, "Plasmonic Mach–Zehnder modulator with > 70 GHz electrical bandwidth demonstrating 89.6 Gbit/s 4-APSK," *Proc. OFC'2015*.

[120] C.-Y. Lin, X. Wang, S. Chakravarty, *et al.*, *Appl. Phys. Lett.*, **97**, 093304-1 (2010).

[121] X. Zhang, A. Hosseini, S. Chakravarty, *et al.*, *Opt. Lett.*, **38**, 4931 (2013).

[122] X. Zhang, A. Hosseini, H. Subbaraman, *et al.*, *IEEE/OSA J. Lightwave Technol.*, **32**, 3774 (2014).

[123] J. Luo, H. Ma, M. Haller, A. K.-Y. Jen, and R. R. Barto, *Chem. Comm.*, **888** (2002).

[124] P. A. Sullivan, B. C. Olbricht, A. J. P. Akelaitis, *et al.*, *J. Mater. Chem.*, **17**, 2899 (2007).

[125] A. F. Tillack, L. E. Johnson, M. Rawal, L. R. Dalton, and B. H. Robinson, *Proc. Mater. Res. Soc.*, **1698**, mrss14–1698-jj08–05 (2014).

[126] G. Ghosh, *Handbook of Thermo-Optic Coefficients of Optical Materials with Applications*, New York, Academic Press (1998).

8 Overview of applications

In this chapter, an overview of applications of electro-optics is presented. Waveguide electro-optic devices are discussed in greater detail in Chapter 9 and applications involving second-order nonlinear optical effects other than electro-optic phenomena (e.g., frequency conversion) are discussed in Chapter 10. Photorefractivity phenomena, materials, and applications are discussed in Chapter 11. Some prototype devices, particularly silicon–organic hybrid (SOH) and plasmonic–organic hybrid (POH) devices, were briefly discussed in Chapter 7.

8.1 Device parameters and materials requirements

Applications of electro-optic materials include telecommunications, computing, defense, medical, and sensing technologies. A wide range of specific devices are critical for these technologies. Electro-optic devices can be divided into three general categories: (1) stripline devices such as Mach–Zehnder interferometers and directional couplers; (2) resonant devices such as ring microresonators, etalons, photonic crystal, and metamaterial devices; and (3) prism-based devices such as spatial light modulators. The relationships between device performance and material performance parameters vary with device type. Some properties, such as acceptable material optical loss, depend on system architectures, as well as individual device architectures. Material requirements, such as thermal and photochemical stability, will depend upon device and system operating conditions, e.g., temperature and optical power levels that are used. Telcordia standards define minimum thermal stability requirements, e.g., long-term operational stability at 85 °C (and 85% humidity). Optical power levels vary with application area, e.g., telecommunication applications currently utilize power levels of 10–20 mW at wavelengths centered around 1.3 or 1.55 μm.

Clearly, material requirements and their relationship to device performance will depend upon application and upon device and system architectures, so general and simple requirements cannot be defined in a meaningful way. However, in the following, we provide insight into the relationship between device performance and materials characteristics for the simple general device structures noted above. A more in-depth discussion relevant to waveguide device architectures is presented in the next chapter. An even broader range of device architectures are relevant to the applications discussed in Chapters 10 and 11.

Stripline devices. Representative stripline devices include Mach–Zehnder interferometers (amplitude modulators), birefringence or phase modulators, and directional couplers (see Figs. 9.6 and 9.8) [1,2]. The relationship between Mach–Zehnder drive voltage (V_π, the voltage required to produce a phase shift of π and thus full wave modulation) and material electro-optic activity, r_{33}, is as follows:

$$V_\pi = \frac{\lambda h}{n^3 r_{33} L \Gamma} \tag{8.1}$$

where λ is the operating wavelength, h is the electrode spacing, n is the index of refraction, L is the electrode (device) length, and Γ is the modal (optical/radiofrequency fields) overlap parameter. If a push–pull Mach–Zehnder interferometer structure is employed, a factor of 2 should be added to the denominator of the V_π equation. In a push–pull Mach–Zehnder structure, the electro-optic material is poled in opposite directions in the two arms of the interferometer so that application of a driving electric field advances the phase in one arm while retarding the phase in the other arm, leading to enhanced interference (see Figs. 7.1 and 7.38).

Critical performance parameters of Mach–Zehnder modulators (and electro-optic devices in general) include bandwidth and optical insertion loss as well as drive voltage. These are highly interdependent; for example, each depends on device length L (see Fig. 7.35). Device bandwidth depends on electrode length through the resistivity of the metal electrodes. For example, gold has a resistivity-defined microwave loss of 0.75 dB (GHz)$^{1/2}$/cm, which, for example, limits device 3 dB$_e$ bandwidth to 90 GHz for a 5-mm electrode. Device insertion loss also depends on device length through material propagation loss. The other factor that defines insertion loss is fiber coupling loss, which for organic electro-optic materials is defined by mode size mismatch between the organic electro-optic waveguide and silica fiber waveguide rather than index of refraction mismatch between the organic material and silica. For example, a material propagation (absorption and scattering) loss of 2 dB/cm and a fiber coupling loss of 0.8 dB/facet leads to a total insertion loss of 2.6 dB for a 5 mm device length. For a single device, maximum acceptable insertion loss is of the order of 6–7 dB for many applications. For a material with an electro-optic activity of 300 pm/V and an electrode spacing h of 8 μm, a 5-mm-long device will have a drive voltage (V_π) of 0.75 V.

For electro-optic modulation (electrical-to-optical signal transduction) a critical parameter is the signal loss (or gain) that is associated with the transduction process. The loss-to-gain transition for the electrical-to-optical-to-electrical transduction process (RF photonics) occurs at a drive voltage of approximately 0.2 V for a Mach–Zehnder device. To achieve gain with an electro-optic material with an activity of 300 pm/V, the device length would have to be increased to approximately 2 cm; however, this would increase insertion loss to 5.6 dB. As can be seen from Fig. 7.35 (or a consideration of the resistivity of gold electrodes), this increase in length would also reduce device bandwidth to approximately 40 GHz. For a 5-mm-length Mach–Zehnder modulator, the material electro-optic activity would have to be increased to above 1000 pm/V to achieve gain. Of course, specific requirements will vary from one application to the next and will change with time.

Table 8.1 A comparison of the performance characteristics of various amplitude modulator types.

Modulator type	Drive voltage (V)	Bandwidth (GHz)	Insertion loss (dB)	Operating range (nm)	Integration
Electro-absorption	1–3	25[c]	>6	5–10	Monolithic
LiNO$_3$ electro-optic	2–5	25	3–6	>50	None
Electro-optic polymer	< 1V [a,b]	>100[d]	<6[e]	>50	Heterogeneous

[a] Y. Shi et al. Science **288**, 119–122 (2000).
[b] Y. Enami et al. Appl. Phys. Lett. **91**, 093505 (2007).
[c] C. Zhang et al. IEEE Phot. Tech. Lett. **11**, 191–193 (1999).
[d] D. Chen et al. Appl. Phys. Lett. **70**, 3335–3337 (1997).
[e] C. T. DeRose et al. Opt. Express **17**, 3316–3319 (2009).

From Eq. (8.1), it is clear that $n^3 r_{33}$, rather than r_{33}, is the material parameter that defines device drive voltage performance. Thus, different organic electro-optic materials should be compared with respect to $n^3 r_{33}$. Comparison of organic and inorganic materials is even more complex because different material characteristics impact bandwidth and drive-voltage performance for organic and inorganic materials. The intrinsic bandwidth of organic electro-optic materials is defined by the phase relaxation time of the quasi-delocalized π-electron system, which is of the order of tens of femtoseconds. Thus, the intrinsic bandwidth of organic electro-optic materials is of the order of tens of terahertz. As noted earlier, such bandwidths have been demonstrated using organic second-order nonlinear optical materials in all-optical modulation experiments (a third-order nonlinear optical phenomenon), in terahertz generation and detection (including terahertz spectroscopy) experiments, and in pulsed experiments. For organic electro-optic devices in practice, the resistivity of metal electrodes defines device bandwidth performance. For the inorganic material lithium niobate, the velocity mismatch of propagating optical and electrical waves plays a significant role in defining device bandwidth performance. This velocity mismatch arises from the dramatic difference between the dielectric permittivity ($\varepsilon = 35$) and index of refraction ($n = 2.2$) of lithium niobate. A high dielectric constant is unfavorable owing to its effect on applied electric (radiofrequency, microwave, and millimeter wave) fields. Thus, the parameter $n^3 r_{33}/\varepsilon$ is often used as a figure-of-merit in comparing electro-optic materials. Given that materials technologies are rapidly evolving, a comparison of different modulator/material types may not be meaningful, but Table 8.1 presents such a comparison.

The equation for V_π of a Mach–Zehnder (MZ) modulator is given by Eq. (8.1). The V_π for a birefringence (BR) or phase modulator can be related to that for a MZ modulator by V_π (BR) $\approx r_{33}/(r_{33} - r_{13}) V_\pi$(MZ). In the limit of small acentric order parameters, the drive voltage of a birefringence or phase modulator is 1.5 times that of a Mach–Zehnder modulator. The minimum drive voltage of a directional coupler (the voltage required to switch the signal from one output to another) is 1.7 times

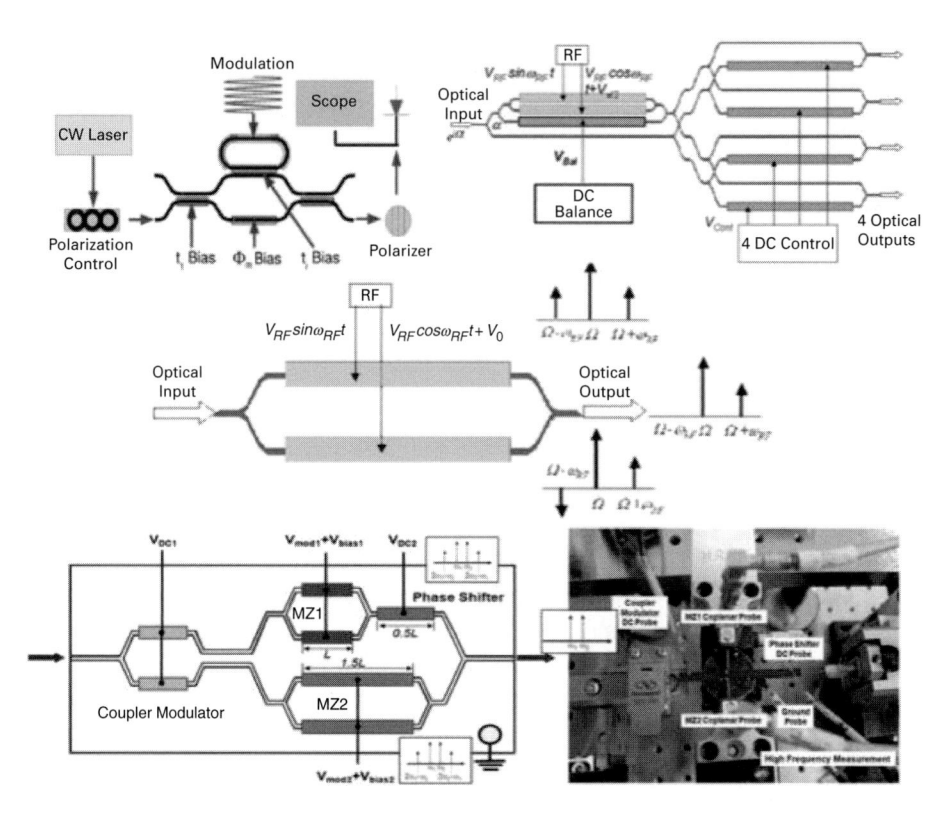

Figure 8.1 A variety of devices based on stripline organic electro-optic modulators (both schematic and actual device (bottom right) are shown [9 to 12]). Upper left: Optical signal processor. Upper right: Optical SSB modulator. Middle: Four-channel RF phase shifter. Bottom: Linearized double parallel MZ modulator.

that of a Mach–Zehnder modulator. It is readily seen that somewhat greater drive voltages are required for phase modulators and directional couplers than for Mach–Zehnder interferometers.

Specialized applications may require additional characteristics such as conformal or flexible material. Stripline devices are frequently used for electrical-to-optical signal transduction, signal routing, and sensing (including electric field sensing) applications. Stripline devices form critical components of more complex systems such as phased array radar systems, ultra-stable oscillators, acoustic spectrum analyzers, optical gyroscopes, high-frequency A/D converters, etc. [2 to 8]. Plasmonics have been incorporated into stripline devices [9,10] leading to sub-λ confinement of light. Some representative device structures are shown in Fig. 8.1 (from the laboratory of H. Fetterman [9 to 12]).

Stripline SOH and POH devices are discussed briefly in Chapter 7. State-of-the-art voltage-length performance values for these device are <0.5 V-mm and 50 V-µm, respectively. Moreover, SOH devices as short as 0.5 mm have been fabricated and POH devices as short as 10 microns have been demonstrated. Such short devices permit

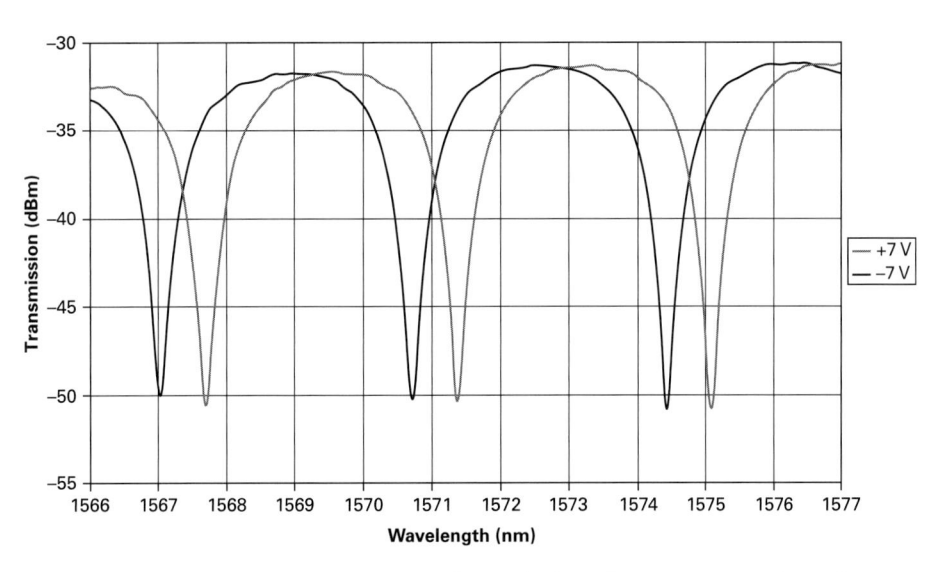

8.1 Device parameters and materials requirements

Figure 8.2 The shifts in bandpass resonances of an early ring microresonator are shown for application of voltages of +7 V and −7 V.

higher propagation losses to be tolerated. Excellent power efficiencies (approximately 1 fJ/bit) and extinction coefficients have been achieved with these device architectures. The footprints of these devices are competitive with resonant devices discussed in the next section.

Resonant devices. For resonant devices, such as ring microresonators or etalons, bandwidth and drive voltage are inter-related through the quality, Q, factor of the device [1]. The Q factor can be viewed as the number of times that the optical wave transits the device. The relevant device parameter for analog resonant devices is the bandwidth-sensitivity factor ($V_{\pi\ \mathrm{equiv}}/\Delta f_{\mathrm{3dBe}}$), which is related to material electro-optic activity (r_{33}) by ($V_{\pi\ \mathrm{equiv}}/\Delta f_{\mathrm{3dBe}}$) = $4\pi h \lambda/[3\sqrt{3})(n_{\mathrm{eff}})2r_{33}c]$, where h is the electrode spacing, λ is the operational wavelength, n_{eff} is the effective index of refraction in the device, c is the speed of light in a vacuum, and $\Delta f_{\mathrm{3dBe}} = c/\lambda Q$. As a representative example, consider a device with the following characteristics: $n_{\mathrm{eff}} = 1.6$, $r_{33} = 300$ pm/V, $h = 6$ μm, and $\lambda = 1.3$ μm, then ($V_{\pi\ \mathrm{equiv}}/\Delta f_{\mathrm{3dBe}}$) = 0.08 V/GHz. In other words, an applied voltage of 0.08 V is required to produce a 1-GHz shift in the bandpass resonance, or an applied voltage of 1 V produces a 12.5-GHz shift. Such a shift would require a resonant device with a Q of 3×10^5. Voltage-induced bandpass shifting for a ring microresonator is shown in Fig. 8.2.

Resonant devices effect modulation by producing a shift in the bandpass of the device (see Section 9.4 and Fig. 9.9). Full wave modulation corresponds to shifting the bandpass by the full width at half maximum. Tuning range can be enhanced exploiting the Vernier effect using coupled multiple ring resonators. The effect is illustrated for a double ring resonator in Fig. 8.3.

The voltage-controlled shifts of the bandpass resonance are shown for an electro-optic etalon structure in Fig. 8.4.

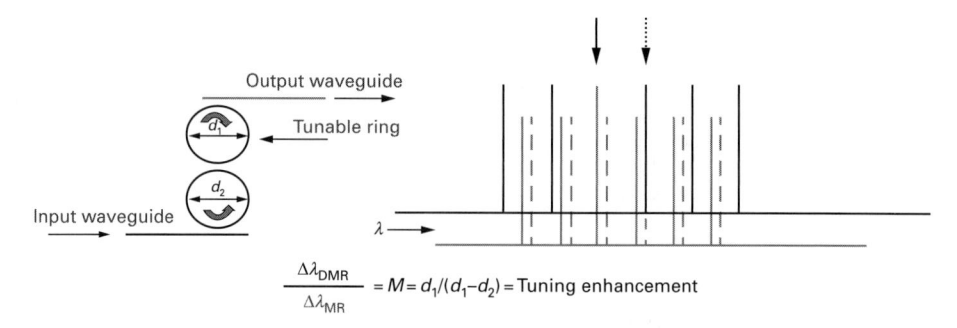

Figure 8.3 Tuning enhancement is shown for a double ring resonator. The black and gray lines indicate the resonances of the two ring resonators. The dashed lines indicate the new resonance positions when an electric field is applied. Note that the resonant wavelengths of the two resonators can be tuned in and out of resonance with each other. Transmission occurs only when both rings are resonant. A tuning enhancement factor, M, of 40 (leading to voltage tuning of 0.04 nm/V), with a side mode suppression of >30 dB, is easily realized for organic electro-optic double ring microresonators (e.g., d_1 = 240 μm and d_2 = 236 μm).

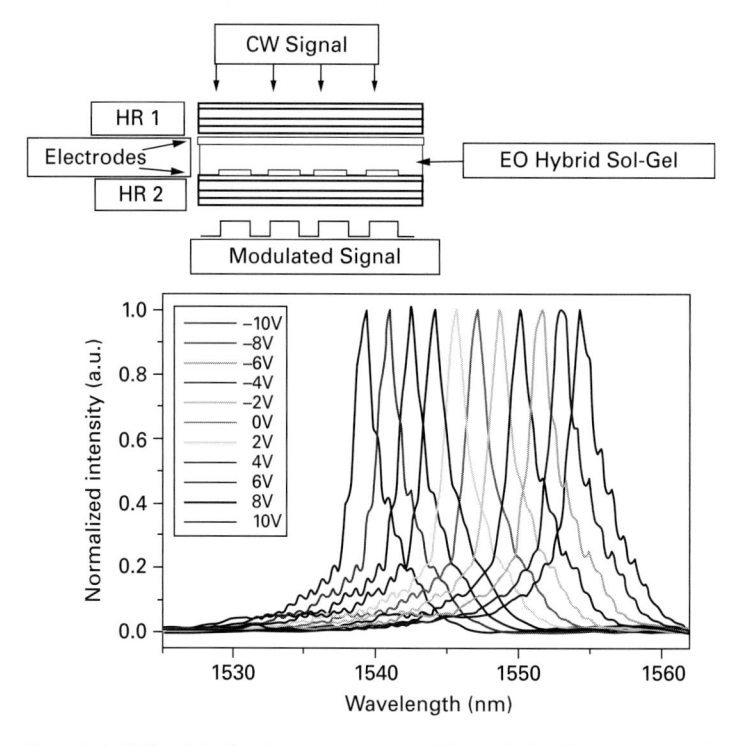

Figure 8.4 Shift of the bandpass resonance with applied voltage for an etalon structure.

A different relationship between bandwidth and drive voltage applies for processing digital data, namely, $V_{10dB} = 3h\lambda B/((n_{eff})^2 r_{33} c)$ where B is the bit rate and c is the speed of light. As an example, consider digital data with a bit rate of 10 Gb/s, n_{eff} = 1.6, r_{33} = 300 pm/V, λ = 1.3 μm, and $Q = 5 \times 10^3$, then an applied voltage of V_{10dB} = 1 V is

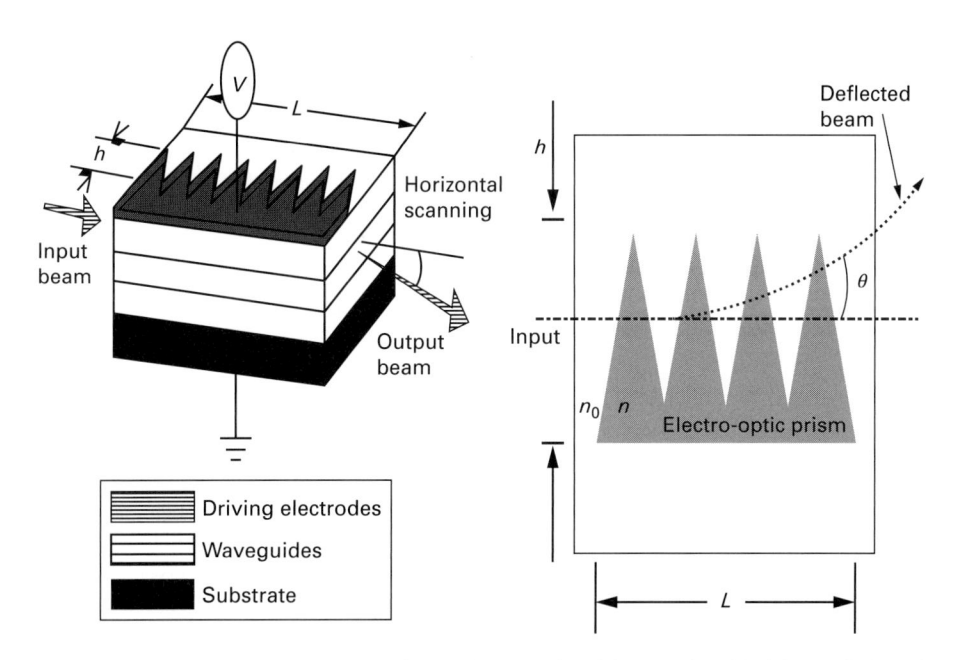

Figure 8.5 A cascaded prism beam steering device. Reprinted from Ref. [16] with permission of the American Chemical Society.

required. Drive voltage requirements for transducing digital data obviously scales with bit rate, so a bit rate of 100 Gb/s would require a drive voltage of 10 V. The preceding two examples (for analog and digital data processing) illustrate the importance of employing a material with a large electro-optic coefficient.

Ring microresonator devices are useful for a number of information processing applications including electrical-to-optical signal transduction, optical signal routing, active (voltage-controlled) wavelength-division multiplexing (WDM), and sensing. Frequently, individual resonators are configured in complex systems architectures to accomplish information technology tasks. Ring resonators do involve issues related to thermal stabilization (avoidance of thermal drift) and to additional losses associated with the bend of ring resonators (i.e., bending losses).

Beam steering devices. The relationship of device performance (beam steering angle) to material electro-optic activity for prism and cascaded prism device structures is given by

$$\theta = (n)^3 r_{33}(VL/dh) \tag{8.3}$$

where L and d are the length and width of the prism or the array of prisms, and V is the voltage applied across an overall thickness h (or the electrode spacing) [13 to 16]. In a cascaded prism device, L is the length of the base of the prism cascade (see Fig. 8.5). Again, it can be seen that smaller devices can be used if EO activity is sufficiently high.

Again, $n^3 r_{33}$ is the relevant material parameter for comparing various organic electro-optic materials. The effective electro-optic activity of organic liquid crystalline materials is orders of magnitude greater than that of organic "electronic" electro-optic materials; thus, liquid crystalline materials dominate spatial light modulation applications that do not require exceptional switching speeds. Currently, electronic organic electro-optic materials are only realistic for small-angle beam steering applications.

Silicon slotted waveguide devices. As illustrated in Fig. 9.27(b and c), light is concentrated in the low-refractive-index slots of silicon slotted waveguide devices. This phenomenon is readily predicted by solution of Maxwell's equations for such structures. The concentration of light immediately results in an enhancement of the efficiency of optical rectification and second-harmonic generation (although the generated second harmonic will not be supported in silicon waveguides in many cases). Optical rectification can occur with input (laser diode) optical powers as low as microwatt levels. The small dimensions of silicon waveguides permit greatly reduced electrode spacing, particularly when doped silicon is used for the electrodes. The consequence can be greatly reduced drive voltage requirements (note that Eq. (8.1) remains operative for calculating performance). Currently, drive voltage performance as low as 0.25 V has been demonstrated for vertical slot waveguides [17].

It should be noted that horizontal slot waveguides (see Fig. 7.37) have also been demonstrated [18].

8.2 Applications

Telecommunications. Currently, the most common application of electro-optic devices is that of single Mach–Zehnder modulators for electrical-to-optical signal transduction. Directional couplers are frequently used to perform signal routing. Ring microresonator structures are convenient for signal transduction, routing, and active wavelength division multiplexing. Resonant devices, such as ring microresonators, are also convenient for sensing applications. This includes devices exploiting surface plasmon resonance (SPR) phenomena.

A natural evolution of device technology involves moving from the use of single discrete devices to the chipscale integration of multiple devices. In Fig. 8.6, a chipscale reconfigurable optical add-drop multiplexer (ROADM) is shown [19].

Chipscale integration is most commonly accomplished on silicon and frequently involves integrating organic electro-optic materials into silicon or plasmonic photonic waveguide circuitry. The high index of refraction of silicon ($n = 3.48$ at $\lambda = 1.55$ μm) has the advantage of permitting reduction in the dimensions of photonic circuitry, which facilitates chipscale integration. Electro-optic devices may also play important roles in optical digital signal processing.

The high optical intensities realized with slotted silicon photonic devices also open the way for additional nonlinear optical applications relevant to telecommunications, including optical rectification and all-optical switching. Early demonstrations of these phenomena illustrate their potential [20,21].

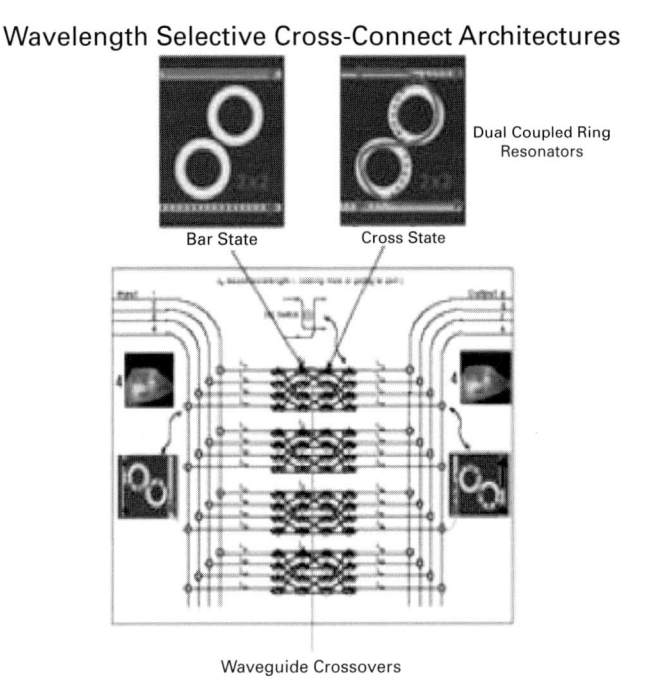

Wavelength Selective Cross-Connect Architectures

Figure 8.6 A 4×4 reconfigurable optical add-drop multiplexer.

Computing. Photonic/electronic integration is being increasingly pursued by computer chip manufacturers to deal with problems of bandwidth, power consumption, and thermal management. Current commercial systems involve off-chip integration, but research and development activities focus increasingly on chipscale integration based on silicon photonics. Such integration exploits the size compatibility between silicon electronics and photonics. Silicon modulators and indium phosphide (InP) electro-absorptive modulators are serious competitors to hybrid organic EO/silicon devices. The main limitations of these competitive technologies include thermal management, optical loss, and the linearity of reponse. As will be discussed in Chapter 12, modulated lasers can be serious competitors for some applications because of their low cost.

Defense and sensing. Increasingly, electrical coaxial cable is being replaced by silica fiber for transmission of information over distances greater than a meter in military planes, ships, and land vehicles. In satellites, the size, weight, and power savings resulting from such replacement are even more important as they translate into serious cost savings. Chipscale integration, discussed above for telecommunication and computing applications, is also relevant for defense systems that utilize these technologies.

Defense also utilizes a variety of sensors based on electro-optic device structures, ranging from electric field sensors including phased array radar to terahertz spectroscopy and imaging. Resonant devices (and particularly slotted silicon resonant devices) are particularly attractive for sensing owing to the enhancement of detector sensitivity

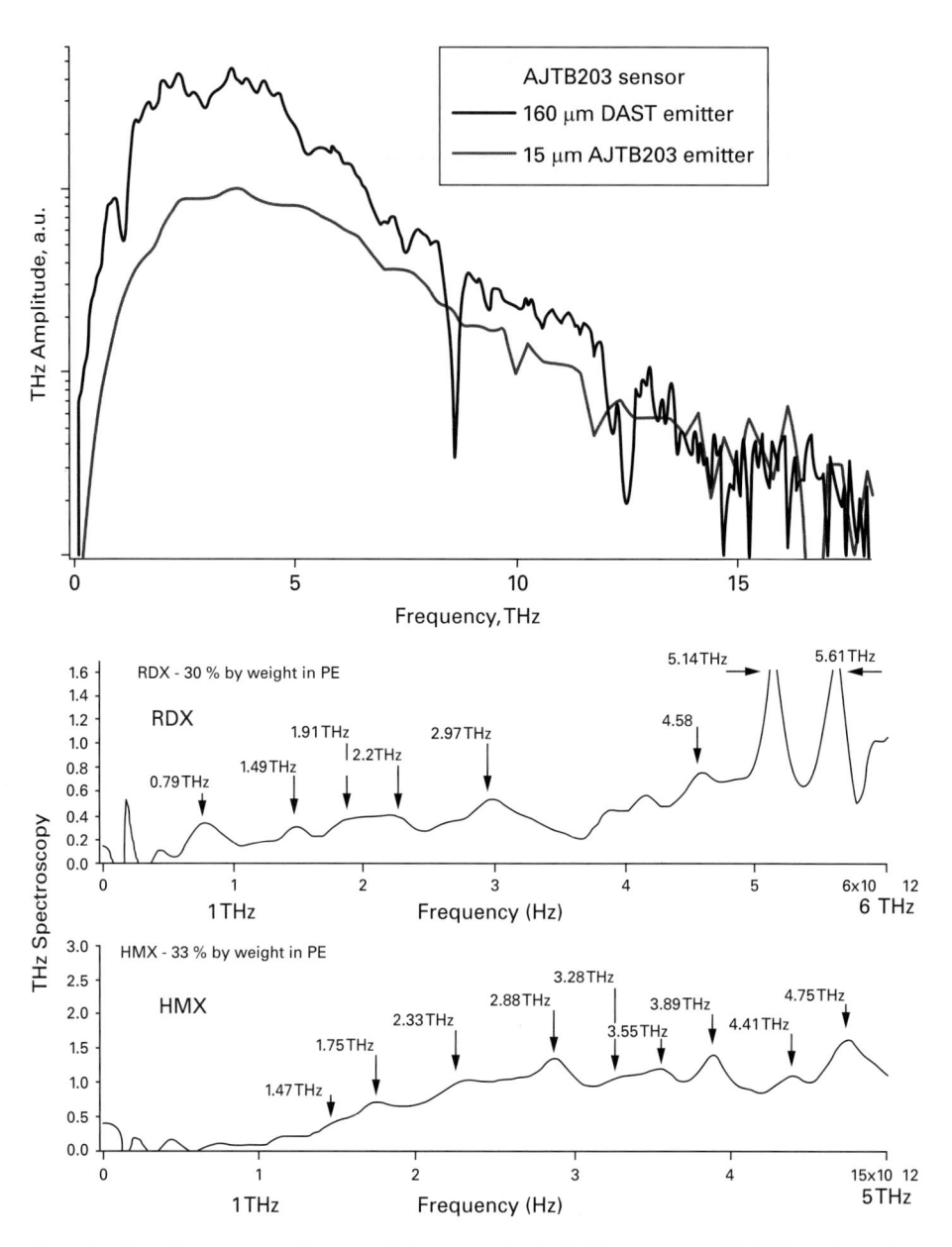

Figure 8.7 Top: Terahertz amplitude for EO polymer (AJTB203) and crystal (DAST) emitters (the AJTB203 polymer is used as a sensor in both cases). The absence of phonon peaks in the polymer spectrum is clearly evident. Bottom: THz spectra for the military-grade explosives RDX (spectrum to 6 THz shown) and HMX (spectrum to 5 THz shown). Figures are courtesy of Professor Michael Hayden (University of Maryland–Baltimore County).

by the Q factor of the resonant devices (see Fig. 9.26 for examples of integration of ring microresonators with optical fiber technology) and owing to concentration of light in silicon slotted waveguide devices. Because atoms or molecules entering the electro-optic device region will produce a change in the index of refraction (which can be

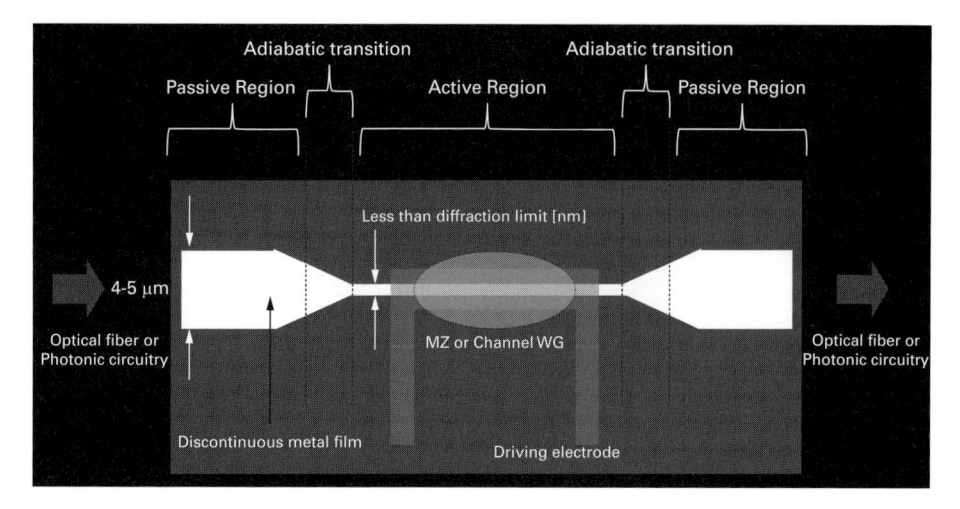

Figure 8.8 A plasmonic nanowire modulator.

quantitatively evaluated and translated to a voltage change exploiting the electro-optic effect), chemical and biochemical sensing can be effected. Changes in dimensions of resonators (particularly organic devices) can be exploited to measure physical properties (e.g., stress, shear).

Two applications of sensor technology that are receiving considerable attention include terahertz sensing/spectroscopy [22] and phased array radar [23]. Terahertz generation and detection using second-order nonlinear optical materials is an example of optical rectification (or difference-frequency generation). Electrically poled macromolecular organic electro-optic materials have an advantage relative to inorganic crystalline materials in that phonon resonances are absent. Thus, operation to 15 THz has been demonstrated, and operation to 30 THz is highly probable [22]. Terahertz radiation has been used in imaging applications and also for terahertz spectroscopy. The example of detection of explosive materials by their terahertz absorption fingerprints is illustrated in Fig. 8.7.

Other defense-related devices include optical gyroscopes, spectrum analyzers, high-frequency oscillators, and A/D converters.

Plasmonic and metamaterial devices. Plasmonic and metamaterial device architectures afford the possibility of new levels of device performance through exploitation of "negative index of refraction" phenomena and resonance effects. Superprisms, superlenses, and cloaking devices are receiving increasing attention. Also, plasmonic and metamaterial effects can be used to concentrate light into nanoscopic dimensions (see Fig. 8.8) permitting considerable reduction in the size of photonic circuit elements [9,10,12]. Plasmonic stripline devices as short as 10 μm have been fabricated as discussed in Chapter 7.

The major problem that has plagued these architectures to the present is that of optical loss. This point is illustrated in Fig. 8.9, where the optical loss associated with plasmonic concentration of light and enhancement of nonlinear optical activity is shown.

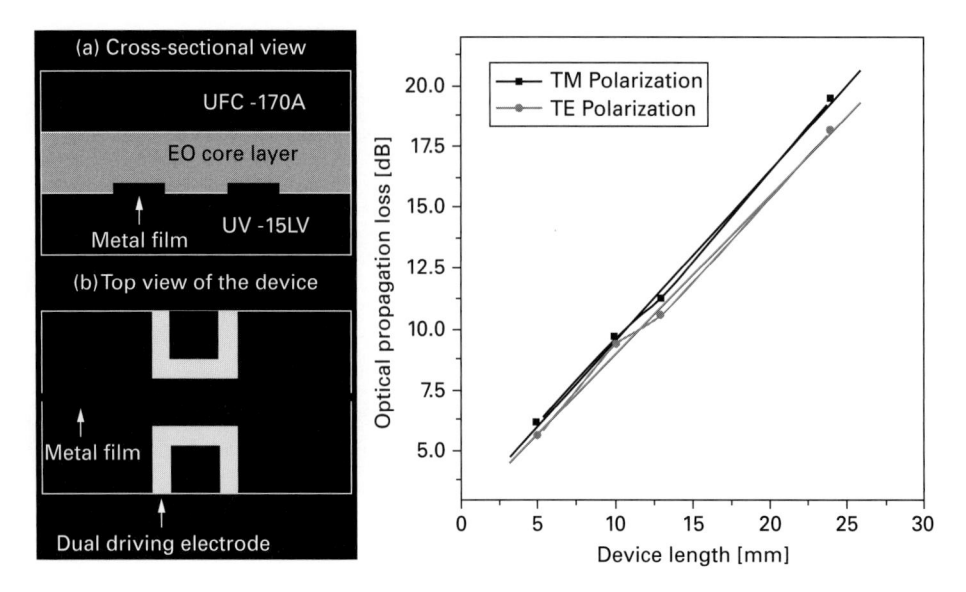

Figure 8.9 The optical loss for TM and TE modes for a simple plasmonic electro-optic modulator. UV −170A and UFC −15LV are UV-curable polymers.

The effects demonstrated at 1550 nm telecommunications wavelengths require an order of magnitude increase in optical loss. Such high optical loss is a major factor in inhibiting practical application. Optical loss can be somewhat reduced by utilizing nanostructured metals, but insertion loss is still typically above 10 dB for a 1-cm device. However, with short plasmonic-organic hybrid (POH), devices as short as 10 μm have been demonstrated with a plasmonic loss of 0.5 dB/micron and an insertion loss of 12.5 dB.

References

[1] W. H. Steier and L. R. Dalton, in *Broadband Optical Modulators: Science, Technology, and Applications*, A. Chen and E. Murphy, Eds., New York, Taylor & Francis (2011), Ch. 9, pp. 221–254.
[2] L. R. Dalton, *J. Phys.: Condens. Matter*, **15**, R897 (2003).
[3] S.-S. Lee, A. H. Udupa, H. Erlig, et al., *IEEE Microw. Guided Wave Lett.*, **9**, 357 (1999).
[4] D. H. Chang, H. Erlig, M.-C. Oh, et al., *IEEE Photon. Tech. Lett.*, **12**, 537 (2000).
[5] A. Yacoubian, V. Chuyanov, S. M. Garner, et al., *IEEE J. Sel. Top. Quantum Electron.*, **6**, 810 (2000).
[6] H. R. Fetterman, D. H. Chang, H. Erlig, et al., *Proc. SPIE*, **4114**, 44 (2000).
[7] J. H. Bechtel, Y. Shi, H. Zhang, et al., *Proc. SPIE*, **4114**, 58 (2000).
[8] H. R. Fetterman, D. Chang, W. H. Steier, et al., *Trends in Optics and Photonics. Integrated Photonics Research. Optical Society of America Tech. Digest*, 146 (2000).
[9] S.-K. Kim, N. Sylvain, S. J. Benight, et al., *Proc. SPIE*, **7754**, 775403 (2010).
[10] S.-K. Kim, H. R. Fetterman, K. Geary, P. Berini, and L. R. Dalton, *Proc. 2001 Conf. Laser and Electro-Optics*, May 1–6, 2011, Baltimore, MD, 1–3 (2011).

[11] S. Kim, W. Liu, Q. Pei, L. R. Dalton, and H. R. Fetterman, *Opt. Express*, **19**, 7865 (2011).
[12] L. R. Dalton, A. K.-Y. Jen, B. H. Robinson, *et al.*, *Proc. SPIE*, **7935**, 793502 (2011).
[13] L. Sun, J.-H. Kim, C.-H. Jang, *et al.*, *Opt. Eng.*, **3950**, 98 (2000).
[14] L. Sun, J.-H. Kim, C.-H. Jang, *et al.*, *Opt. Eng.*, **40**, 1217 (2001).
[15] J.-H. Kim, L. Sun, C.-H. Jang, *et al.*, *Proc. SPIE*, **4279**, 37 (2001).
[16] L. R. Dalton, P. A. Sullivan, and D. H. Bale, *Chem. Rev.*, **110**, 25 (2010).
[17] T. Baehr-Jones, B. Penkov, J. Huang, *et al.*, *Appl. Phys. Lett.*, **92**, 163303 (2008).
[18] H. Figi, D. H. Bale, A. Szep, L. R. Dalton, and A. Chen, *J. Opt. Soc. Am. B*, **28**, 2291 (2011).
[19] J. Takayesu, M. Hochberg, T. Baehr-Jones, *et al.*, *IEEE J. Lightwave Technol.*, **27**, 440 (2008).
[20] T. Baehr-Jones, M. Hochberg, G. Wang, *et al.*, *Opt. Express*, **13**, 5216 (2005).
[21] M. Hochberg, T. Baehr-Jones, G. Wang, *et al.*, *Nat. Mater.*, **5**, 703 (2006).
[22] C. V. McLauglin, L. M. Hayden, B. Polishak, *et al.*, *Appl. Phys. Lett.*, **92**, 151107-1 (2008).
[23] R. S. Kim, A. Szep, N. G. Usechak, *et al.*, *Proc. SPIE*, **8259**, 82590B1 (2012).

9 Organic electro-optic waveguides, switches, and modulators

9.1 Light propagation in optical waveguides

Light as an electromagnetic wave experiences diffraction when propagating in free space or in a homogeneous medium. This is because, in all practical cases, beams of light are laterally limited and not infinite. This limits the propagation distances for free-space optical communications, as well as the interaction lengths with the applied electric fields for electro-optic applications and with other optical fields for other nonlinear optical applications. The smallest possible diffraction is experienced by a beam with a Gaussian intensity profile of the following form (see also Fig. 9.1(a))

$$I(\rho, z) = I_0 \left(\frac{w_0}{w(z)}\right)^2 \exp\left(-\frac{2\rho^2}{w^2(z)}\right), \tag{9.1}$$

where $w(z)$ is the radius of the beam propagating along the Cartesian axis z with the radial coordinate $\rho = \sqrt{x^2 + y^2}$, w_0 is the waist radius (minimal radius of the beam chosen at $z = 0$, i.e. in the focus) with the following propagation parameters

$$w(z) = w_0 \sqrt{1 + \left(\frac{z}{z_0}\right)^2}; \quad z_0 = \frac{\pi w_0^2}{\lambda}, \tag{9.2}$$

which can be derived from the wave equation (see, for example, Ref. [1]). At the radial distance of the beam radius $\rho = w(z)$, the intensity drops by a factor of $1/e^2 \sim 0.135$, i.e. most of the beam power remains within a circle of radius $w(z)$.

When a beam is focused to a waist diameter of $2w_0$, it will diffract so that for large $z \gg z_0$ its radius is increasing linearly with z at an angle $\theta = w_0/z_0$ (see Fig. 9.1(a)). The Rayleigh length z_0 (also called the depth of focus) specifies the propagation distance, in which the beam has a radius smaller than $\sqrt{2}w_0$. If the beam is strongly focused to a waist comparable to the wavelength $w_0 \sim \lambda$, the beam diffracts very strongly with $z_0 \sim \pi \lambda$ and $\theta > 15°$.

The efficiencies of nonlinear optical processes scale with the intensity of the pump beams; see Eq. (2.25). The peak intensity can be related to the beam power P as

$$I_0 = \frac{2P}{\pi w_0^2}. \tag{9.3}$$

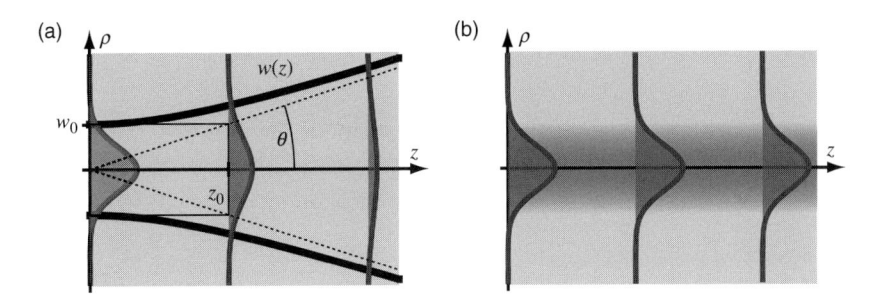

Figure 9.1 (a) Propagation of a Gaussian beam with waist radius w_0 in a homogeneous material. The beam diverges as $\theta \approx w_0/z_0$, where the Rayleigh range $z_0 = \pi w_0^2/\lambda$. (b) Propagation of a waveguide mode in a waveguide.

We usually have a constant beam power available, so we can greatly increase the efficiency of nonlinear optical processes by focusing the beam. However, the efficiency also scales quadratically with the interaction length L; see Eq. (2.25). If the beam diverges within this length, the efficiency will drop, so we are limited to interaction distances of about $L \sim z_0$. Keeping a high beam intensity over much longer interaction distances, as in Fig. 9.1(b), can increase efficiencies of nonlinear optical processes by more than two orders of magnitude.

In electro-optic applications, the half-wave voltage needed to drive the devices scales with d/l (see Eq. (3.23)) – but owing to diffraction, the propagation distance l is considerably limited in homogeneous materials. For example, choosing $l = 1$ cm with $l < \pi w_0^2/\lambda$ or $l < \pi d^2/(4\lambda)$, the distance between the electrodes should be considerably larger than $\sqrt{4\lambda l/\pi} \sim 100\,\mu$m for $\lambda \sim 1\,\mu$m. Without diffraction, we could focus the beam down to a waist diameter of the order of the light wavelength, keeping the electrodes a few μm apart, which could therefore reduce the half-wave voltage by two orders of magnitude. Indeed, typical half-wave voltages for bulk $LiNbO_3$ optical modulators are of the order of 1000 V, while good waveguide $LiNbO_3$ modulators reach half-wave voltages below 10 V.

The possibility of guiding light in optical waveguides has enabled optical communications, as well as nonlinear optical interactions and electro-optic effects, at very moderate optical-power levels and voltages. Optical waveguides are basic units of all integrated optics, allowing guided-wave propagation in micrometer-size cross-sections without experiencing diffraction (Fig. 9.1(b)).

Optical waveguides require non-homogeneous media with properly designed refractive-index spatial profiles. Most common are dielectric waveguides that consist of a core region with dimensions comparable to light wavelength, which has a higher refractive index than the surrounding cladding region (Fig. 9.2), resulting in light confinement in the region of the higher refractive index. This confinement is simply based on the total internal reflection principle at the boundaries. The refractive index can also be continuously changed from the core to the cladding. By choosing the refractive index of the core and of the cladding region, we can tune the effective refractive index (i.e. phase velocity) of the light propagating in a waveguide. Several optical modes with

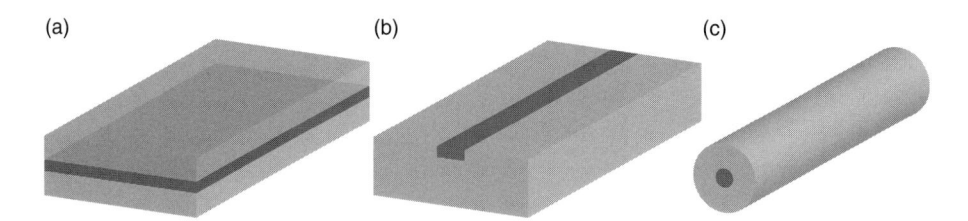

Figure 9.2 1D optical waveguide or slab waveguide (a) and 2D waveguides: strip (b), and fiber (c).

different phase velocity may propagate in a waveguide, depending on the dimensions of the core region and the refractive-index profile. Waveguides can be planar or one-dimensional (see Fig. 9.2(a)), allowing spatial confinement only in one lateral dimension, or two-dimensional, laterally completely confined, with a rectangular/trapezoidal (Fig. 9.2(b)) or a circular/oval (Fig. 9.2(c)) cross-section.

For organic electro-optic materials, a configuration such as that shown in Fig. 9.2(b) is most convenient from the fabrication point of view. In some cases, a planar configuration is used if diffraction in the lateral plane is not a problem. To use the electro-optic effect most efficiently, the core material, in which most light is concentrated, is made of an electro-optic material, while the cladding may be a non-active material such as a passive polymer. This is also a most convenient geometry from the point of view of materials properties, since active electro-optic organic materials usually have higher refractive indices than non-active ones, owing to their high electronic polarizability. More recently, however, a need has emerged to integrate organic materials in more compact photonic devices such as those based on silicon waveguides. Silicon has a refractive index much higher than common organic materials, and therefore different device configurations are required – these are discussed later in this chapter. Based on the high light confinement possible in silicon nanophotonic waveguides, efficiencies exceeding the state-of-the-art of conventional waveguides by at least one order of magnitude have been demonstrated.

9.1.1 Guided modes of slab waveguides

The 1D waveguide structure of the slab waveguide (Fig. 9.2(a)) is particularly simple to analyze (see, for example, Refs. [1 to 4]). The results contain all the necessary components to understand propagation in more complicated 2D waveguiding structures, which are nowadays commonly analyzed numerically. We will consider the configuration sketched in Fig. 9.3: for simplicity we will assume isotropic materials in a configuration symmetric around $x = 0$, i.e., consisting of only two materials with different refractive indices $n_1 > n_2$. We will assume no variation of geometry or field distribution in the y direction, i.e., all derivatives $\partial/\partial y$ will be 0. This simplification allows us to decompose the field in two types of modes with a well-defined polarization: transverse-electric (TE) modes with the electric field vector parallel to y, and transverse-magnetic (TM) modes with the magnetic field vector parallel to y. This is not the case

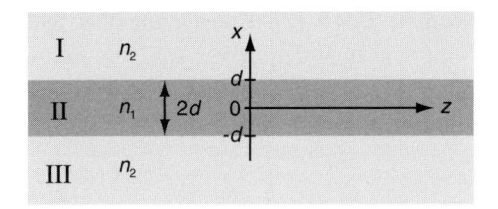

Figure 9.3 Cross-sectional view of a planar (1D, slab) dielectric waveguide of thickness 2d, $n_1 > n_2$.

for 2D waveguides, whose modes are in general hybrid, but are in most practical cases very close to pure TE/TM modes.

We can employ the common wave equation, which for isotropic materials has a particularly simple form

$$\nabla^2 \mathbf{E}(\mathbf{r},t) - \frac{n^2(\mathbf{r})}{c^2} \frac{\partial^2 \mathbf{E}(\mathbf{r},t)}{\partial t^2} = 0, \tag{9.4}$$

where $c = 1/\sqrt{\varepsilon_0 \mu_0}$ is the speed of light in vacuum, $\mathbf{E}(\mathbf{r},t)$ the electric field of the optical wave, and $n(\mathbf{r})$ the spatially varying refractive index of the (composite) material:

$$n(\mathbf{r}) = \begin{cases} n_1 & |x| < d \\ n_2 & |x| > d. \end{cases} \tag{9.5}$$

Note that for strongly anisotropic materials a more complex analysis may be necessary for precise evaluation of propagation parameters [5 to 7]. The electric field vector can be written as $\mathbf{E}(\mathbf{r},t) = \mathbf{E}(x,z)e^{i\omega t}$, where ω is the angular frequency of light that does not change in the material if the material is linear. This gives the following reduced wave equation:

$$\frac{\partial^2 \mathbf{E}(x,z)}{\partial x^2} + \frac{\partial^2 \mathbf{E}(x,z)}{\partial z^2} + k_0^2 n^2(x) \mathbf{E}(x,z) = 0; \quad k_0 = \omega/c. \tag{9.6}$$

Since light propagates along the chosen direction z, we can consider the following spatial dependence

$$\mathbf{E}(x,z) = \mathbf{E}(x) e^{-i\beta z}, \tag{9.7}$$

where $\beta = k_0 n_{\text{eff}}$ is the propagation constant depending on the effective index n_{eff} of the guided wave (or its phase velocity c/n_{eff}). n_{eff} depends on waveguide geometry and refractive indices n_1 and n_2. The spatial dependence (Eq. 9.7) further reduces the wave equation (9.6) into

$$\frac{d^2 \mathbf{E}(x)}{dx^2} + (k_0^2 n^2(x) - \beta^2) \mathbf{E}(x) = 0. \tag{9.8}$$

To obtain the solution of this equation, which gives the propagation constant β and the electric field profile $\mathbf{E}(x)$, we will discuss the solutions for the TE and TM light polarization separately.

TE modes are obtained by setting $E_z = 0$ [2]. Considering this and Maxwell's equations for an isotropic case with vanishing derivatives $\partial/\partial y$, the following electric and magnetic field components are not zero: E_y, H_x, and H_z. The wave equation in the TE case becomes

$$\frac{d^2 E_y(x)}{dx^2} + (k_0^2 n^2(x) - \beta^2) E_y(x) = 0. \tag{9.9}$$

With the help of Maxwell's equations, the magnetic field components H_x and H_z can be related to the electric field E_y as

$$H_x(x) = -\frac{\beta}{\omega \mu_0} E_y(x), \quad H_z(x) = \frac{i}{\omega \mu_0} \frac{dE_y(x)}{dx} \tag{9.10}$$

Considering the refractive index profile (Eq. (9.5)), the solution for the field profile differs inside (region II of Fig. 9.3) and outside (regions I and III) of the waveguide core region. We can further simplify the analysis by considering even and odd modes separately.

Even guided TE modes have the following mode profile inside region II ($|x| < d$)

$$E_y^{II}(x) = A \cos k_x x \tag{9.11}$$

with

$$k_x^2 = k_0^2 n_1^2 - \beta^2 > 0, \tag{9.12}$$

resulting in, considering (Eq. 9.10)

$$H_x^{II}(x) = -\frac{\beta}{\omega \mu_0} A \cos k_x x, \quad H_z^{II}(x) = -\frac{i k_x}{\omega \mu_0} A \sin k_x x. \tag{9.13}$$

Outside the slab, inside regions I and III ($|x| > d$), the field for guided modes decays exponentially

$$E_y^{I,III}(x) = B e^{-\kappa_x(|x|-d)} \tag{9.14}$$

with

$$\kappa_x^2 = \beta^2 - k_0^2 n_2^2 > 0, \tag{9.15}$$

resulting in, considering (Eq. 9.10)

$$H_x^{II}(x) = -\frac{\beta}{\omega \mu_0} B e^{-\kappa_x(|x|-d)}, \quad H_z^{II}(x) = -\frac{i \kappa_x}{\omega \mu_0} \frac{x}{|x|} B e^{-\kappa_x(|x|-d)}. \tag{9.16}$$

Boundary conditions for the electromagnetic waves imply that the components of **E** and **H** fields parallel to the boundaries should be continuous, i.e. $E_y^I(d) = E_y^{II}(d)$ and $H_z^I(d) = H_z^{II}(d)$, resulting in

$$A \cos k_x d = B \tag{9.17}$$

$$A \sin k_x d = \frac{\kappa_x}{k_x} B \tag{9.18}$$

Dividing the above equations gives the eigenvalue equation

$$\tan k_x d = \frac{\kappa_x}{k_x}. \tag{9.19}$$

Using Eqs. (9.12) and (9.15), κ_x can be related to k_x as $\kappa_x = \sqrt{k_0^2(n_1^2 - n_2^2) - k_x^2}$; therefore the eigenvalue equation is the transcendental equation for k_x

$$\tan k_x d = \frac{\sqrt{k_0^2(n_1^2 - n_2^2) - k_x^2}}{k_x}, \tag{9.20}$$

which can be solved numerically for a given dielectric contrast $\Delta n^2 = n_1^2 - n_2^2$, thickness d, and light angular frequency $\omega = k_0 c$.

Odd guided TE modes have the following mode profiles inside region II ($|x| < d$), and inside regions I and III ($|x| > d$)

$$E_y^{II}(x) = A \sin k_x x, \quad E_y^{I,III}(x) = \frac{x}{|x|} B e^{-\kappa_x(|x|-d)} \tag{9.21}$$

From boundary conditions we can obtain the following eigenvalue equation for this case

$$\tan k_x d = -\frac{k_x}{\kappa_x} = -\frac{k_x}{\sqrt{k_0^2(n_1^2 - n_2^2) - k_x^2}}. \tag{9.22}$$

Figure 9.4(a) shows a graphical solution of the transcendental eigenvalue equations for even and odd TE modes obtained by plotting the left-hand side and the right-hand side of Eqs. (9.20) and (9.22).

Considering Eq. (9.12) and the solutions of the transcendental equations we obtain the propagation constants β_m and the effective refractive indices $n_{\text{eff},m}$ of different guided modes m, which have values between n_1 and n_2 ($n_2 < n_{\text{eff}} < n_1$, see Fig. 9.4(b)).

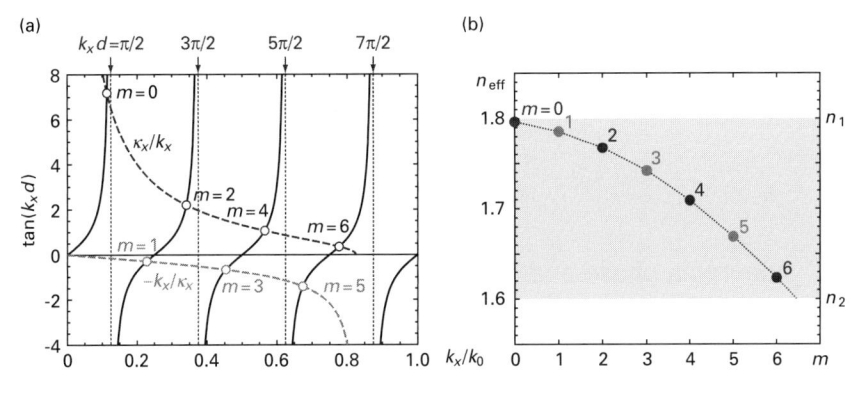

Figure 9.4 (a) Graphical solution of the eigenvalue equations (9.20) and (9.22) for TE modes and light wavelength $\lambda_0 = 2\pi/k_0 = 1.0\,\mu\text{m}$, refractive indices $n_1 = 1.8$ and $n_2 = 1.6$, and slab thickness $2d = 4\,\mu\text{m}$. (b) The corresponding effective refractive indices n_{eff} of the seven guided modes $m = 0$ to 6 existing in this particular waveguide configuration. The number of guided modes supported by a waveguide depends on the waveguide dimensions, refractive indices, and light wavelength.

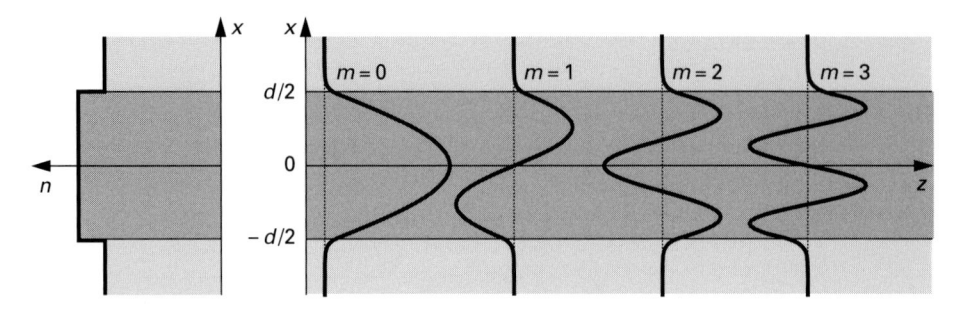

Figure 9.5 Optical field profiles of the first four waveguiding modes $m = 0,1,2,3$ in a waveguide (right) with the refractive index profile that is shown on the left.

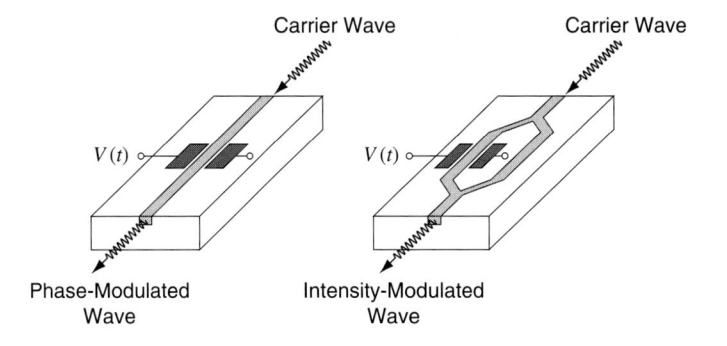

Figure 9.6 An integrated-optic phase modulator (left) and a Mach–Zehnder intensity modulator (right) based on the electro-optic effect.

The corresponding optical field profiles of the first four waveguiding modes are illustrated in Fig. 9.5.

TM modes are obtained by setting $H_z = 0$. Considering this and the Maxwell's equations for an isotropic case with vanishing derivatives $\partial/\partial y$, the following electric and magnetic field components are not zero: H_y, E_x, and E_z. We can find a solution in an analogous way as for the TE modes. The resulting eigenvalue equations for even and odd TM modes are

$$\tan k_x d = \frac{n_1^2}{n_2^2} \frac{\kappa_x}{k_x} \text{ (even TM)}, \quad \tan k_x d = \frac{n_2^2}{n_1^2} \frac{k_x}{\kappa_x} \text{ (odd TM)}. \tag{9.23}$$

9.2 Integrated phase and amplitude electro-optic modulators

Figure 9.6 shows the simplest schemes of integrated-optics phase and amplitude modulators.

Their operation principle is very similar to the one in bulk materials discussed in Section 3.4. Using a phase modulator, the phase ϕ of the optical field $\tilde{E}(t)$ will change with the applied voltage $V(t) = V_m \sin(\omega_m t)$ as

$$\Delta\phi = -\pi \frac{V(t)}{V_\pi}; \quad \tilde{E}(t) = A \cos\left(\omega t + \pi \frac{V_m}{V_\pi} \sin(\omega_m t)\right). \tag{9.24}$$

The principle of the Mach–Zehnder amplitude (intensity) modulator is based on the interference between the beam experiencing phase modulation and the reference beam. At the first Y-branch, a guided optical mode is divided into two guided modes, one of which passes through a phase electro-optic modulator. These two guided beams are combined again after the second Y-branch. If the branches divide the beam with intensity I_0 into two equal parts, the outgoing signal beam intensity is, as in the case of bulk intensity modulator (see Eq. (3.22)), equal to

$$I_{\text{out}} = I_{\text{in}} \cos^2\left(\frac{\Delta\phi}{2}\right) = I_{\text{in}} \cos^2\left(\frac{\pi}{2}\frac{V}{V_\pi}\right), \tag{9.25}$$

where I_{in} is the input beam intensity. We can choose the working point at the point of the maximum slope dT/dV at $T(V) = 0.5$ by using an appropriate bias voltage to obtain a linear intensity modulator. In this case the modulated transmitted light intensity is directly proportional to the modulation of the electric signal $V(t)$, as also illustrated in Fig. 3.4.

9.3 Optical coupling between waveguides

Optical power can be coupled in and out of a waveguide by using several optical coupling approaches:

Prism coupling: By using a prism of a refractive index higher than that of the cladding material, light can be coupled into waveguides owing to the evanescent field at the total internal reflection within the prism. For multimode waveguides, this method allows determination of the effective indices of various waveguide modes, which depend on the particular coupling angle within the prism. This method is only applicable to planar waveguides and is often also used to precisely measure the refractive index of various materials that can be deposited in a thin-film form on a lower-index substrate with a known refractive index, since the optical modes (i.e. their effective refractive indices) are very sensitive to small changes of the refractive index between the material and the cladding. Micro-prisms or mirrors can also be directly integrated into planar photonic circuits to realize coupling or sharp bends.

End-fire coupling: Optical power from a free-space propagating beam can be coupled into and out of a waveguide/fiber by using strong focusing, usually by microscope objectives or so-called collimating lenses. Coupling is optimized when the beam is focused so that its profile and polarization correspond to guided mode(s). The efficiency is limited by reflection losses at the input surface and mode mismatching, both being particularly critical for strongly confined waveguide modes in waveguides with a high refractive index and a high index contrast between the core and the cladding.

Fiber coupling: Similar to end-fire coupling, but the light is coupled directly from an optical fiber to the waveguide.

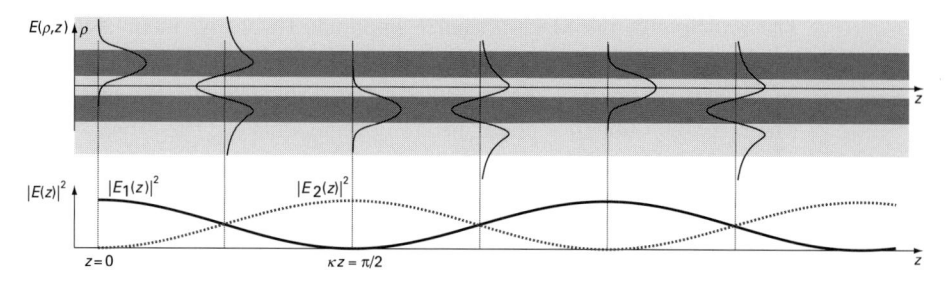

Figure 9.7 Optical coupling between two parallel single-mode waveguides with spatially overlapping fields of their optical modes. When propagating, the optical power will move from one waveguide to another and back. The graph below illustrates the corresponding power in each waveguide as a function of the propagation direction z for the case of identical waveguides.

> **Grating coupling:** This method is used either with free-space propagating beam or beam from an optical fiber. At the input of a waveguide, a refractive-index grating can be fabricated, which is used for coupling light directly from above into the waveguide.
>
> **Waveguide tapers:** Coupling from one waveguide mode to another can be integrated into the same plane by using a tapered waveguide, which serves as a mode converter. It can be combined with other coupling schemes, e.g., when light from an optical fiber with a larger beam diameter is to be coupled into a small-core high-index-contrast waveguide.

Beside the above coupling schemes, light can be coupled between adjacent waveguides from one waveguide to another. This approach can be also used for various optical functionalities in integrated optics. The basic principle of this coupling is described in the following paragraphs.

We consider two single-mode waveguides with different propagation constants ($\beta = k_0 n_{\text{eff}}$) β_1 and β_2, which are the propagation constants of isolated waveguides. If these waveguides are placed so close together that both optical modes overlap, the optical power can be coupled from one waveguide to another and back. Such a situation is illustrated in Fig. 9.7.

To obtain the optical mode profile in such a two-waveguide structure, we can rigorously solve the wave equation similarly as for a single-waveguide structure. However, if the coupling is weak, we can proceed in a slowly varying amplitude approximation assuming that the electric field amplitude $A_i(z)$ in each waveguide ($i = 1$ or 2) is only slowly changing, resulting in the following electric field as a function of the propagation direction z

$$a_1(z) = A_1(z)e^{i\beta_1 z}, \tag{9.26}$$

$$a_2(z) = A_2(z)e^{i\beta_2 z}. \tag{9.27}$$

Note that for a single waveguide the amplitude is not changing as a function of z, but has only a phase factor as in Eq. (9.7). For weak coupling we can assume that the lateral field amplitudes do not change. The derivative of the electric field in one waveguide as a function of z will in first approximation contain a factor proportional to the field in the other waveguide (see e.g. Refs. [1 to 4])

$$\frac{da_1}{dz} = i\beta_1 a_1 + i\kappa_1 a_2, \quad (9.28)$$

$$\frac{da_2}{dz} = i\beta_2 a_2 + i\kappa_2 a_1, \quad (9.29)$$

where κ_i is the (real) coupling constant, which depends on the spatial overlap between both optical modes (for an exact expression see e.g. Ref. [2]). The conservation of energy implies that $\kappa_1 = \kappa_2 = \kappa$. Assuming a boundary condition that at $z = 0$ all optical power is in the first waveguide, i.e. $a_1(0) = a_0$ and $a_2(0) = 0$, the solution of the above differential equations is

$$a_1(z) = a_0 e^{i\beta z}\left(\cos \gamma z - \frac{i\Delta\beta}{2\gamma}\sin \gamma z\right), \quad (9.30)$$

$$a_2(z) = a_0 e^{i\beta z}\frac{i\kappa}{\gamma}\sin \gamma z, \quad (9.31)$$

where $\beta = (\beta_1 + \beta_2)/2$, $\Delta\beta = \beta_2 - \beta_1$, and $\gamma^2 = (\Delta\beta/2)^2 + \kappa^2$. The above solution shows a simple periodic exchange of power between two waveguides, which is illustrated in Fig. 9.7 for the case of identical waveguides with $\beta_1 = \beta_2$, leading to a complete power exchange and the following simple solution

$$a_1(z) = a_0 e^{i\beta z} \cos \kappa z, \quad (9.32)$$

$$a_2(z) = a_0 e^{i\beta z} \sin \kappa z. \quad (9.33)$$

In integrated optics, optical power often has to be transferred to other waveguides, which can be done by using a coupler with a length within which a complete transfer to the other waveguide is achieved, $L = \pi/(2\kappa)$ (see Figs. 9.7 and 9.8). Such couplers can also be used to split a signal from one to two waveguides, as needed for example for Mach–Zehnder modulators – in this case the coupler length of $L = \pi/(4\kappa)$ is optimal.

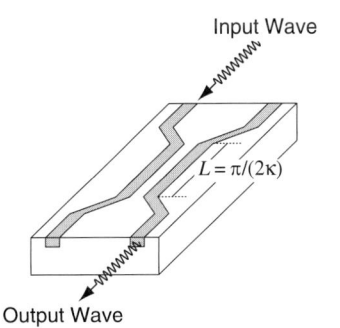

Figure 9.8 Directional coupler: for a certain interaction distance L, light can be completely coupled from one waveguide to the other. By introducing a phase mismatch $\Delta\beta = 2\sqrt{3}\kappa$ in this configuration, for example by applying an electric field to one of the two waveguides, light can be switched from one waveguide to the other.

Waveguide couplers can also be used as switching elements if made out of electro-optic materials. For example, a coupler with $\Delta\beta=0$ and interaction length $L = \pi/(2\kappa)$ will transmit all input power from the first waveguide into the second waveguide, as shown in Fig. 9.8.

When introducing a propagation constant mismatch $\Delta\beta$ by applying an electric field and thus changing the effective index of one waveguide (or applying a field of opposite sign to both waveguides), the transmission into the second waveguide will change according to Eq. (9.31) and will reach zero for $\kappa L = \pi$, which corresponds to $\Delta\beta = 2\sqrt{3}\kappa$ – in this case all light will exit from the first waveguide. Directional couplers can be also used as electro-optic modulators: by applying a modulated voltage, the transmitted light is modulated in intensity, in a similar way as in a Mach–Zehnder interferometer.

9.4 Microring resonators

Mach–Zehnder integrated-optic elements are CMOS compatible and allow for high speed (>40 GHz) modulation in a traveling-wave configuration. However, present devices based on LiNbO$_3$ are still characterized by relatively large size (lengths of the order of 1 cm) and by relatively high power consumption (several watts). Compact-size and low-power devices may be possible with organic and hybrid electro-optic materials. A promising structure is a microring resonator that allows for long electrical and optical interaction lengths on a very small scale (of the order of 10–100 µm), allowing for low power consumption and very large scale integration (VLSI) [4, 8 to 16].

The basic design of a microresonator is shown in Fig. 9.9(a). The device consists of two straight waveguides that are coupled by a microring waveguide.

Light guided in the input waveguide is partially coupled into the microring if the ring is close enough (laterally or on top of) the input waveguide, so that their optical modes overlap. A very small coupling constant κ is sufficient – the ideal coupling constant depends on the losses inside the ring, as discussed below. At resonator optical frequencies the intensity of the light in the ring increases markedly, and is coupled out into the drop-port waveguide. The free-spectral range (FSR) is the wavelength separation between the adjacent transmission peaks. If the ring is made of an electro-optic material, we can tune its refractive index and therefore also the resonance condition by applying an external electric field E. Figure 9.9(c) illustrates a result of such a filtering by observing light transmission through the drop port while scanning the wavelength. One can see sharp increases of the intensity when the resonance condition is satisfied. The resonance condition is satisfied if after one round trip in the ring the light is in phase with the light coupled into the microring, i.e.

$$\phi = kL = \frac{2\pi}{\lambda_0} nL \quad \text{round-trip phase change} \tag{9.34}$$

$$\phi = m \cdot 2\pi \,(\text{at resonance}) \Rightarrow \lambda_m = \frac{nL}{m} \quad \text{resonance wavelength} \tag{9.35}$$

9.4 Microring resonators

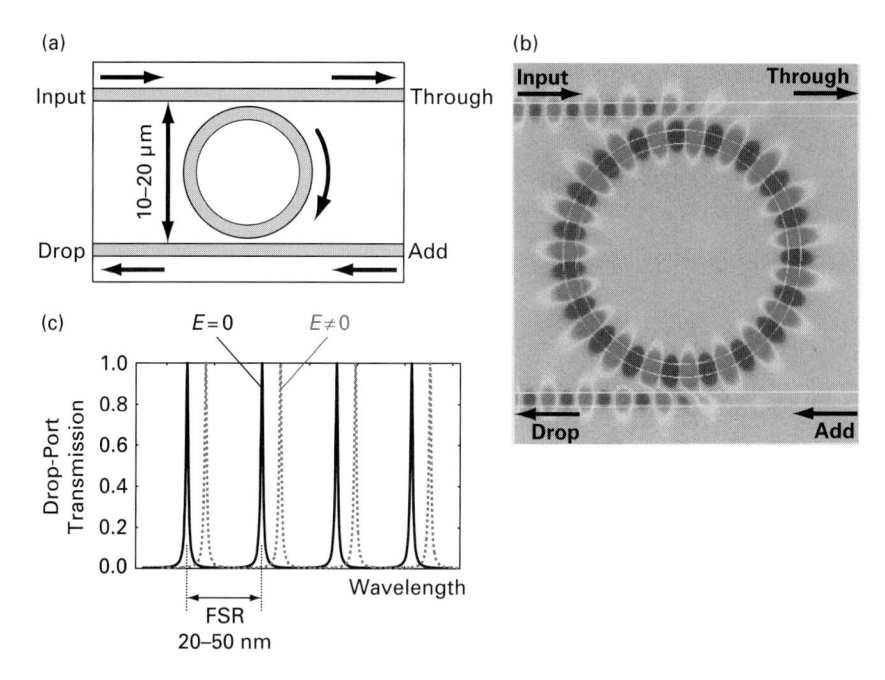

Figure 9.9 (a) A basic scheme of a single-mode VLSI microring filter viewed from the top. (b) Illustration of the optical field in a microring at resonance with $m = 19$. (c) Drop-port transmission as a function of light wavelength and upon switching an external field E on or off.

where $L = 2\pi R$ is the round-trip propagation length in a ring of a radius R, and n is the effective mode index. At $\lambda \sim 1.55$ µm, for a ring of a diameter $2R = 50$ µm and an effective index $n = 2$, the integer m is about 200 and resonances are separated by FSR ~ 8 nm. The resonance condition is quite sensitive and depends on the geometry, the wavelength, and the material refractive index. Changing the refractive index by means of the electro-optic effect, we can effectively tune the resonance condition. Based on microresonators, several types of high-speed photonic elements can be constructed, including optical switches, modulators, wavelength multiplexers, and filters.

9.4.1 Transmission of a microring resonator

The simplest microresonator configuration consists of only one straight waveguide coupled to a ring waveguide, as shown in Fig. 9.10(a). The input optical field with an amplitude A is transmitted through the coupler region with an amplitude transmission factor τ and partially coupled into the ring with a coupling constant $i\kappa$, where κ and τ are real if the propagation constants, i.e. the effective indices, are equal in both waveguides.

The relation between the optical field amplitudes is therefore

$$B = \tau A + i\kappa a, \tag{9.36}$$

$$b = i\kappa A + \tau a, \tag{9.37}$$

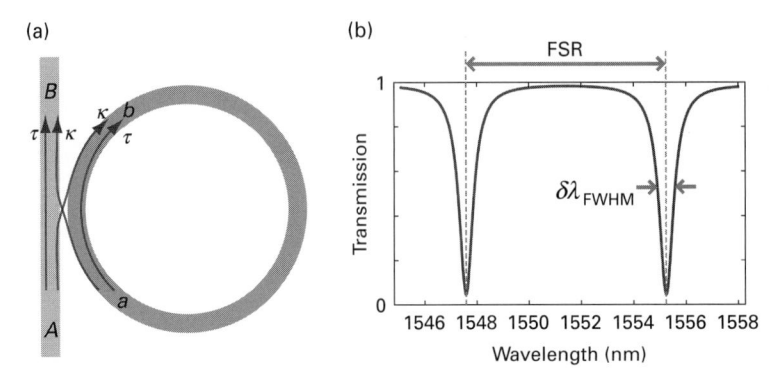

Figure 9.10 (a) Microring resonator based on evanescent optical mode coupling between a straight and a ring waveguide. (b) Transmission function according to Eq. (9.41).

$$a = b(\xi e^{i\phi}), \tag{9.38}$$

$$\xi = e^{-\alpha L/2}, \tag{9.39}$$

where ξ is the round-trip (amplitude) loss factor and α the (intensity) loss constant, which includes the absorption, scattering, and bending losses. The optical field amplitudes A, B, a, and b are defined in Fig. 9.10(a).

Solving equations (9.36–9.38) and considering that $\kappa^2 + \tau^2 = 1$ for lossless coupling we obtain

$$\frac{B}{A} = \frac{\tau - \xi e^{i\phi}}{1 - \xi \tau e^{i\phi}}, \tag{9.40}$$

leading to the following intensity transmission function

$$T = \left|\frac{B}{A}\right|^2 = \frac{(\xi - \tau)^2 + 4\xi\tau \sin^2 \frac{\phi}{2}}{(1 - \xi\tau)^2 + 4\xi\tau \sin^2 \frac{\phi}{2}} = 1 - \frac{(1 - \xi^2)(1 - \tau^2)}{(1 - \xi\tau)^2 + 4\xi\tau \sin^2 \frac{\phi}{2}} \tag{9.41}$$

The obtained transmission function is analogous to the transmission of a Fabry–Perot resonator and is shown in Fig. 9.10(b). As for Fabry–Perot resonators, we can define the following quantities that can be derived from the transmission function (9.41):

- Free spectral range (FSR) (the spacing between two adjacent resonance wavelengths):

$$\text{FSR} = \lambda_m - \lambda_{m+1} \cong \frac{\lambda_m^2}{Ln_g}, \tag{9.42}$$

where the group effective index $n_g \cong n$ for negligible dispersion $\partial n/\partial \lambda$; otherwise, $n_g = n - (\partial n/\partial \lambda)\lambda_m$.

- Spectral width of the resonances at full-width half-maximum (FWHM):

$$\delta\lambda_{\text{FWHM}} = \frac{\lambda_m^2}{\pi Ln_g} \frac{1 - \xi\tau}{\sqrt{(\xi\tau)}}. \tag{9.43}$$

- Resonator finesse:

$$F = \frac{\text{FSR}}{\delta\lambda_{\text{FWHM}}} = \pi \frac{\sqrt{(\xi\tau)}}{1-\xi\tau}. \tag{9.44}$$

- Quality factor:

$$Q = \frac{\lambda}{\delta\lambda_{\text{FWHM}}} = F\frac{\lambda}{\text{FSR}}. \tag{9.45}$$

Minima of the transmission function (9.41) occur when the resonance condition (9.35) is satisfied, leading to the following transmission

$$T_{\min} = \frac{(\xi-\tau)^2}{(1-\xi\tau)^2}, \tag{9.46}$$

which reaches its minimum for the case of "critical coupling" or that $\xi = \tau$, i.e. for the perfect balance between the coupling coefficient and the round-trip loss factor. Using the quantities introduced above, we can write the transmission function (9.41) in the following form

$$T = 1 - \frac{T_0}{1 + \left(\frac{2F}{\pi}\right)^2 \sin^2\frac{\phi}{2}}, \tag{9.47}$$

where $T_0 = 1 - T_{\min}$ is equal to 1 for the case of critical coupling.

9.4.2 Electro-optic microresonator filter

If microresonators are made out of electro-optic media, their resonance wavelengths can be shifted by applying an electric field E, as illustrated in Fig. 9.9(c). To obtain the magnitude of this shift $\Delta_E\lambda_m = \lambda_m(E) - \lambda_m$ as a function of E, we consider the resonance condition (9.35) for the mth resonance

$$m = \frac{n(\lambda_m)L}{\lambda_m} = \frac{n(\lambda_m(E))L}{\lambda_m(E)}, \quad n(\lambda_m(E)) = n(\lambda_m) + \frac{\partial n}{\partial \lambda}\Delta_E\lambda_m + \Delta_E n, \tag{9.48}$$

where we have considered the change of the effective index at the shifted wavelength due to dispersion $\partial n/\partial \lambda$ and due to the electro-optic effect $n(E)$. This leads to

$$\Delta_E\lambda_m = \lambda_m \frac{\Delta_E n}{n_g(\lambda_m)}, \tag{9.49}$$

i.e. the relative resonance shift $\Delta\lambda/\lambda$ is, neglecting dispersion, equal to the relative change of the effective index $\Delta n/n$, and does not depend on the Q factor, finesse, and resonator size.

9.4.3 Electro-optic modulation in microring resonators

Microring resonators are also very interesting for micrometer (micron)-size electro-optic modulators. The principle of operation is illustrated in Fig. 9.11 in comparison to Mach–Zehnder integrated modulators as a function of the round-trip phase change ϕ

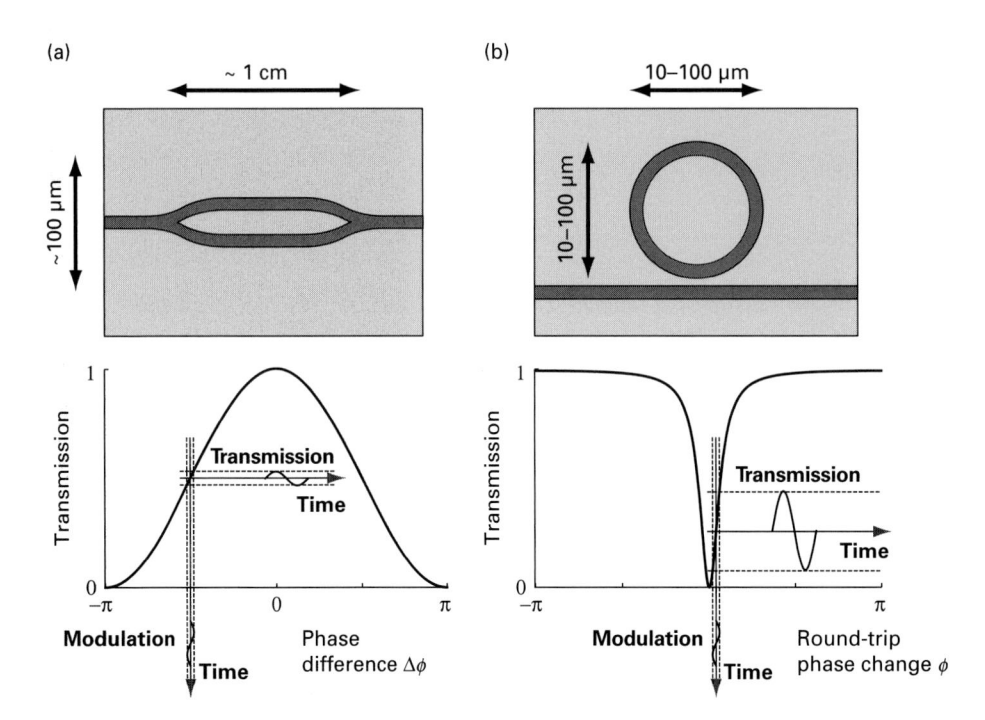

Figure 9.11 Electro-optic modulation using a microring resonator (right) and a Mach–Zehnder modulator (left). The transmission function for the microresonator is plotted according to (9.47) for $F = 15$, $T_0 = 1$ and for the Mach–Zehnder modulator according to $T = \cos^2(\Delta\phi/2)$.

in the microresonator, and as a function of the phase difference $\Delta\phi$ experienced in the arms of the Mach–Zehnder interferometer. Beside the advantage of the much smaller size of the microresonators, one can expect from the figure that the same applied field amplitude will result in a much larger intensity modulation amplitude compared with Mach–Zehnder devices.

The transmission of the Mach–Zehnder interferometer is given by Eq. (9.25) as $T = \cos^2(\Delta\phi/2) = \cos^2(\pi/2 \cdot V/V_\pi)$. Figure 9.11 shows that the modulation sensitivity depends on the maximum slope of the transmission function

$$\left.\frac{dT}{dV}\right|_{max} = \frac{\pi}{2}\frac{1}{V_\pi} \quad \text{(Mach–Zehnder)} \qquad (9.50)$$

For microring resonators we can therefore define an equivalent half-wave voltage V_π^{eq} as

$$V_\pi^{eq} = \frac{\pi}{2}\left(\left.\frac{dT}{dV}\right|_{max}\right)^{-1} = \frac{\pi}{2}\left(\left.\frac{dT}{d\phi}\frac{d\phi}{dV}\right|_{max}\right)^{-1} \qquad (9.51)$$

One can derive that the maximum slope of the transmission function (9.47) is [9]

$$\left.\frac{dT}{d\phi}\right|_{max} = \frac{3\sqrt{3}}{8}\frac{T_0}{\pi}F \approx \frac{F}{5} \qquad (9.52)$$

Considering

$$\phi = \frac{2\pi}{\lambda} nL \quad \text{and} \quad \Delta_E \phi = \frac{2\pi}{\lambda}\left(-\frac{n^3 r}{2}\frac{V}{d}\right)L = -\pi\frac{V}{V_\pi} \tag{9.53}$$

we obtain

$$V_\pi^{eq} = \frac{5}{2F} V_\pi, \tag{9.54}$$

i.e. the voltage required to achieve the same intensity modulation amplitude is reduced by a factor of $2F/5$ in microring modulators, which may be substantial considering that F is typically around 20 or more [9].

9.5 Light propagation in periodic media: photonic crystals

Another possibility to reduce the size of optical elements is to use periodically structured media. The term "photonic crystals" usually refers to regular arrays of materials with different refractive index. Figure 9.12 shows simple examples of such arrays made of two different materials periodically structured in one (1D), two (2D) or three (3D) dimensions. The spatial period of photonic crystals, the lattice constant a, is in the order of the wavelength of the electromagnetic waves; i.e., 1 μm or less for visible light.

The evaluation of various possible configurations for efficient light propagation and modulation in photonic crystals can be performed numerically using commercially available programs. To understand the basic principles of light propagation in photonic crystals, we here consider the simplest case of a 1D photonic crystal made of two different isotropic materials alternating along the x axis, which is also parallel to the propagation direction or the wave vector **k** (see Fig. 9.13).

The derivation is analogous to the derivation of the electronic states in ordinary crystals assuming non-interacting valence electrons in a periodic potential of lattice ions (see e.g. [17,18]). We are looking for solutions of the wave equation (9.4) in one dimension

$$\frac{c^2}{\varepsilon(x)} \frac{\partial^2 E}{\partial x^2} = \frac{\partial^2 E}{\partial t^2}. \tag{9.55}$$

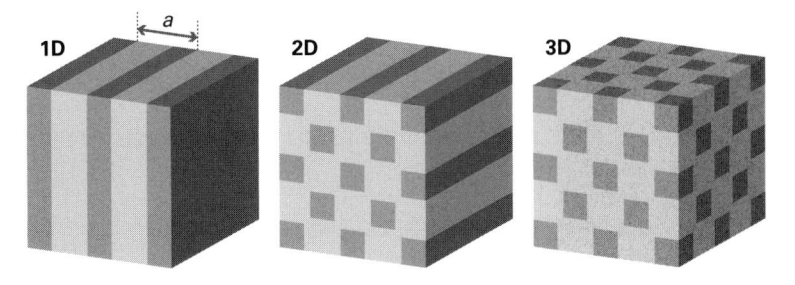

Figure 9.12 Schematic illustration of 1D, 2D, and 3D photonic crystals.

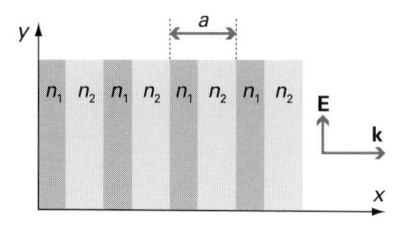

Figure 9.13 Cross-sectional view of a 1D photonic crystal with a lattice constant a.

Since $\varepsilon(x)$ is periodic, $\varepsilon^{-1}(x)$ is also periodic and can be therefore expanded in a Fourier series

$$\varepsilon^{-1}(x) = \sum_{m=-\infty}^{\infty} \alpha_m e^{-i\frac{2\pi m}{a}x}; \quad m \ldots \text{integer}. \tag{9.56}$$

Since $\varepsilon(x)$ is a real quantity, $\alpha_{-m} = \alpha_m^*$.

It can be shown that the Bloch theorem is valid for electromagnetic waves in photonic crystals, exactly as it holds for electronic eigenstates in ordinary crystals. Electric field eigenmodes are characterized by a wavenumber k and can be expressed as

$$E(x,t) \equiv E_k(x,t) = e^{i(\omega_k t - kx)} u_k(x), \tag{9.57}$$

where ω_k is the angular frequency of the eigenwave k and $u_k(x)$ presents a periodic function

$$u_k(x+a) = u_k(x) \Rightarrow u_k(x) = \sum_{m=-\infty}^{\infty} E_m e^{-i\frac{2\pi m}{a}x} \Rightarrow \tag{9.58}$$

$$E_k(x,t) = \sum_{m=-\infty}^{\infty} E_m e^{i\omega_k t - i\left(k + \frac{2\pi m}{a}\right)x}. \tag{9.59}$$

We can further simplify the discussion by assuming a simple cosine modulation of the refractive index $n(x) = \bar{n} + \Delta n \cos(2\pi x/a)$ or that in the Fourier expansion (9.56) only the components with $m = 0$ and $m = \pm 1$ are dominant

$$\varepsilon^{-1}(x) = \alpha_0 + \alpha_1 e^{-i\frac{2\pi}{a}x} + \alpha_{-1} e^{i\frac{2\pi}{a}x} \tag{9.60}$$

Considering the derivatives

$$\frac{\partial^2 E_k(x,t)}{\partial x^2} = \sum_m -\left(k + \frac{2\pi m}{a}\right)^2 E_m e^{i\omega_k t - i\left(k + \frac{2\pi m}{a}\right)x}$$

$$\frac{\partial^2 E_k(x,t)}{\partial x^2} = -\omega_k^2 E_k(x,t)$$

9.5 Light propagation in periodic media: photonic crystals

we can rewrite the wave equation (9.55) as

$$\left(\alpha_0 + \alpha_1 e^{-i\frac{2\pi}{a}x} + \alpha_{-1}e^{i\frac{2\pi}{a}x}\right) \cdot \sum_m \left(k + \frac{2\pi m}{a}\right)^2 E_m e^{i\omega_k t - i\left(k+\frac{2\pi m}{a}\right)x} = \frac{\omega_k^2}{c^2} \sum_m E_m e^{i\omega_k t - i\left(k+\frac{2\pi m}{a}\right)x},$$
(9.61)

which gives, considering the coefficients at $e^{i\omega_k t - i\left(k+\frac{2\pi m}{a}\right)x}$,

$$\left(\frac{\omega_k^2}{c^2} - \alpha_0 \left(k + \frac{2\pi m}{a}\right)^2\right) E_m = \alpha_1 \left(k + \frac{2\pi(m-1)}{a}\right)^2 E_{m-1} + \alpha_{-1}\left(k + \frac{2\pi(m+1)}{a}\right)^2 E_{m+1},$$
(9.62)

e.g., for $m = 0$ and $m = -1$

$$m = 0: \quad \left(\frac{\omega_k^2}{c^2} - \alpha_0 k^2\right) E_0 = \alpha_1 \left(k - \frac{2\pi}{a}\right)^2 E_{-1} + \alpha_{-1}\left(k + \frac{2\pi}{a}\right)^2 E_1$$

$$m = -1: \quad \left(\frac{\omega_k^2}{c^2} - \alpha_0 \left(k - \frac{2\pi}{a}\right)^2\right) E_{-1} = \alpha_1 \left(k - \frac{4\pi}{a}\right)^2 E_{-2} + \alpha_{-1}k^2 E_0. \quad (9.63)$$

The relation between ω_k and k in absence of refractive index modulation ($\alpha_1 = \alpha_{-1} = 0$) is $(\omega^0)_k^2 = \alpha_0 c^2 k^2$ – the well-known relation for homogeneous materials (dotted lines in Fig. 9.14).

We can first consider the solution for $k \approx \pi/a$. In this case $(\omega^0)_k^2 \approx (\omega^0)_{k-2\pi/a}^2$. In the expansion (9.59) we may consider only the first-order modulation terms (as for the dielectric function); therefore, for $k \approx \pi/a$, the terms with $m = 0$ and $m = -1$ will be dominant. The relations (9.63) then give

$$(\omega_k^2 - \alpha_0 c^2 k^2) E_0 - \alpha_1 c^2 (k - 2\pi/a)^2 E_{-1} = 0$$

$$-\alpha_{-1} c^2 k^2 E_0 + (\omega_k^2 - \alpha_0 c^2 (k - 2\pi/a)^2) E_{-1} = 0. \quad (9.64)$$

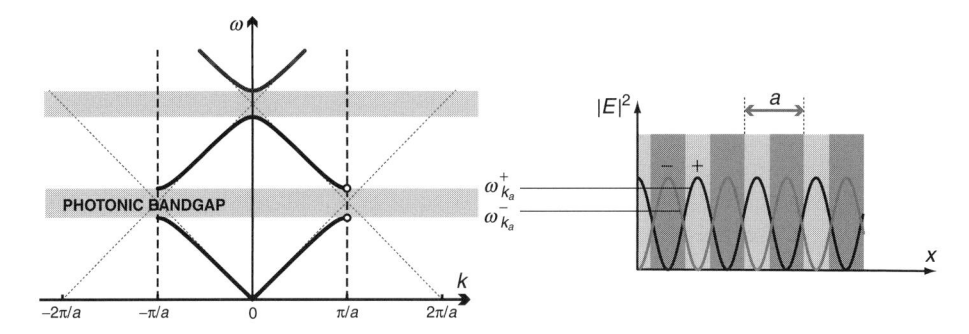

Figure 9.14 Dispersion relation for a 1D photonic crystal (solid lines). The dashed vertical lines denote the boundaries of the first Brillouin zone. The dotted lines denote the dispersion for a uniform material $\omega = kc/\bar{n}$. The solutions are folded into the first Brillouin zone. The solutions for $|E|^2$ for the wave vector $k=k_a=\pi/a$ at the edge of the first Brillouin zone are illustrated on the right; darker shading represents a higher-refractive-index material.

To obtain a non-trivial solution of the above equations we should set the determinant of the coefficients to zero. For $k = k_a = \pi/a$ this leads to

$$(\omega_{k_a}^2 - \alpha_0 c^2 k_a^2)^2 - |\alpha_1|^2 c^4 k_a^4 = 0$$

$$\omega_{k_a}^{\pm} = (ck_a)\sqrt{\alpha_0 \pm |\alpha_1|}, \qquad (9.65)$$

i.e. at this point we have, instead of one solution $\omega_k^0 = ck\sqrt{\alpha_0}$ as for a homogeneous material, two non-equal solutions $\omega_{k_a}^+$ and $\omega_{k_a}^-$ for the angular frequency ω_{k_a}. The full dispersion relation $\omega_k(k)$ for a general k is illustrated in Fig. 9.14.

No optical mode may propagate in the following angular-frequency interval

$$(c\pi/a)\sqrt{\alpha_0 - |\alpha_1|} < \omega < (c\pi/a)\sqrt{\alpha_0 + |\alpha_1|} \qquad (9.66)$$

which is known as the photonic bandgap, in analogy with the energy bandgap for electrons in a periodic potential. The solutions for the wave vectors that differ by $2\pi/a$ are regarded as the same, and therefore we may limit the discussion to the first Brillouin zone $[-\pi/a, +\pi/a]$. Other photonic bandgaps appear at the places where the two dispersion lines for the homogeneous material $\omega = c_{\bar{n}}k$ cross (see Fig. 9.14).

We can insert the solution (9.65) for $k = k_a = \pi/a$ into the coupled equations (9.64) to get the ratio between the coefficients E_0 and E_{-1}, which we can then consider in (9.59). In this way, we obtain that both solutions present standing waves that are relatively shifted by $\pi/2$, i.e.

$$\omega_{k_a}^+ = (ck_a)\sqrt{\alpha_0 + |\alpha_1|} \; : \; E_{\pi/a}^+(x) \propto \cos\frac{\pi}{a}x \qquad (9.67)$$

$$\omega_{k_a}^- = (ck_a)\sqrt{\alpha_0 - |\alpha_1|} \; : \; E_{\pi/a}^-(x) \propto \sin\frac{\pi}{a}x \qquad (9.68)$$

if $x = 0$ is set to the center of the lower-index material (so that $\varepsilon^{-1}(x) = \alpha_0 + |\Delta\alpha|\cos 2\pi x/a$). We can now intuitively understand the difference between the effective refractive index of both waves with the same wave vector k. As shown in Fig. 9.14, one of the waves has intensity maxima in the regions of the higher index and the other in the regions of the lower index; they therefore experience a different effective index, although their wavenumber is the same.

A well-known example of a photonic crystal in nature is opal, a mineral that does not have a crystalline structure, but consists of densely packed aggregates of silica (SiO_2) spheres. One can make very similar artificial structures by self-assembly of colloidal particles (e.g. sub-micron-diameter silica or PMMA spheres). Another possibility is to fabricate periodic structures in dielectric materials by using deep UV, electron-beam, or focused ion beam lithography.

Introducing defects in photonic crystals, such as waveguides or point-like cavities, may result in localized electromagnetic states inside the photonic crystals, equivalent to electronic dopant states in ordinary crystals. If a defect is formed such that it supports a mode that is within the photonic bandgap, this mode is forbidden from propagating in the bulk of the photonic crystal. Since photonic crystals are perfect "optical insulators", one can confine light without losses around sharp bends and within very small (wavelength-scale) cavities. Also, since the group velocity $\partial\omega/\partial k$ becomes very small close to the photonic bandgap (see Fig. 9.14), light may be slowed down significantly using photonic crystals, which may significantly enhance nonlinear optical and electro-optic effects [19 to 23]. Therefore, very interesting opportunities for bandgap tuning, switching, active filtering, and modulation are possible by exploiting electro-optic and general nonlinear optical effects in photonic crystals.

9.6 Single-crystalline organic waveguides and modulators

9.6.1 Overview of the structuring techniques for organic crystals

As for general optical waveguides, we need to structure organic crystals with a sub-micron precision so that a suitable refractive index contrast for optical waveguiding is achieved. Although for optical waveguiding even very small refractive index changes of the order of $\Delta n \sim 10^{-3}$ may be sufficient, for small waveguides needed for large-scale integration, as well as for reducing the half-wave voltage of EO modulators, larger index contrast is desired. For example, to fabricate microring resonators with a small radius below 10 μm, the refractive index of the guiding medium should be larger than that of the surrounding medium by at least $\Delta n \sim 0.5$ to avoid high losses. Organic EO crystals have a relatively high refractive index compared with poled polymers, which can reach the values of inorganic ferroelectric crystals such as $LiNbO_3$ ($n \sim 2.2$). This basically allows very efficient high-index contrast waveguiding with respect to substrate materials such as SiO_2. Some of the organic EO crystals may be also strongly anisotropic with birefringence as high as $\Delta n > 0.5$ at non-resonant wavelengths, which should be taken into account when designing the waveguides. The particular best orientation of the waveguides with respect to the propagation direction and the optical and electric field orientations depends on the particular tensor properties of the material; see examples in Section 9.6.2. Organic crystals are also suitable as active cladding materials to high-index silicon photonic passive waveguides ($n_{Si} \sim 3.5$), which can result in very compact EO modulators with high figures-of-merit, if organic crystals can be oriented in a suitable way; examples are discussed in Section 9.6.2.

The main challenges in building integrated EO modulators based on organic crystals are related to their processing possibilities: the organic crystal should be deposited on appropriate substrate materials and in a desired orientation to achieve planar light confinement, and then structured with an appropriate technique to achieve horizontal light confinement. In the following section, we describe several approaches and techniques that were developed to fabricate optical waveguides in organic EO crystals. The techniques of photolithography, photostructuring (including photobleaching and femtosecond laser ablation), ion implantation, electron-beam irradiation, and direct deposition into pre-structured inorganic templates are schematically presented in Table 9.1 with some of their main features.

Table 9.1 Structuring methods investigated for organic single-crystalline waveguides with second-order nonlinear optical activity. More details are given in Section 9.6.2. Δn is estimated assuming SiO_2 substrates (where applicable).

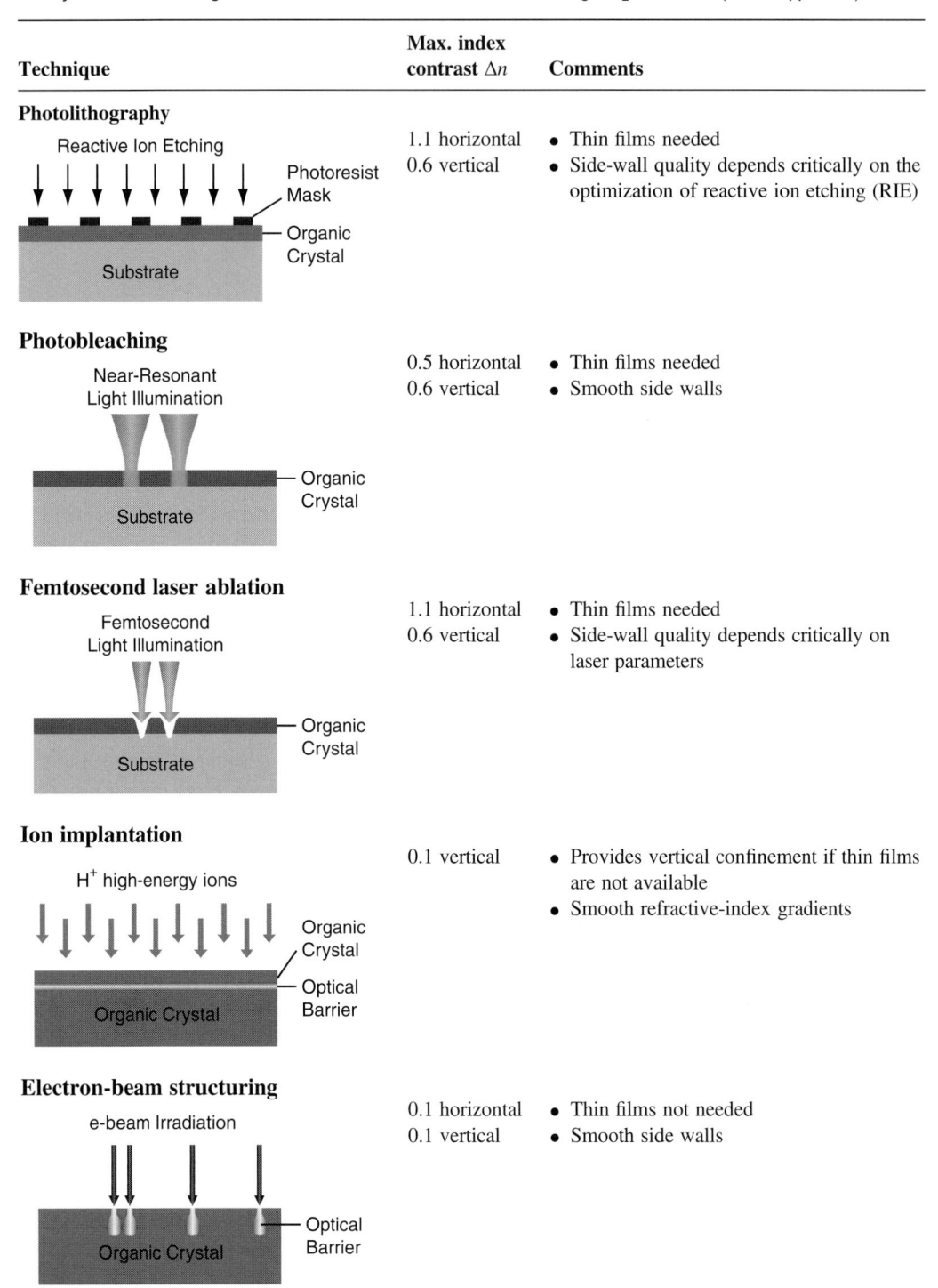

Technique	Max. index contrast Δn	Comments
Photolithography Reactive Ion Etching	1.1 horizontal 0.6 vertical	• Thin films needed • Side-wall quality depends critically on the optimization of reactive ion etching (RIE)
Photobleaching Near-Resonant Light Illumination	0.5 horizontal 0.6 vertical	• Thin films needed • Smooth side walls
Femtosecond laser ablation Femtosecond Light Illumination	1.1 horizontal 0.6 vertical	• Thin films needed • Side-wall quality depends critically on laser parameters
Ion implantation H^+ high-energy ions	0.1 vertical	• Provides vertical confinement if thin films are not available • Smooth refractive-index gradients
Electron-beam structuring e-beam Irradiation	0.1 horizontal 0.1 vertical	• Thin films not needed • Smooth side walls

Table 9.1 (*cont.*)

Technique	Max. index contrast Δn	Comments
Epitaxial growth	0.6 horizontal 0.6 vertical	• Very versatile • Structuring performed only in standard inorganic substrates (SiO$_2$, Si) • Easy electrode or other material integration • Limited by crystallization properties of organic material

9.6.2 Examples of modulators and fabrication techniques

Waveguide fabrication by photolithography. Standard photolithographic microfabrication techniques have been developed for semiconductors and can also be used for several inorganic nonlinear optical crystals and for organic polymers (macromolecular materials), but mostly cannot be straightforwardly applied to organic crystals. This is because many organic crystals are incompatible with common photoresist solvents that will generally etch the surfaces of the crystals. Therefore, specific photolithographic procedures have been developed for some of the organic crystals depending on their particular physico-chemical properties.

A few of the organic crystals are insoluble in common organic solvents; for these, standard photolithography and oxygen reactive ion etching may be used to fabricate channel waveguiding structures, as demonstrated for MNBA [27]. Channel waveguides were also produced in (−)MBANP, which is soluble in organic solvents, by using a special water-soluble inorganic photoresist heteropolyacid [28]. Neither of these techniques, however, can be used for highly nonlinear organic salts such as DAST, which are soluble in both water and standard organic solvents used for photolithography. A special photolithographic technique has been developed to produce channel waveguides in DAST by Kaino *et al.* [29, 30], where a protective PMMA layer was used to prevent DAST crystals from being dissolved in the photoresist solution and developer. An alternative photolithographic process for DAST employed the standard photoresist SU8 (MicroChem) for the structuring and a LOR-B5 (MicroChem) lift-off resist as a protective cladding [31].

Photolithographic structuring of another promising highly nonlinear EO crystal, phenolic polyene OH1, is less complex than for DAST crystals, since OH1 is not soluble in water. Also, since thin films of OH1 can be fabricated directly on inorganic substrates, the structuring is analogous to structuring of inorganic materials. A photolithographic technique optimized for the fabrication of wire optical waveguides in OH1 thin films is schematically depicted in Fig. 9.15(a–d). OH1 thin films on glass substrates were first covered with a water-soluble PVA (polyvinyl alcohol) layer to protect OH1 from organic solvents. Standard photoresist SU8 was deposited on

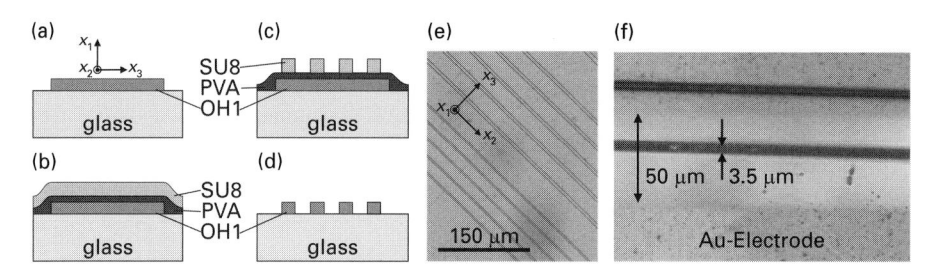

Figure 9.15 (a–d) Processing steps for the fabrication of OH1 wire waveguides on glass substrates. (e) Microscope image of single crystalline OH1 wires on glass. (f) OH1 waveguide with deposited gold electrodes for EO modulation experiments. Reprinted with permission from Ref. [32].

top and structured so that the wires were perpendicular to the x_3 polar axis of the film (Fig. 9.15(c)). The waveguide pattern was transferred to the OH1 film by optimizing the etching process [32] and OH1 wires as shown in Fig. 9.15(d and e) were obtained. Gold electrodes were vapor-deposited on the sides of the waveguides by using a simple shadow mask, resulting in a relatively large electrode separation of 50 μm, as shown in Fig. 9.15(f).

The resulting waveguides with a height of 3.5 μm exhibited propagation losses of 2 dB/cm, 9 dB/cm, and 17 dB/cm for waveguide widths of 7.6, 5.4, and 3.4 μm, respectively, at 980 nm. The losses additionally increase in smaller waveguides because of the very high core-cladding index contrast of $\Delta n = 1.23$. For phase-modulation measurements, a straight-waveguide sample with cross-sectional dimensions of 3.4×3.5 μm^2, shown in Fig. 9.15(f), was used. The measured half-wave voltage × interaction length product $V_\pi L$ resulting from this configuration is 8.4 V-cm and 28 V-cm at 632.8 nm and 852 nm, respectively [32]. With an optimized electrode configuration with 1 μm spacing between the electrode and the waveguide, the half-wave voltage × length product $V_\pi L$ is expected to be reduced from 8.4 V-cm, as obtained in these first devices, to 0.3 V-cm in the optimized case. This technique is therefore promising to achieve high-index-contrast sub-1-V half-wave voltage, organic EO modulators with highly stable chromophore orientation.

Photostructured optical waveguides. An alternative technique for waveguide patterning is photobleaching, which is often used also for microstructuring of polymer-based devices. Photobleaching refers to the change of the chemical composition of the molecules after high-intensity light exposure in the resonant wavelength regime, which leads to a decrease of the refractive index that can be used for confining light laterally. The refractive indices of DAST crystals can be reduced by photobleaching using light within the absorption band of DAST from 260 nm to 700 nm [33,34]. The depth range of photobleaching can be varied between 0.2 μm and 2.6 μm by selecting a suitable wavelength [34]. The refractive index change for light polarization along the polar axis x_1 is relatively very high, about $\Delta n = 0.5$ at 1.55 μm, which is important for large-scale integration. Photobleaching was used to produce channel waveguides in thin DAST samples by Kaino *et al.* [30] with a UV resin used as an undercladding, leading to propagation losses of about 11 dB/cm.

9.6 Single-crystalline organic waveguides

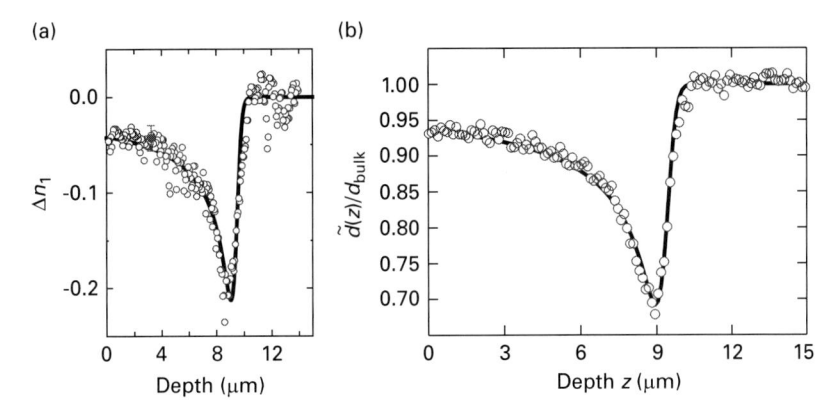

Figure 9.16 (a) Measured refractive index change Δn_1 as a function of the implantation depth at $\lambda = 633$ nm. (b) Normalized susceptibility profile at the fundamental wavelength $\lambda = 1176$ nm. The waveguides were produced in DAST by 1 MeV H$^+$ ion implantation with a fluence of $\phi = 1.25 \times 10^{14}$ ions/cm^2 at an angle of 60°. Reprinted with permission from (a) Ref. [36] and (b) Ref. [37].

Another possibility for direct structuring of organic crystals by using light is femtosecond laser ablation, which was investigated for the patterning of DAST surfaces [35]. An ideal fluence range for optimal, almost damage-free ablation was determined at the wavelengths of 775 nm, 600 nm, and 550 nm with a pulse width of about 170 fs. The profiles of the grooves reveal that the ablation of ridge waveguides using fs lasers in DAST is a promising technique alternative to photolithography for the realization of optical waveguides in organic crystals [35].

Ion-implanted waveguides. Since the growth of organic single crystalline thin films with accurate thickness control needed for integrated optics is very challenging for some materials, the ion-implantation approach to produce planar waveguides in bulk organic crystals was developed [36]. The refractive index in DAST decreases because of electronic excitations induced by ion irradiation, in contrast to inorganic materials, for which the major refractive index changes are due to ion-induced nuclear displacements.

In H$^+$ ion-implanted DAST, the electronic loss curve has a peak at the end of the ion range, which results in an optical barrier suitable for optical waveguiding. The measured profile of the refractive index n_1 at $\lambda = 633$ nm for an implantation fluence of $\phi = 1.25 \times 10^{14}$ ions/cm^2 is shown in Fig. 9.16(a). A maximal peak refractive index change of around $\Delta n = -0.2$ at 633 nm and $\Delta n = -0.1$ at 810 nm was measured [36].

To use ion-implanted waveguides for active integrated photonic devices, it is of major importance to maintain the high NLO and EO properties in the waveguide core region. The nonlinear optical coefficient in the waveguide core region is preserved by more than 90% after implantation, as shown in Fig. 9.16(b) [37].

Waveguiding in ion-implanted waveguides was demonstrated by using the conventional end-fire light coupling. The light was polarized parallel to x_1, which is the most interesting configuration for EO modulation in DAST. The estimated transmission losses were around 10 dB/cm at $\lambda = 1550$ nm [36]. Electro-optic phase modulation in

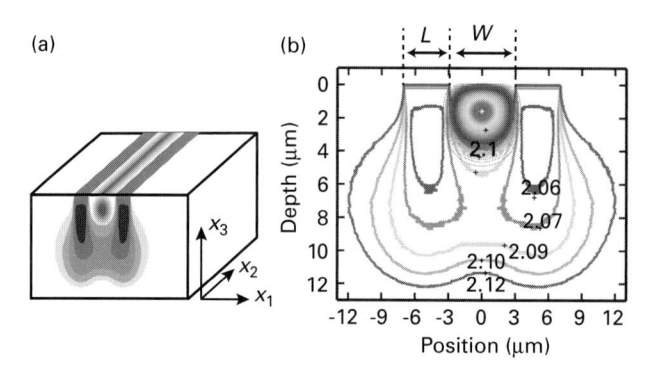

Figure 9.17 (a) Concept of channel waveguide patterning in organic crystals: Two lines spaced by the waveguide core width are exposed by e-beam (gray stripes, deposited energy). An unexposed region surrounded by an exposed area with lowered refractive index is created, forming a waveguide (gray middle stripe). (b) Calculated 2D profile of the refractive index n_1 in DAST at a wavelength of $\lambda = 1.55$ μm for an electron fluence of $\phi = 2.6$ mC/cm^2, a line width $L = 4$ μm, and a waveguide core width of $W = 6$ μm. The corresponding intensity profile of the first-order guided mode with calculated losses of below 0.1 dB/cm is also illustrated. Reprinted with permission from Ref. [38].

ion-implanted waveguides was also demonstrated [37]. The EO coefficient in H$^+$-implanted waveguides measured at $\lambda = 1550$ nm was $r_{111} = 42 \pm 10$ pm/V, which is about 10% lower compared with the bulk value of 47 ± 8 pm/V, in agreement with the nonlinear susceptibility profile shown in Fig. 9.16(b). For the target devices in DAST crystals with an electrode distance below 5 μm, the half-wave voltage × interaction length products for ion-implanted waveguides are expected to be below 1.7 V-cm.

Electron-beam induced waveguides. Electron-beam (e-beam) irradiation permits patterning of EO channel waveguiding structures in bulk organic NLO crystals even without the necessity for thin-film technology [38]. The electrons of the writing beam are scattered in the material and therefore the beam is widened in the target material. This circumstance can be exploited to directly write channel waveguides in bulk crystals by exposing two lines separated by the waveguide width as depicted in Fig. 9.17(a). The advantage of this configuration is that the waveguide core is mostly in the virgin material, in which the nonlinear and EO properties are the same as in the bulk material.

DAST crystals were exposed with an e-beam system providing an electron energy of 30 keV. The refractive index was reduced by e-beam exposure. For a fluence of $\phi = 2.6$ mC/cm^2, a maximal refractive index reduction of $\Delta n = -0.3$ at $\lambda = 633$ nm was achieved [38]. The two-dimensional refractive index cross-section after e-beam irradiation is shown in Fig. 9.17(b).

Channel waveguides and Mach–Zehnder modulators were produced by e-beam irradiation in DAST, using various fluences and waveguide dimensions. The $x_1 x_3$ end-faces were subsequently polished and waveguiding demonstrated by using standard end-fire coupling [38]. Electrodes were patterned subsequent to the e-beam exposure. The EO modulation measurements were performed in the experimental configuration shown in Fig. 9.18(b). The applied modulation voltage (amplitude of 10 V, lower curve)

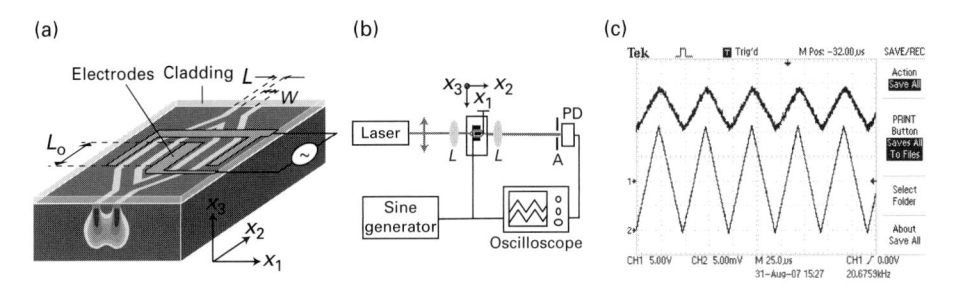

Figure 9.18 (a) Mach–Zehnder modulator geometry in DAST with in-plane electrodes exploiting the EO coefficient r_{111}. (b) Set-up for the EO modulation experiment with a photodiode (PD), aperture (A) and lens (L). (c) The applied modulation voltage with an amplitude of 10 V (lower curve) and the detected signal with the photodiode (upper curve). Reprinted with permission from Ref. [38].

and the measured modulated signal (upper curve) at the output of the Mach–Zehnder device are shown in Fig. 9.18(c) for a waveguide width of $W = 4$ μm. The amplitude of the modulation was about 20% of the output signal. In this first demonstration, the half-wave voltage was higher than 10 V, since the modulator dimensions and the electrode arrangement were not optimized. The electrode spacing was relatively wide at 20 μm, and the effective optical–electric interaction length was only $L_o = 0.85$ mm long.

Nevertheless, this technique seems very promising for structuring DAST and also other organic crystals. It allows for direct and single-exposure step structuring of channel waveguides in bulk crystals with particularly smooth side walls. The depth and the lateral size can be precisely tuned in the range of 0–12 μm needed for integrated optics. The refractive index contrast of about 0.1 at 1.55 μm allows for structuring curved structures with a radius of 100 μm and above. Compared with the ridge channel waveguides fabricated by photolithography or femtosecond laser ablation, edge polishing for in- and out-coupling of light is much less demanding. Furthermore, there is no need for additional processing in order to achieve confinement in the vertical direction as required when using photolithography, photobleaching or laser ablation.

Graphoepitaxially grown waveguides. An interesting method for producing DAST crystalline waveguides by graphoepitaxial melt growth was proposed by Geis *et al.* [39]. Melt growth of DAST single crystals is limited owing to the thermal decomposition of DAST molecules above the melting point. Nevertheless, it was shown that in nitrogen atmosphere the molten DAST is relatively stable for about 200–500 seconds, which was enough for a fast growth of DAST onto a structured substrate.

Microstructures with waveguide forms and crossed gratings to seed the growth were etched into an oxidized silicon substrate. The substrate was coated with a polycrystalline film of DAST, and then heated above the melting point for a short time. The resulting crystals presented a much higher quality than without the grating-like microstructure, and reasonable orientation with the c axis normal to the substrate to within approximately ±4° (see Fig. 9.19(a)). The graphoepitaxially grown waveguides with the desired orientation (b axis parallel to the waveguide) determined by grating patterning exhibited optical losses below 10 dB/cm. Electro-optic modulation was also

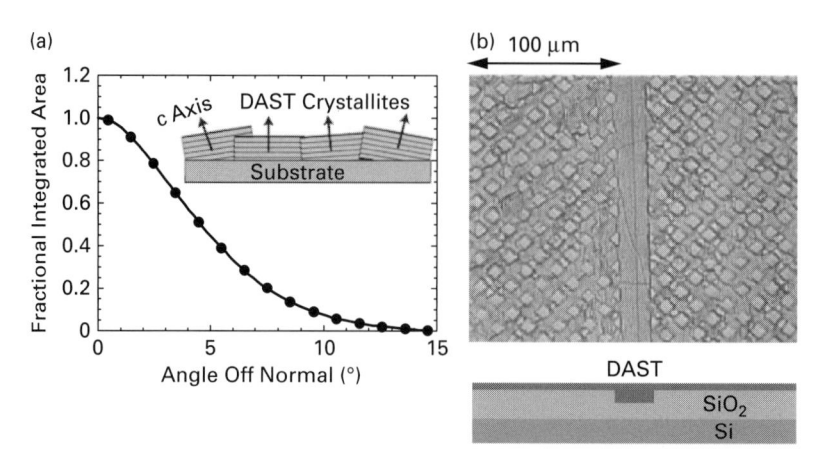

Figure 9.19 (a) Distribution of the c axis of DAST crystallites grown graphoepitaxially as a function of the angle from the structure normal. (b) Optical micrograph of DAST waveguide 4 μm deep and 15 μm wide using polarized light. In-plane variation of 5° or more is visible as a color variation (figure above). Schematic cross-section (figure below). Reprinted with permission from Ref. [39].

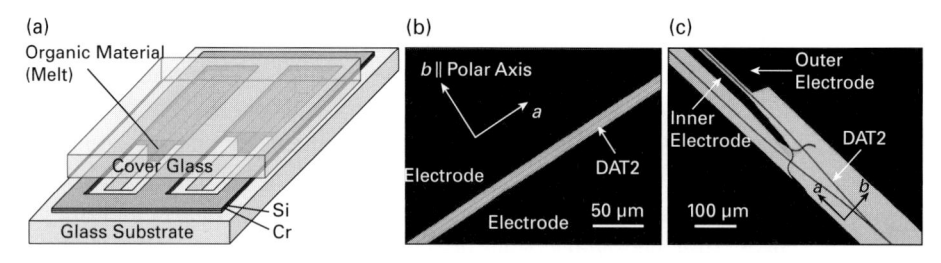

Figure 9.20 (a) Schematic of melt capillary growth inside electrode-equipped pre-structured microchannels. Transmission microscope image between crossed polarizers of a DAT2-based phase (b) and Mach–Zehnder (c) EO modulator. Reprinted with permission from (a,b) Ref. [40] and (c) Ref. [41].

demonstrated in a Mach–Zehnder geometry using ∼100-nm TiN or Cr conductive coating of the grating substrate prior to DAST growth [39].

Melt capillary grown waveguides inside microchannels. Another direct growth of organic waveguides from melt developed recently is based on single-crystalline growth inside microfluidic-like templates. In this case, materials that are stable at the melting temperature were chosen. The growth can be therefore slower and highly controlled, which is beneficial for the quality and single crystallinity of the grown structures. The direction of the melt flow and gradients determines the crystalline orientation, since most of the organic EO crystals tend to grow preferentially along one direction. Figure 9.20(a) illustrates the basic principle of this technique. Borosilicate glass substrates were structured so that the waveguiding structures formed channels. Chromium electrodes were deposited on the sides with a thin Si layer on top, which was used for anodic bonding of the cover glass [40]. Finally the organic material was placed at the edges of the cover glass and heated up to the melting temperature, at which the melt started to flow into the channels by the capillary force. On cooling down in presence of a temperature gradient, single crystalline wires grew inside the channels. Single crystalline wires, several millimeters long, with a width of several microns, and a thickness from several microns down to 30 nm or below (see also Fig. 6.18) have been obtained by this method, using melt-processable materials COANP and DAT2.

By using this melt capillary growth technique, both phase and amplitude EO modulators were fabricated in DAT2, as shown in Fig. 9.20(b) and (c). These waveguides exhibit a high refractive index contrast of $\Delta n = 0.54$ between the DAT2 material and the surrounding glass. The EO modulation was measured for both TE and TM modes, which allowed the EO coefficients of DAT2 to be estimated: $r_{12} = 7.4 \pm 0.4$ pm/V and $r_{22} = 6.7 \pm 0.4$ pm/V at 1.55 µm. The half-wave voltages of the fabricated structures were $V_\pi L = 78 \pm 2$ Vcm for TE-modes and $V_\pi L = 60 \pm 1$ Vcm for TM-modes at 1.55 µm, limited by the EO coefficients of the DAT2 material. Novel materials with state-of-the-art EO coefficients and a possibility for melt growth are therefore greatly desired to improve the figures-of-merit for this technique.

9.6.3 Single-crystalline organic microresonators

Single-crystalline organic EO microring-resonator filters and modulators were demonstrated by using the melt capillary growth technique [42]. In this case organic material COANP with very good melt-crystallization properties and a moderate EO coefficient $r_{33} = 15 \pm 2$ pm/V at 633 nm was employed. A top view transmission microscope image of a COANP crystal grown in a microring resonator channel waveguide is depicted in Fig. 9.21(a). Very high single-crystalline quality of these waveguides was confirmed by

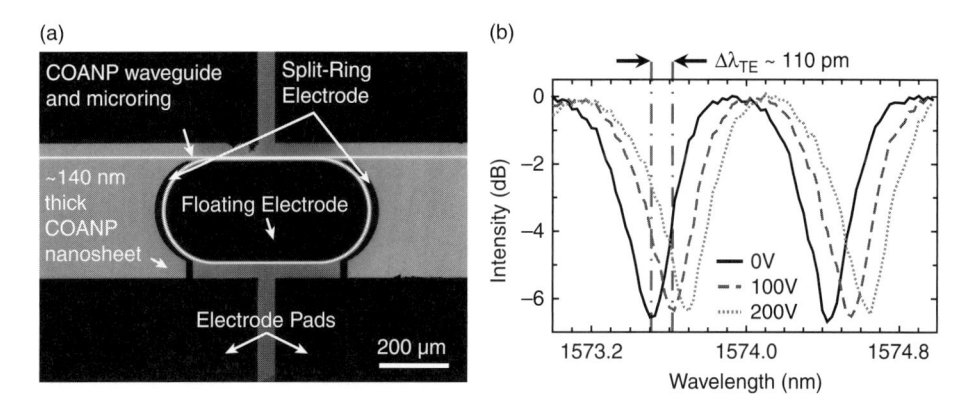

Figure 9.21 (a) Transmission microscope image between crossed polarizers of a COANP waveguide with a racetrack microring resonator grown by the melt capillary method in prefabricated channels. (b) Resonance curve of a TE mode at a wavelength around 1.574 µm (solid line); the dashed and dotted line are the corresponding electro-optically shifted curves by applying 100 V and 200 V voltage to the device electrodes. Reprinted with permission from Ref. [42].

optical waveguiding characterization. Typical devices fabricated showed almost perfectly symmetric high extinction ratio resonance peaks of about 10 dB, ring losses $\alpha = 12\pm0.3$ dB/cm, and a finesse $F = 6.2\pm0.2$. The measured TE spectrum of the racetrack resonator shown in Fig. 9.21(a) showed a $\Delta\lambda = 110$ pm-shift in response to an applied voltage of 100 V, corresponding to a frequency tunability of 0.11 GHz/V, which is comparable to the value reported for ion-sliced $LiNbO_3$ microring resonators [15]. A great improvement in performance is expected if materials with state-of-the-art EO figures-of-merit ($n^3 r$ of DAST or OH1 is more than one order of magnitude higher than for COANP) and higher index contrast (Δn with respect to borosilicate is at 1.55 μm about 0.15 for COANP and almost 0.7 for OH1 and DAST) can be used for melt growth.

9.7 Polymer waveguides and modulators

As already mentioned, one of the significant advantages of macromolecular (polymer and dendrimer) electro-optic materials is the range of processing options that can be employed and the compatibility of macromolecular materials with a diverse array of materials (silicon, inorganic oxide glasses, polymers, metals, etc.). Mass production techniques such as nano-imprint lithography can be used to stamp out complex electro-optic circuitry, and conformal/flexible devices can be fabricated in a straightforward manner (see Section 7.12). Material properties such as index of refraction, dielectric permittivity, optical loss, and birefringence can be tuned over wide ranges. Although early values for index of refraction and birefringence were significantly less than correponding values for organic crystalline materials, the high number densities of current macromolecular materials yield index of refraction values approaching those of crystalline materials. In like manner, as acentric order is increased in electrically poled macromolecular materials, birefringence increases. With electrically poled macromolecular thin films, the problems of growth anisotropy, microdomain formation, and crystal orientation of crystalline materials are avoided. Electrically poled organic macromolecular electro-optic materials are used with all types of devices discussed in the next section (Section 9.8 – see particularly Fig. 9.27) that either exploit modulation of light propagating in the core waveguide (Fig. 9.27(a)) or modulation of the evanescent light propagating above a silicon core (Fig. 9.27(a) and (b)). However, electrically poled organic macromolecular electro-optic materials can be divided into two categories according to the type of device into which they are incorporated: (1) All-organic electro-optic devices typically have structures consisting of bottom cladding, active core, top cladding (see Figs. 9.26(a) and 7.1(bottom)). (2) Hybrid organic/silicon photonic devices (discussed in Section 9.8.2) typically have structures shown in Fig. 9.27(a) and (b). In this section, we focus on the former. With all-organic (or organic cladding–core–cladding) device structures, the index of refraction of the core waveguide must be greater than the claddings to ensure propagation of light in the core. Values of refractive index for commonly employed cladding materials are given in Table 7.4, and the refractive indices of most organic electro-optic materials (1.6 to 2.2) are greater than these cladding materials owing to the presence of π-electron chromophores. Ideally, the cladding materials should have low optical loss. Although this is frequently assumed to be the case, Table 7.4 illustrates that this is not necessarily a good assumption. However, the optical loss of the cladding material is normally tolerated because it is only the evanescent field that is propagating in the cladding. The refractive index contrast between core and cladding is critical in defining the confinement of light in the core and also in determining the thickness of cladding required to prevent significant light losses from interaction with the metal (e.g., gold) electrodes. Of course, to minimize drive voltages it is important to keep electrode spacings as small as possible and this means minimizing the thicknesses of cladding layers. The relative electrical conductivity of core and cladding layers is important at poling temperatures. Ideally, the conductivity of the cladding layers should be higher than that of the core so that the poling field is efficiently dropped across the core. This is a requirement that is hard to realize because of the presence of π-electron chromophores in the core. Moreover, it is difficult to simultaneously achieve high conductivity and low optical loss in a material. One alternative approach to the traditional gold electrode/cladding/core/cladding/gold electrode architecture is to employ "transparent" conducting electrodes such as indium tin oxide (ITO) or doped silicon. However, such electrodes do contribute to optical loss and the increased resistivity relative to gold leads to bandwidth limitations at high operating frequencies. A more recent approach to the problem of conductance of core organic macromolecular electro-optic materials at poling temperatures is to employ charge blocking layers such as titanium dioxide (TiO_2) as discussed in Section 7.12. Use of charge blocking layers can lead to significant improvements in poling-induced electro-optic activity.

Unlike crystalline materials, spin casting is employed to prepare thin films of most electrically poled macromolecular materials, although low-molecular-weight dendrimer materials can also be processed by vapor phase deposition. The cladding–core–cladding structure of all-organic waveguide devices provides vertical confinement of light. Lateral confinement of light for macromolecular electro-optic waveguides is most commonly achieved by reactive ion etching (using an oxygen plasma), by photobleaching, or by electric field poling [43 to 51] although electron-beam etching has been employed in a small number of cases. Electro-optic waveguides have also been induced by stress [52]. Singlet oxygen chemistry plays a role in the first two processes and the index of refraction in the region surrounding the electro-optic core is reduced by decomposition of chromophores. Reactive ion etching (RIE) results in ablation of material surrounding the core and is most commonly used to etch a ridge which will be surrounded by cladding material subsequently deposited by spin casting. Photobleaching simply reduces the index of the region surrounding the core by eliminating π-electron conjugation without loss of material. Photobleaching has also been used to fine-tune the performance of waveguide devices such as Mach–Zehnder modulators (see Fig. 9.22) [43,53 to 57].

For poling-induced waveguides, the poling electrodes are patterned over the waveguides. The alignment of chromophores by the poling process increases the index of refraction for the TM polarization [51].

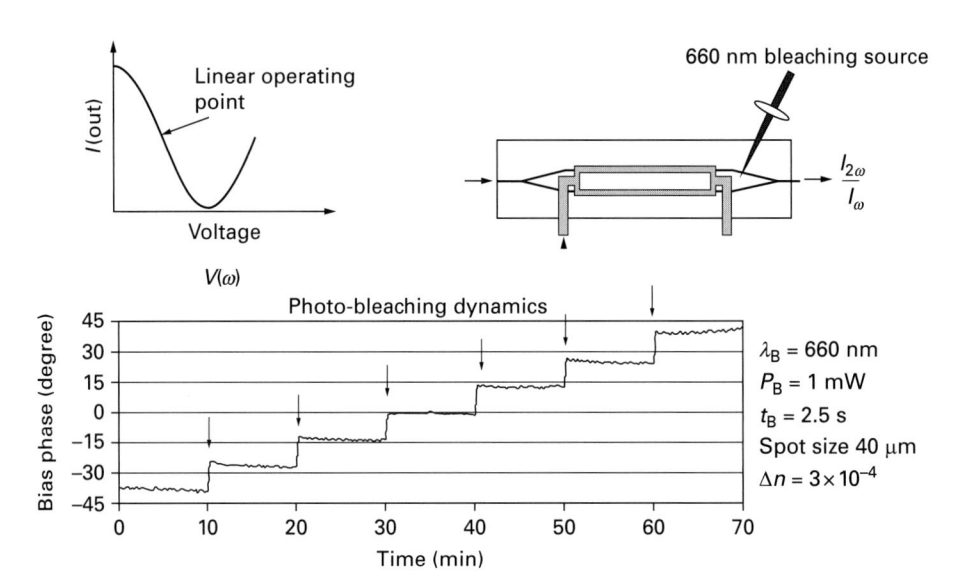

Figure 9.22 The use of photobleaching to fine-tune the balance of the arms of a Mach–Zehnder modulator.

As noted in Chapter 7, optical loss in electrically poled organic macromolecular materials can be kept to below 2 dB/cm while realizing electro-optic coefficients in single thin-film devices as high as 450 pm/V. In multilayer devices (such as triple stack, cladding–core–cladding devices discussed above), the same materials typically yield electro-optic coefficients in the range 100–250 pm/V, which is reasonable given the increased electrode spacing and the problems of the relative conductance of core and cladding materials discussed in the preceding paragraphs. One way around this is to deposit the top cladding layer after using corona poling to induce electro-optic activity in the core layer. An example of fabrication of a buried channel electro-optic waveguide employing RIE and corona poling is shown in Fig. 9.23.

However, in practical terms, corona poling and co-planar electrode poling have not competed effectively with poling through triple stack device architectures. Also, parallel plate poling with the top electrode directly on top of the electro-optic layer (followed by removal of the poling electrode, deposition of a top cladding layer, and deposition of the upper drive electrodes) has been utilized; however, this approach has not proved attractive for production of devices. The parallel plate poling of a triple stack device is popular because it yields an operational device without further processing.

Coupling loss arising from mode size and shape mismatch to silica fibers (or to laser modes) can be problematic for end-fire or fiber coupling discussed in Section 9.2. Mode transformers (taper coupling) have been the approach of choice for reducing coupling losses to a few tenths of a dB [58 to 63]. Various masking techniques, such as utilization of a shadow mask or a grayscale mask, can be employed to create vertical transitions and thus mode transformers such as shown in Fig. 9.24.

9.7 Polymer waveguides and modulators

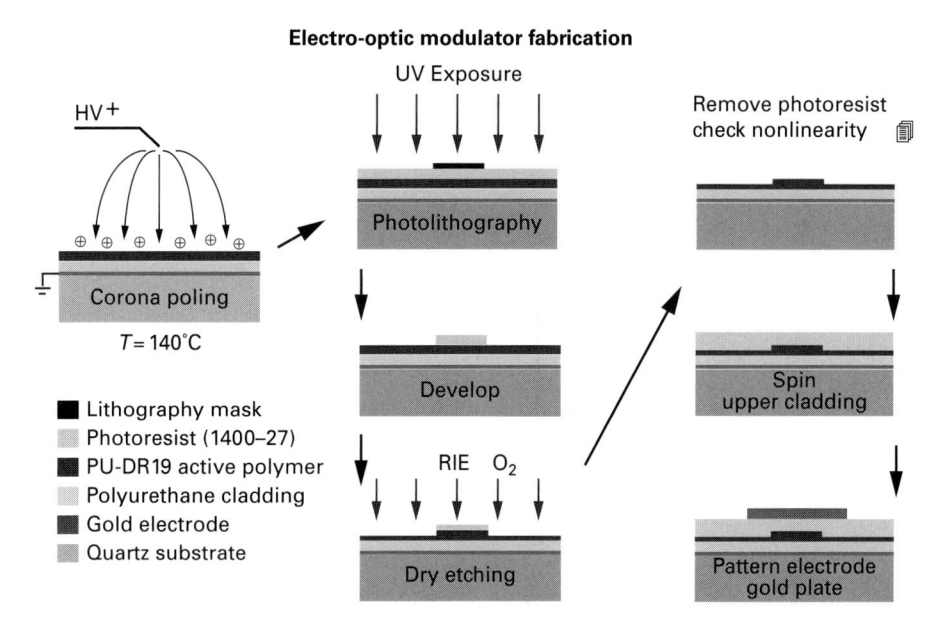

Figure 9.23 Fabrication of a buried channel electro-optic waveguide modulator by corona poling and reactive ion etching. Reprinted with permission from [46].

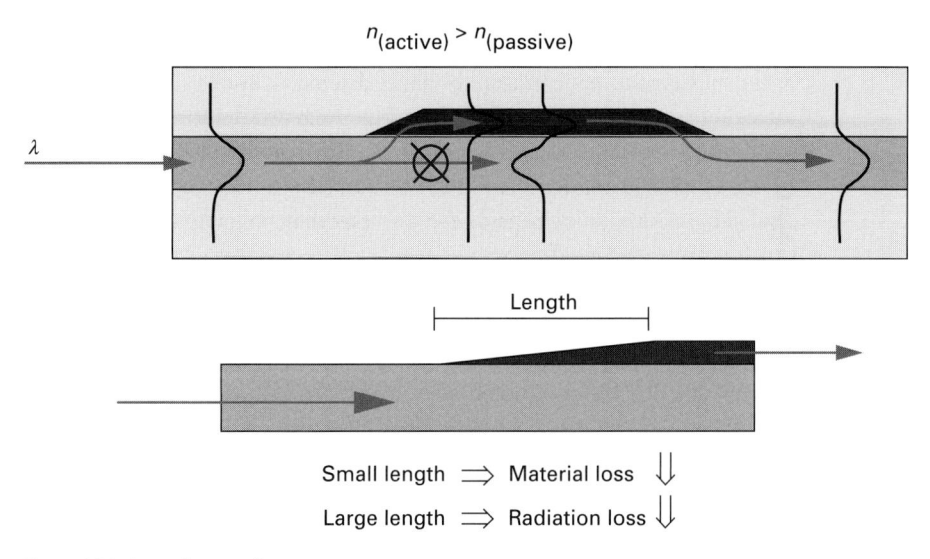

Figure 9.24 A mode transfomer.

All-organic (triple stack) push–pull Mach–Zehnder modulators have exhibited drive voltages as low as 0.2 V (at 1.55 μm). Insertion loss of less than 7 dB and operational bandwidths of >40 GHz have also been demonstrated for low-drive-voltage modulators. It is interesting that these numbers are very comparable to values observed for organic EO/silicon photonic hybrid devices. The poorer poling efficiencies currently realized for hybrid devices offset gains from smaller electrode spacings. Bandwidths

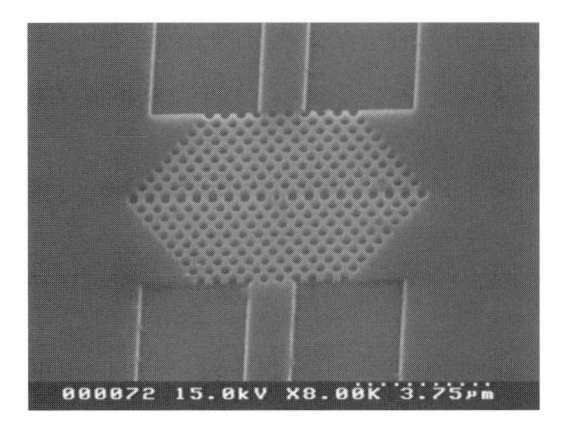

Figure 9.25 A photonic crystal structure fabricated by nano-imprint lithography.

in excess of 100 GHz have been demonstrated for all-organic devices but not accompanied by low drive voltages [64,65]. Trade-offs between drive voltage, bandwidth, and insertion loss have been discussed in Chapter 7 (see particularly Fig. 7.34). All-organic devices based on poled-polymer electro-optic materials have been commercialized by GigOptix and have satisfied Telcordia standards. As discussed in Chapter 7, lattice hardening to improve thermal and photochemical stability is important for practical devices. Photochemical stability can also be improved by the packaging of devices.

Organic electro-optic materials have also been used to fabricate photonic crystal waveguide structures such as shown in Fig. 9.25, which illustrates a structure fabricated using nano-imprint lithography [66]. For other examples, the reader is referred to the work of Eich and co-workers [24 to 26]. Active photonic crystal devices are also being created by incorporating organic electro-optic materials into silicon photonic structures, which will be discussed in Section 9.8.2.

Ring micro-resonator and photonic crystal structures have been employed for sensing, including the sensing of electromagnetic radiation, chemical and biochemical analytes, temperature, and strain [67 to 73]. Such structures are frequently passive sensors and so will not be discussed here other than to note the various schemes that have been used for enhanced performance (see Fig. 9.26).

In this section, we have focused on the simplest waveguide device structures. These structures are frequently modified to achieve more sophisticated functions or to achieve improved performance, e.g., improved linearity, suppression of nonlinear intermodulation distortion, or greater spur free dynamic range (SFDR) for Mach–Zehnder modulators [74]. Macromolecular (e.g., polymeric) electro-optic waveguides have also been integrated with both continuous and nanostructured gold (and silver) metals to exploit plasmonic effects [75]. Sub-wavelength confinement of guided light has been demonstrated as well as a variety of simple (amplitude and phase modulators) and more complex device structures. Both insulator–metal–insulator (IMI) and metal–insulator–metal (MIM) plasmonic structures have been investigated. Optical loss is a critical issue with plasmonic, photonic crystal, and metamaterial device architectures. For plasmonic

9.8 Silicon–organic hybrid waveguides

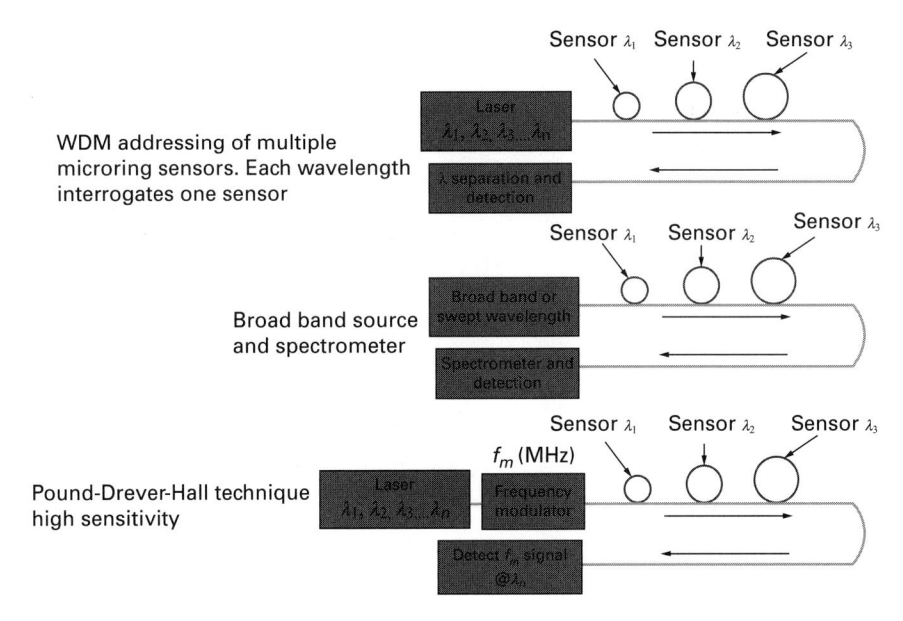

Figure 9.26 Microring resonator-based techniques for detection of multiple analytes and their spatial patterns.

waveguide devices, the use of nanostructured metals permits utilization of long-range surface modes (LRSM) rather than the more lossy long-range surface plasmon polaritons (LRSPP). This permits compact devices structures with acceptable optical loss to be realized, although the performance of these devices has yet to equal the performance of hybrid organic electro-optic material/silicon photonic waveguide devices, to be discussed in the next section.

9.8 Silicon–organic hybrid waveguides

9.8.1 Silicon photonics and silicon–organic hybrid concept

In the past few years, silicon photonics has expanded as one of the most promising platforms for future on-chip very-large-scale photonic integration. This is because of its wide compatibility with present CMOS processing, as well as silicon's high refractive index with a high index contrast with respect to SiO_2 substrates, allowing for very small waveguide cross-sections of only about 200×450 nm^2 [76]. Efficient electro-optic modulation in silicon waveguides is possible based on the free-carrier dispersion effect, with the EO bandwidth currently limited to about 30 GHz [77]. However, silicon does not exhibit an intrinsic electro-optic Pockels effect, which is often a desired modulation process, in particular for high-speed modulation and low power modulation. Another possibility to functionalize silicon waveguides is to integrate EO cladding materials. This is very challenging for most EO materials, owing to their limited processing possibilities within silicon structures.

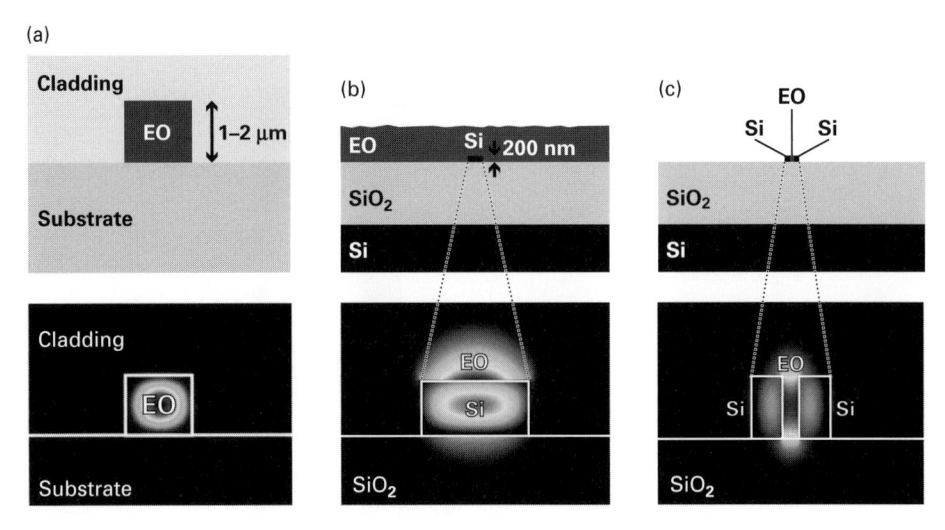

Figure 9.27 Schematic of electro-optic waveguides based on organic crystals: (a) conventional wire waveguides with electro-optic core material, (b) silicon waveguides with electro-optic cladding, (c) silicon slot waveguides with embedded electro-optic material. The bottom figures show the 2D intensity profile of the fundamental modes in the structures above: TM mode (or very similarly TE) for structure (a), TM mode for structure (b), and TE mode for structure (c).

Figure 9.27 illustrates three basic possibilities for tuning the effective index of guided modes by means of the electro-optic effect. Type (a) is the conventional geometry, in which the waveguide core material is electro-optically active. This modulation scheme is used most often, for example in commercial LiNbO$_3$ high-speed modulators. The other two types, (b) and (c), can be directly combined with the highly advanced silicon-on-insulator (SOI) technology. Cladding modulation is the scheme used for type (b), where an electro-optic refractive index change in the cladding material results in a change of the effective index, corresponding to the guided mode in the SOI waveguide structure. Type (c) is a non-conventional waveguide structure, in which the modal intensity is enhanced in the sub-100-nm wide slot consisting of the active material embedded in a silicon wire single-mode waveguide [78,79]. In both waveguide types (b) and (c), the silicon structures can be used also as electrodes, provided they are adequately conductive (i.e., appropriately doped). Especially for the type (c) modulators, this is an extremely interesting approach, since it leads to high electric fields for low applied switching voltages, as demonstrated e.g. for poled polymer/silicon hybrid structures [80 to 82], as well as modulation speeds exceeding those of silicon alone [83,84].

Figure 9.27 also shows the simulated 2D intensity distribution for these waveguide structures with a typical refractive index of an organic electro-optic cladding material ($n \approx 1.6$). For conventional optical waveguides the electric field of the fundamental optical mode is concentrated in the core waveguide structure, which is due to the core sizes with dimensions comparable to the optical wavelength or larger.

For silicon waveguides of type (b) and (c), a large amount of the optical field is in the cladding material. This is because of a high refractive index contrast between the silicon core ($n = 3.48$) and the cladding, allowing much smaller waveguides and therefore a high integration density of photonic devices. Despite the high index contrast, both waveguide structures guide a large part of the modal intensity in the cladding material. This is a consequence of the boundary conditions of the electromagnetic field, which requires that the normal component of the electric field $E_\perp$ behaves as

$$\varepsilon_{\text{core}} E_{\perp,\text{core}} = \varepsilon_{\text{clad}} E_{\perp,\text{clad}} \Rightarrow \frac{E_{\perp,\text{clad}}}{E_{\perp,\text{core}}} = \frac{n_{\text{core}}^2}{n_{\text{clad}}^2} > 1 \qquad (9.69)$$

at the interface of two materials with different refractive index n_{core} and n_{clad}, i.e. the electric-field amplitude can be considerably enhanced on the side of the lower-index material [78]. Relation (9.69) is apparent for the TM mode in Fig. 9.27(b) and the TE mode in (c).

Very promising results of hybrid organic–silicon integration have been obtained by using poled polymers, as described in Section 9.8.2, since polymers can be relatively easily processed, but a high poling efficiency and long-term stability in sub-micron structures required for silicon hybrid modulators is difficult to achieve for polymers. First results on hybrid integration of organic EO crystalline materials that do not need a poling procedure with silicon waveguides have also been demonstrated and are presented in Section 9.8.3.

9.8.2 Polymeric organic/silicon hybrid modulators

Electrically poled macromolecular (polymer and dendrimer) materials have been incorporated into a variety of silicon photonic stripline, ring microresonator, and photonic crystal structures including slow wave structures [80 to 82,85 to 93]. A variety of vertical and horizontal slot structures have been investigated as have silicon nanowire structures. As already noted, the structures shown in 9.27(b) and (c) have been investigated. The low conductivity of doped silicon has made realization of good poling efficiency very difficult; frequently, macromolecular materials exhibiting electro-optic coefficients of several hundred picometers/volt when poled as thin films exhibit electro-optic activities of less than 100 pm/V when poled in silicon photonic device structures. However, even with poor poling efficiencies, the smaller electrode separation of silicon photonic waveguides permits drive voltages as small as 0.2 V to be realized. Modulation to tens of gigahertz has been demonstrated. The concentration of optical fields in the reduced dimensions of silicon photonic waveguides permits optical rectification and all-optical modulation to be demonstrated to frequencies as high as 10 THz.

Coupling of light into and out of silicon photonic circuits has proven to be a challenge, but a variety of techniques including use of AWGs and tapers has permited realization of devices with insertion loss of less than 10 dB.

9.8.3 Crystalline silicon–organic hybrid modulators

A cladding modulation scheme was chosen for first hybrid modulators employing OH1 single crystals, in which the waveguide core is the high-index-contrast silicon-on-insulator (SOI) waveguide and the cladding an active EO material of a lower refractive index, as shown in Fig. 9.22(b). An EO refractive index change in the cladding material results in a change of the effective index of the guided mode in the SOI core waveguide. OH1 was chosen as the active cladding material for Mach–Zehnder SOI modulators owing to its favorable thin-film growth properties [94]. Since the OH1 crystalline thin films grow with their x_1-axis perpendicular to the substrate, the optimal alignment is the one with the polar x_3-axis perpendicular to the electrodes on both sides of the waveguides, which results in the employment of the EO coefficient r_{113} and r_{333} for TM and TE modes, respectively.

The process steps of the fabrication of integrated Mach–Zehnder (MZ) type modulators are depicted in Fig. 9.28. Silicon waveguides, MZ-structures, and Cr electrodes were patterned using standard optical lithography. On top of these structures, large-domain, single-crystalline OH1 was grown to a thickness of 2–5 µm, as illustrated in Fig. 9.28(c). After thin-film growth, the device was covered with a 2-µm-thick polyvinyl alcohol (PVA) protection layer. The structures were characterized by a standard end-fire coupling set-up with single-mode laser light in the wavelength range 1530–1610 nm. The modulation signal detected by the photodiode upon applying a sinusoidal voltage with 10 V amplitude was visualized on an oscilloscope [95].

The cladding modulation efficiencies for these structures were also determined numerically [95]. The results for the experimental device configuration agree well with the measured values, indicating that the EO coefficients of bulk OH1 crystals are preserved in thin films deposited on silicon. MZ modulators based on wire waveguides with OH1 as EO cladding can lead to half-wave voltages $V_{\pi,\text{ideal}} \cdot L$ below 1 V-cm. These results show that the cladding modulation scheme is superior with respect to conventional electro-optically active channel waveguides, owing to the stronger light confinement and possible considerable reduction of the gap between the electrodes.

It has already been noted that horizontal as well as vertical slot silicon photonic modulators have been fabricated. Horizontal slot waveguides afford a convenient deposition surface for the growth of organic electro-optic crystalline materials or deposition of very thin (∼25 nm) films of organic electro-optic materials by sequential-synthesis/self-assembly techniques. Recently, horizontal slot modulators have been demonstrated with crystalline materials [96].

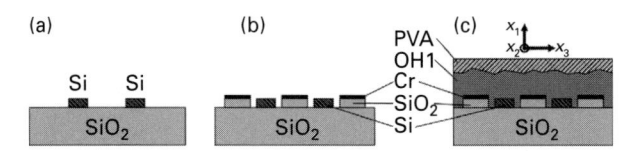

Figure 9.28 Process steps for the fabrication of an electro-optically active MZ interferometer based on the SOI waveguide technology. (a) Waveguide patterning using optical lithography. (b) Second lithography step for Cr electrode patterning. (c) Growth of the electro-optic single crystalline OH1 thin film, and spin-coating of the PVA protection layer. Reprinted with permission from Ref. [95].

References

[1] B. E. A. Saleh and M. C. Teich, *Fundamentals of Photonics*, New York, John Wiley & Sons, Inc. (1991).
[2] D. Marcuse, *Light Transmission Optics*, New York, Van Nostrand (1972).
[3] A. Yariv and P. Yeh, *Optical Waves in Crystals*, New York, John Wiley & Sons (1984).
[4] K. Okamoto, *Fundamentals of Optical Waveguides*, San Diego, Academic Press (2000).
[5] D. Marcuse, *IEEE J. Quantum Electron.*, **14**, 736 (1978).
[6] D. Marcuse and I. P. Kaminow, *IEEE J. Quantum Electron.*, **15**, 92 (1979).
[7] A. Knoesen, T.K. Gaylord, and M.G. Moharam, *J. Lightwave Technol.*, **6**, 1083 (1988).
[8] B. E. Little, S. T. Chu, H. A. Haus, J. Foresi, and J. P. Laine, *J. Lightwave Technol.*, **15**, 998 (1997).
[9] I. L. Gheorma and R. M. Osgood, *IEEE Photonics Technol. Lett.*, **14**, 795 (2002).
[10] P. Rabiei, W. H. Steier, C. Zhang, and L. R. Dalton, *IEEE J. Lightwave Technol.*, **20**, 1968 (2002).
[11] V. Van, T. A. Ibrahim, P. P. Absil, *et al.*, *IEEE J. Sel. Top. Quantum Electron.*, **8**, 705 (2002).
[12] K. J. Vahala, *Nature*, **424**, 839 (2003).
[13] Q. F. Xu, B. Schmidt, S. Pradhan, and M. Lipson, *Nature*, **435**, 325 (2005).
[14] H. Tazawa, Y. H. Kuo, I. Dunayevskiy, *et al.*, *IEEE J. Lightwave Technol.*, **24**, 3514 (2006).
[15] A. Guarino, G. Poberaj, D. Rezzonico, R. Degl'Innocenti, and P. Gunter, *Nat. Photonics*, **1**, 407 (2007)
[16] D. Rezzonico, M. Jazbinsek, A. Guarino, O. P. Kwon, and P. Gunter, *Opt. Express*, **16**, 613 (2008).
[17] J. Joannopoulos, S. Johnson, and J. W. and D. Meade, *Photonic Crystals: Molding the Flow of Light*, Princeton, NJ, Princeton University Press (2008).
[18] K. Sakoda, *Optical Properties of Photonic Crystals*, Berlin/Heidelberg, Springer (2001).
[19] Y. Dumeige, P. Vidakovic, S. Sauvage, *et al.*, *Appl. Phys. Lett.*, **78**, 3021 (2001).
[20] M. Soljacic, S. G. Johnson, S. H. Fan, *et al.*, *J. Opt. Soc. Am. B*, **19**, 2052 (2002).
[21] M. Soljacic and J. D. Joannopoulos, *Nature Mater.*, **3**, 211 (2004).
[22] Y. A. Vlasov, M. O'Boyle, H. F. Hamann, and S. J. McNab, *Nature*, **438**, 65 (2005).
[23] M. Roussey, F. I. Baida, and M. P. Bernal, *J. Opt. Soc. Am. B*, **24**, 1416 (2007).
[24] C. Liguda, G. Bottger, A. Kuligk, *et al.*, *Appl. Phys. Lett.*, **78**, 2434 (2001).
[25] G. Bottger, M. Schmidt, M. Eich, R. Boucher, and U. Hubner, *J. Appl. Phys.*, **98**, 103101 (2005).
[26] M. Schmidt, M. Eich, U. Huebner, and R. Boucher, *Appl. Phys. Lett.*, **87**, 121110 (2005).
[27] S. Fukuda and T. Gotoh, *Laser Kenkyu*, **21**, 1134 (1993).
[28] K. Tsuda, T. Kondo, F. Saito, T. Kudo, and R. Ito, *Jpn. J. Appl. Phys. 2*, **31**, L134 (1992).
[29] K. Takayama, K. Komatsu, and T. Kaino, *Jpn. J. Appl. Phys. 1*, **40**, 5149 (2001).
[30] T. Kaino, B. Cai, and K. Takayama, *Adv. Funct. Mater.*, **12**, 599 (2002).
[31] M. Jazbinsek, L. Mutter, and P. Gunter, *IEEE J. Sel. Top. Quantum Electron.*, **14**, 1298 (2008).
[32] C. Hunziker, S. J. Kwon, H. Figi, M. Jazbinsek, and P. Gunter, *Opt. Express*, **16**, 15903 (2008).
[33] B. Cai, K. Komatsu, and T. Kaino, *Opt. Mater.*, **21**, 525 (2003).
[34] L. Mutter, M. Jazbinsek, M. Zgonik, *et al.*, *J. Appl. Phys.*, **94**, 1356 (2003).
[35] P. Dittrich, R. Bartlome, G. Montemezzani, and P. Gunter, *Appl. Surf. Sci.*, **220**, 88 (2003).

[36] L. Mutter, A. Guarino, M. Jazbinsek, *et al.*, *Opt. Express*, **15**, 629 (2007).
[37] L. Mutter, M. Jazbinsek, C. Herzog, and P. Gunter, *Opt. Express*. **16**, 731 (2008).
[38] L. Mutter, M. Koechlin, M. Jazbinsek, and P. Gunter, *Opt. Express*, **15**, 16828 (2007).
[39] W. Geis, R. Sinta, W. Mowers, *et al.*, *Appl. Phys. Lett.*, **84**, 3729 (2004).
[40] H. Figi, M. Jazbinsek, C. Hunziker, M. Koechlin, and P. Gunter, *Opt. Express*, **16**, 11310 (2008).
[41] H. Figi, M. Jazbinsek, C. Hunziker, M. Koechlin, and P. Gunter, *Proc. SPIE*, **7599**, 75991N-1 (2010).
[42] H. Figi, M. Jazbinsek, C. Hunziker, M. Koechlin, and P. Gunter, *J. Opt. Soc. Am. B*, **26**, 1103 (2009).
[43] W. H. Steier and L. R. Dalton, In *Broadband Optical Modulators: Science, Technology, and Applications*, A. Chen and E. Murphy, eds., New York, Taylor & Francis (2011), Ch. 9, pp. 221–254.
[44] R. Lytel, G. F. Lipscomb, J. T. Kenney, and E. S. Binkley, In *Polymers for Lightwave and Integrated Optics*, L. A. Hornak, Ed., New York, Marcel Dekker (1992), pp. 433–372.
[45] T. A. Skotheim and J. R. Reynolds, eds., *Handbook of Conducting Polymers, 3rd Edition: Conjugated Polymers, Theory, Synthesis, Properties, and Characterization*, Boca Raton, FL, CRC Press (2007).
[46] L. R. Dalton, A. W. Harper, A. S. Ren, *et al.*, *Ind. Eng. Chem. Res.*, **38**, 8 (1999).
[47] W. H. Steier, A. Chen, S.-S. Lee, *et al.*, *Chem. Phys.*, **245**, 487 (1999).
[48] L. R. Dalton, in *Advances in Polymer Science*, Vol. 158, Heidelberg, Springer-Verlag (2001), pp. 1–86.
[49] L. R. Dalton, *J. Phys.: Condens. Mater.*, **15**, R897 (2003).
[50] S. K. Kim, K. Geary, H. R. Fetterman, *et al.*, *Electron. Lett.*, **39**, 1321 (2003).
[51] S. Kim, K. Geary, D. Chang, *et al.*, *Electron. Lett.*, **39**, 721 (2003).
[52] S. K. Kim, K. Geary, W. Yuan, *et al.*, *Electron. Lett.*, **40**, 866 (2004).
[53] A. Chen, V. Chuyanov, F. I. Marti-Carrera, *et al.*, *IEEE Photon. Tech. Lett.*, **9**, 1499 (1997).
[54] A. Chen, V. Chuyanov, F. I. Marti-Carrera, *et al.*, *Proc. SPIE*, **3147**, 268 (1997).
[55] S. S. Lee, S. Garner, A. Chen, *et al.*, *Appl. Phys. Lett.*, **73**, 3052 (1998).
[56] J. K. S. Poon, Y. Huang, G. T. Poloczi, *et al.*, *Opt. Lett.*, **29**, 2584 (2004).
[57] Y. Huang, J, K. S. Poon, W. Liang, *et al.*, *Appl. Phys. Lett.*, **87**, 071108 (2005).
[58] A. Chen, V. Chuyanov, F. I. Marti-Carrera, *et al.*, *Proc. SPIE*, **3005**, 65 (1997).
[59] A. Chen, V. Chuyanov, F. I. Marti-Carrera, *et al.*, *Opt. Eng.*, **39**, 1507 (2000).
[60] M. C. Oh, C. Zhang, H. J. Lee, W. H. Steier, and H. R. Fetterman, *IEEE Photon. Technol. Lett.*, **14**, 1121 (2002).
[61] D. Y. Zang, G. Shu, T. Downing, *et al.*, *Proc. SPIE*, **4991**, 601 (2003).
[62] D. H. Chang, T. Azfar, S. K. Kim, *et al.*, *Opt. Lett.*, **28**, 869 (2003).
[63] Y. Enami, G. Meredith, N. Peyghambarian, M. Kawazu, and A. K. Y. Jen, *Appl. Phys. Lett.*, **82**, 490 (2003).
[64] M. Lee, H. E. Katz, C. Erben, *et al.*, *Science*, **298**, 1401 (2002).
[65] D. Chen, H. R. Fetterman, A. Chen, *et al.*, *Appl. Phys. Lett.*, **70**, 3335 (1997).
[66] K. A. Firestone, P. J. Reid, R. Lawson, S.-H. Jang, and L. R. Dalton, *Inorg. Chem. Acta*, **357**, 3957 (2004).
[67] A. Chen, H. Sun, A. Pyayt, *et al.*, *IEEE Sensors*, 735 (2005).
[68] H. Sun, A. Pyayt, J. Luo, *et al.*, *Proc. SPIE*, **6117**, 6117131 (2006).
[69] H. Sun, A. Pyayt, J. Luo, *et al.*, *IEEE Sensors J.*, **7**, 515 (2007).
[70] A. Pyayt, X. Zhang, J. Luo, *et al.*, *Proc. SPIE*, **6556**, 65561D1 (2007).

[71] A. Chen, H. Sun, A. Pyayt, *et al.*, *IEEE J. Sel. Top. Quantum Electron.*, **14**, 1281 (2008).
[72] H. Sun, A. Chen, and L. R. Dalton, *IEEE Photon. J.*, **1**, 48 (2009).
[72] H. Sun, A. Chen, and L. R. Dalton, *Opt. Express*, **17**, 10731 (2009).
[73] B. Bhola, H. C. Song, H. Tazawa, and W. H. Steier, *IEEE Photon. Tech. Lett.*, **17**, 867 (2005).
[74] S. Kim, W. Liu, Q. Pei, L. R. Dalton, and H. R. Fetterman, *Opt. Express*, **19**, 7865 (2011).
[75] L. R. Dalton, A. K.-Y. Jen, B. H. Robinson, *et al.*, *Proc. SPIE*, **7935**, 7035021 (2011).
[76] P. Dumon, G. Priem, L. R. Nunes, *et al.*, *Jpn. J. Appl. Phys. 1*, **45**, 6589 (2006).
[77] L. Liao, A. Liu, D. Rubin, *et al.*, *Electron. Lett.*, **43**, 1196 (2007).
[78] V. R. Almeida, Q. F. Xu, C. A. Barrios, and M. Lipson, *Opt. Lett.*, **29**, 1209 (2004).
[79] J. Leuthold, W. Freude, J. M. Brosi, *et al.*, *Proc. IEEE*, **97**, 1304 (2009).
[80] T. Baehr-Jones, B. Penkov, J. Q. Huang, *et al.*, *Appl. Phys. Lett,*, **92**, 163303 (2008).
[81] J. M. Brosi, C. Koos, L. C. Andreani, *et al.*, *Opt. Express*, **16**, 4177 (2008).
[82] R. Ding, T. Baehr-Jones, W. J. Kim, *et al.*, *IEEE J. Lightwave Technol.*, **29**, 1112 (2011).
[83] J. H. Wulbern, S. Prorok, J. Hampe, *et al.*, *Opt. Lett.*, **35**, 2753 (2010).
[84] L. Alloatti, D. Korn, R. Palmer, *et al.*, *Opt. Express*, **19**, 11841 (2011).
[85] B. Maune, R. Lawson, C. Gunn, A. Scherer, and L. R. Dalton, *Appl. Phys. Lett.*, **83**, 4689 (2003).
[86] T. Baehr-Jones, M. Hochberg, G. Wang, *et al.*, *Opt. Express*, **13**, 5216 (2005).
[87] M. Hochberg, T. Baehr-Jones, G. Wang, *et al.*, *Nature Mater.*, **5**, 703 (2006).
[88] J. Takayesu, M. Hochberg, T. Baehr-Jones, *et al.*, *IEEE J. Lightwave Technol.* **27**, 440 (2008).
[89] B. A. Block, T. R. Younkin, R. Reshotko, *et al.*, *Opt. Express*, **16**, 18326 (2008).
[90] R. Ding, T. Baehr-Jones, Y. Liu, *et al.*, *Opt. Express*, **18**, 15618 (2010).
[91] M. Gould, T. Baehr-Jones, R. Ding, *et al.*, *Opt. Express*, **19**, 3952 (2011).
[92] C.-Y. Lin, X. Wang, S. Chakravarty, *et al.*, *Appl. Phys. Lett.*, **97**, 093304-1 (2010).
[93] X. Wang, C.-Y. Lin, S. Chakravarty, *et al.*, *Opt. Lett.*, **36**, 882 (2011).
[94] S. J. Kwon, C. Hunziker, O. P. Kwon, M. Jazbinsek, and P. Gunter, *Cryst. Growth Des.*, **9**, 2512 (2009).
[95] M. Jazbinsek, C. Hunziker, S. J. Kwon, *et al.*, *Proc. SPIE*, **7599**, 75990K–1 (2010).
[96] H. Figi, D. H. Bale, A. Szep, L. R. Dalton, and A. T. Chen, *J. Opt. Soc. Am. B*, **28**, 2291 (2011).

10 Nonlinear optical infrared and terahertz frequency conversion

10.1 Nonlinear optical frequency conversion

As discussed in Chapter 2, there are many different possibilities for converting optical frequencies to other frequencies or even static fields in second-order nonlinear optical materials, such as sum- and difference-frequency generation, including second-harmonic generation and optical rectification (see Fig. 2.2). The big advantage of organic nonlinear optical materials compared with inorganic ones is high nonlinear optical figures-of-merit, reflecting their almost purely electronic response to external fields, as discussed in Section 3.2. Because of this, the best organic materials exhibit considerably higher second-order nonlinear optical susceptibilities compared with the best inorganic materials. This is illustrated in Fig. 10.1, which shows figures-of-merit for second-harmonic generation d^2/n^3 versus transparency range for various organic and inorganic crystals. One can clearly see that the nonlinear optical figures-of-merit of organic materials can be several orders of magnitude higher than in the best inorganic materials. This makes organic materials extremely attractive for nonlinear optical applications.

In the 1980s, the most attractive frequency conversion applications included frequency doubling because of the above advantages and because of the interest in generating blue or green coherent light by using widely available (near-)infrared laser sources, such as diode lasers [1], Ti:sapphire lasers, and Nd:YAG lasers. The early organic materials considered were transparent in the visible. Later on, organic materials with much higher nonlinear optical susceptibilities were developed, but these are no longer transparent in the visible (see Fig. 10.1) and therefore second-harmonic generation with these materials is of limited applicability. At present, the most attractive frequency-conversion applications with organic materials include infrared and far-infrared light generation, as well as generation of electromagnetic waves in the THz frequency range. In this section we mainly describe infrared frequency conversion possibilities with the best organic crystals, such as DAST, DSTMS, and OH1 (see Chapter 6 for details on these materials), and in the next section we look at THz generation with these and other organic materials.

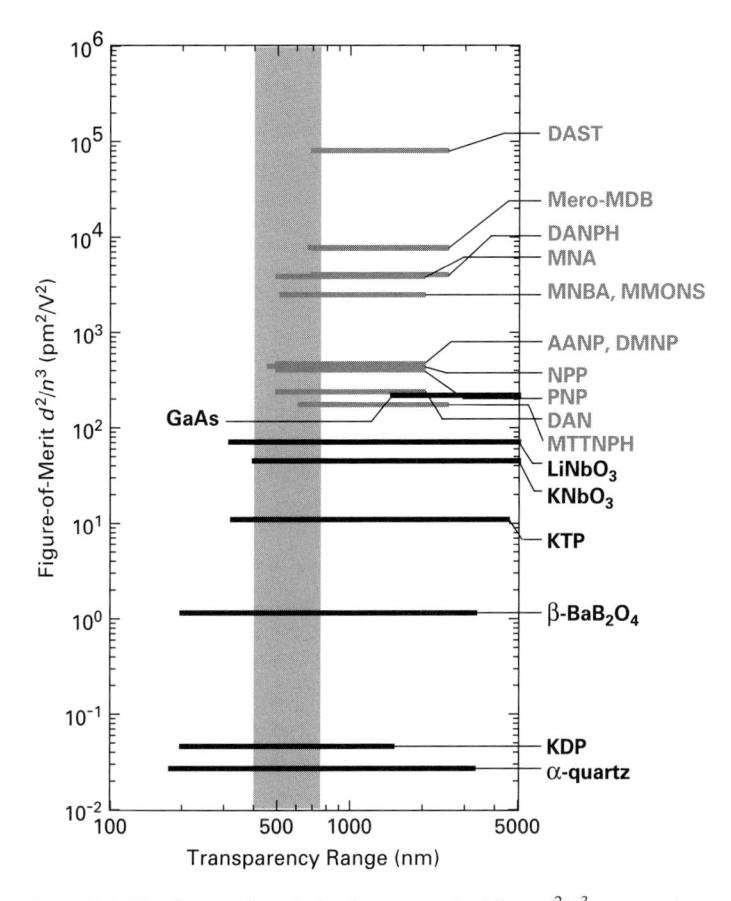

Figure 10.1 The figure-of-merit for frequency doubling, d^2/n^3, versus the transparency range of various organic and inorganic crystals. The shaded area indicates the visible spectral range. Inorganic crystals are indicated in black, organic crystals in gray. It is clearly seen that organic crystals are much superior in terms of the figure-of-merit. Reprinted with permission from [2].

10.1.1 Phase-matched parametric interaction in organic crystals

Beside a high second-order nonlinear optical coefficient d_{ijk}, the most important condition that should be satisfied in order to obtain high conversion efficiency is phase matching (see also Section 2.2.3). This means that beside energy conservation requiring

$$\omega_3 = \omega_1 + \omega_2, \tag{10.1}$$

when mixing waves with three frequencies in nonlinear materials, momentum conservation (or phase matching) must also be fulfilled:

$$\mathbf{k}_{\omega_3} = \mathbf{k}_{\omega_1} + \mathbf{k}_{\omega_2}, \tag{10.2}$$

where $\mathbf{k}_{\omega_i}$ denotes the wave vector at frequency ω_i. In the simplest case of collinear interactions, all wave vectors are parallel, which leads to the following phase-matching condition

$$n(\omega_3)\omega_3 = n(\omega_1)\omega_1 + n(\omega_2)\omega_2. \qquad (10.3)$$

The organic crystal DAST has relatively large off-diagonal nonlinear optical elements d_{122} and d_{212} with $n_1 > n_2 > n_3$, which makes parametric generation of type I and type II possible.

The nonlinear optical polarization $\mathbf{P}^{\omega_3}$ for sum-frequency generation from the fundamental waves $\mathbf{E}^{\omega_{1,2}}$ is described by

$$P_i^{\omega_3} = 2\varepsilon_0 d_{ijk}^{(\omega_3,\omega_1,\omega_2)} E_j^{\omega_1} E_k^{\omega_2}, \qquad (10.4)$$

where d_{ijk} are the corresponding nonlinear optical tensor elements. For general directions of the wave vector and the polarizations in the crystal, the projection of the induced polarization at frequency ω_3 along the direction of the electric field of the generated wave at ω_3 can be written as

$$\left|\mathbf{P}^{\omega_3}\right| = 2\varepsilon_0 d_{\text{eff}} \left|\mathbf{E}^{\omega_1}\right|\left|\mathbf{E}^{\omega_2}\right|, \qquad (10.5)$$

with

$$d_{\text{eff}} = d_{ijk}^{(\omega_3,\omega_1,\omega_2)} \cos\left(\alpha_i^{\omega_3}\right) \cos\left(\alpha_j^{\omega_1}\right) \cos\left(\alpha_k^{\omega_2}\right), \qquad (10.6)$$

where α_i^{ω} is the angle between the electric field vector at frequency ω and the axis i of the Cartesian coordinate system in which d_{ijk} is given [3].

Possible configurations calculated from the known refractive-index dispersion for different propagation directions for type I (the photons corresponding to ω_1 and ω_2 are of the same linear polarization) sum-frequency generation in DAST using Eqs. (10.1) and (10.3) are shown in Fig. 10.2(a) [4]. For type I sum-frequency generation, the pump beam propagates in the x_1–x_3-plane with an angle of θ between 79° and 85°. For type I configuration the polarization of the pump beam lies in the x_1–x_3-plane, whereas the signal and idler waves (λ_1 and λ_2) are polarized along the x_2-axis of the crystal. The effective coefficient for type I phase-matched sum-frequency generation is in the range of $d_{\text{eff}} \approx 2$ pm/V.

Possible configurations for different propagation directions for type II (the photons corresponding to ω_2 and ω_2 are of the orthogonal linear polarization) phase-matched sum-frequency generation are shown in Fig. 10.2(b) [4]. When a DAST crystal is rotated around its x_2-axis (angle tuning), light can be generated over a wide range of wavelengths (from 1 to 3 μm) by optical parametric oscillation using a single pump wavelength. A further possibility is to use a tunable pump laser source in the near infrared region of the spectrum (e.g. Ti:sapphire laser), where the pump beam propagates in the x_1–x_3-plane with angles of θ between 0° and 60°. For the type II configurations, the pump beam is polarized along the x_2-axis of the crystal. The effective coefficient for type II phase-matched sum-frequency generation has been estimated to range from $d_{\text{eff}} \approx 5$ up to 50 pm/V or more, depending on the exact configuration, as shown in Fig. 10.2(b).

Phase-matching curves for optical parametric generation have been also evaluated for DSTMS (the derivative crystal of DAST; see Chapter 6), which has analogous crystal structure and similar dispersion characteristics to DAST; however, even the small differences are enough to modify the phase-matching conditions considerably.

10.1 Nonlinear optical frequency conversion

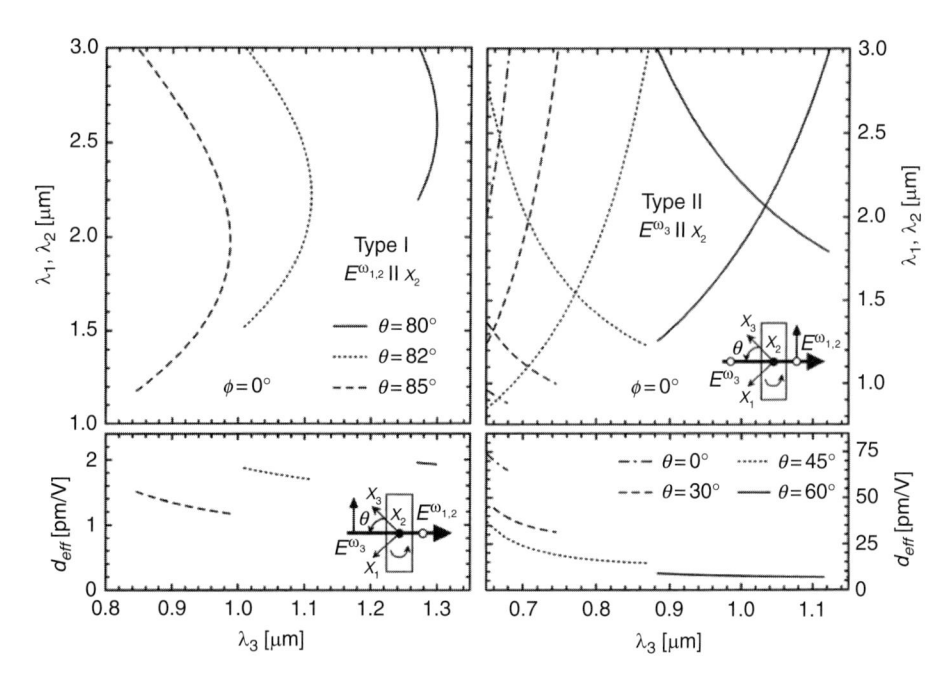

Figure 10.2 Calculated phase-matching curves for sum-frequency generation of type I, i.e. the interacting photons with λ_1 and λ_2 have the same linear polarization (a), and type II, i.e. the interacting photons with λ_1 and λ_2 have orthogonal linear polarization (b), in a DAST crystal. The beams propagate in the x_1–x_3-plane and at different values of θ. For type I phase-matching the polarization of signal and idler waves (λ_1 and λ_2) are along the x_2-axis, whereas for type II the polarization of the pump beam is along the x_2-axis (reprinted with permission from Ref. [4]).

In Fig. 10.3(a1) and (b1), the calculated phase-matching curves for parametric light generation of type I (the generated photons at λ_1 and λ_2 are of the same linear polarization) and II (the generated photons are of orthogonal linear polarization) as functions of the internal angle θ are shown for several wavelengths corresponding to widely available high-power solid-state and diode laser sources. We can see that these sources are all well suited as pump wavelengths for parametric generation of light in DSTMS.

The corresponding effective nonlinear optical coefficients d_{eff} were calculated with Eq. (10.6) taking into account the dispersion of the nonlinear optical coefficient by applying the quantum-mechanical two-level model [5], which is discussed in Chapter 4. Light in the wavelength range from 1 to 2.2 μm can be generated with type I phase matching by tuning the angle θ in a narrow range from 68° to 70.5°. Above 2.2 μm, material absorption hinders the efficient generation of light.

Type II phase matching offers the advantage of higher effective nonlinear optical coefficients, ranging from about 10 to 40 pm/V compared with type I phase matching with 6–10 pm/V, whereas tuning of the angle θ in a broader range is required to access the whole wavelength range from 1 to 2.2 μm. Note that the refractive index dispersion in the far infrared and THz wavelength range has been neglected for these calculations.

Compared with inorganic crystals LiNbO$_3$ with a nonlinear optical figure-of-merit of FOM $= d_{\text{eff}}^2/n^3 \approx 3 \text{ pm}^2/\text{V}^2$ and beta barium borate (BBO) with FOM $< 1 \text{ pm}^2/\text{V}^2$,

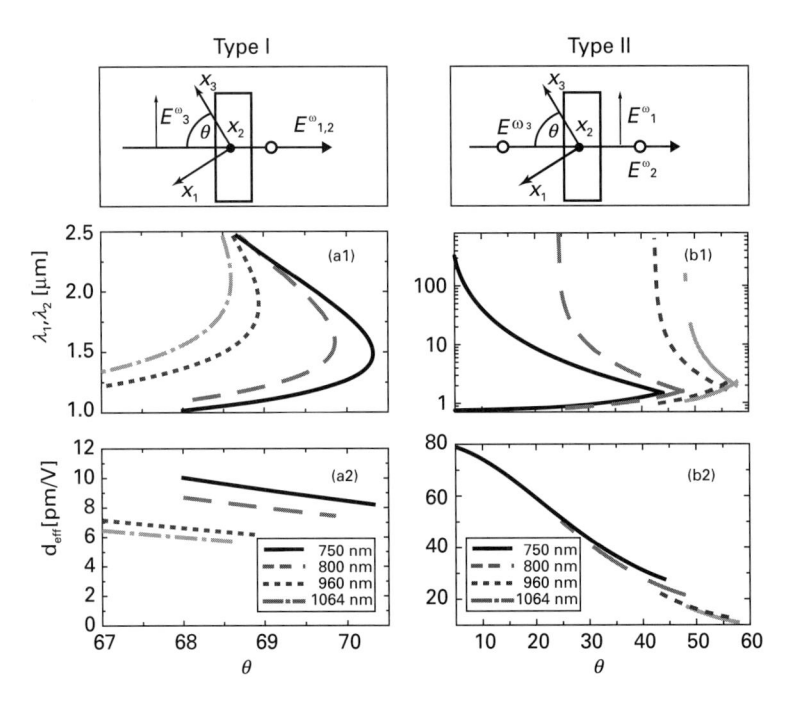

Figure 10.3 Phase-matching curves for parametric light generation of (a1) type I (the interacting photons with λ_1 and λ_2 have the same linear polarization); and (b1) type II (the interacting photons with λ_1 and λ_2 have orthogonal linear polarization) as functions of the internal tuning angle θ for pumping wavelength of 750 nm (solid curve), 800 nm (dashed curve), 960 nm (dotted curve) and 1064 nm (dashed-dotted curve). The corresponding effective nonlinear optical coefficients d_{eff} are depicted in (a2) and (b2). The orientation of the samples and the polarizations of the interacting electric fields are depicted above the graphs (reprinted with permission from Ref. [6]).

which are often used for parametric light generation, DSTMS shows a two orders of magnitude higher figure-of-merit FOM $\approx$ 160 pm^2/V^2 for the generation of $\lambda_1 = 1.5$ µm by type II phase matching pumped with $\lambda_3 = 750$ nm.

10.2 Terahertz-wave generation with organic nonlinear optical materials

The terahertz range of the electromagnetic spectrum is situated between high-frequency electronics (microwaves) and long-wavelength photonics (infrared light). In terms of frequency ν, it is roughly defined to be between 0.1 and 10 THz (see Fig. 10.4). Alternatively, an electromagnetic wave can be described by its vacuum wavelength $\lambda = c/\nu$, wavenumber $\tilde{\nu} = 1/\lambda$, or photon energy $E_{\text{photon}} = h\nu$, where c is the speed of light in vacuum and h is the Planck constant. Figure 10.4 shows the corresponding scales using these parameters for the THz and far-IR frequency range.

Several techniques have been used to generate electromagnetic waves below and above this so-called "THz frequency gap". Figure 10.5 shows the rough power levels of

10.2 Terahertz-wave generation

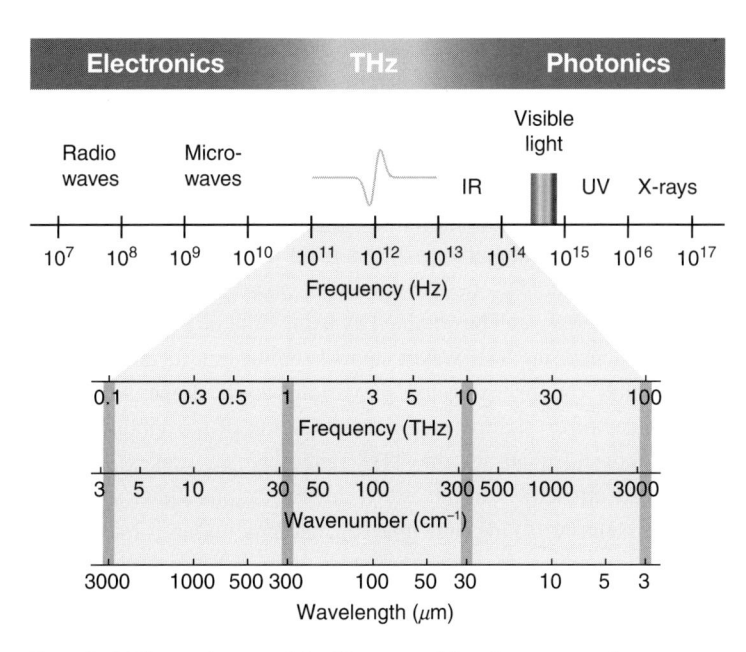

Figure 10.4 The terahertz and far-IR range of the electromagnetic spectrum.

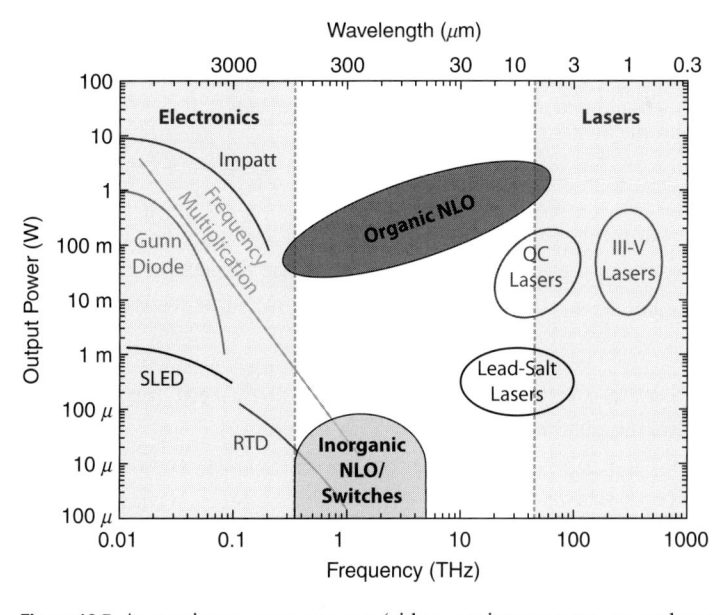

Figure 10.5 Approximate output power (either continuous wave or peak power) as a function of frequency (wavelength) of electromagnetic radiation in the THz range and around when using different generation techniques. Here, Impatt stands for impact ionization avalanche transit-time diodes, RTD for resonant tunneling diodes, QC lasers for quantum cascade lasers, and SLED for superlattice electronic devices. Adapted and updated from Ref. [8].

some of them. Electronic techniques can be used to generate waves with frequencies (mainly by electronic frequency multiplication of lower frequency sources) up to about 0.5 THz. From 0.3 to 3 THz Auston switches [7] are very popular sources. Nonlinear optical techniques (optical rectification and difference-frequency generation) can be used to cover the frequency range between 0.3 to 50 THz and quantum-cascade lasers between roughly 20 THz and 100 THz. Owing to the larger nonlinear optical susceptibilities and velocity matching between THz and optical pump waves of organic materials compared with the inorganic ones, much larger power levels, limited by the damage thresholds of the materials, can be obtained by using organic materials as THz generators.

The interest in generating THz waves stems from the unique interactions of these rays with matter, which can be exploited in various applications. For example, optical phonon resonances of crystalline materials and some of the vibrational and rotational excitations of molecules are in the THz range, which makes THz radiation very interesting for spectroscopy and material identification. Other applications include non-destructive material testing and imaging, various research material investigations such as carrier dynamics in semiconductors with sub-picosecond time resolution, medical diagnostics, and pharmaceutical characterization.

10.2.1 THz-wave generation by difference-frequency mixing

The very same difference-frequency generation (DFG) mechanism as described in the previous section to generate infrared light by using shorter-wavelength pump light can also be used to generate light in the THz frequency range. In this case we need a pump source consisting of two frequencies ω_1 and ω_2 that are very close to each other, so that their difference frequency lies in the THz range: $\omega_{THz} = \omega_1 - \omega_2$. The phase-matching condition should be also satisfied

$$\Delta \mathbf{k} = \mathbf{k}_{THz} - (\mathbf{k}_1 - \mathbf{k}_2). \tag{10.7}$$

For collinear difference-frequency generation and assuming that the optical frequencies are close together, so that in first approximation the dispersion in the optical range can be considered as $n_2 = n_1 + (\partial n / \partial \lambda)\Delta\lambda$, $\Delta\lambda = \lambda_2 - \lambda_1$, this leads to

$$\Delta k = \frac{\omega_{THz}}{c}(n_{THz} - n_g) \tag{10.8}$$

and the following coherence length for THz generation

$$l_c = \frac{\lambda_{THz}}{2(n_{THz} - n_g)}, \tag{10.9}$$

where $n_g = n - (\partial n / \partial \lambda)\lambda$ is the group index of the optical wave. The equation (10.9) is valid for a relatively small dispersion in the optical range, i.e. up to several THz if we use infrared pump light. For larger THz frequencies, it should be calculated as

$$l_c = \frac{1}{2}\left(\frac{n_{THz}}{\lambda_{THz}} - \frac{n_1}{\lambda_1} + \frac{n_2}{\lambda_2}\right)^{-1}. \tag{10.10}$$

10.2 Terahertz-wave generation

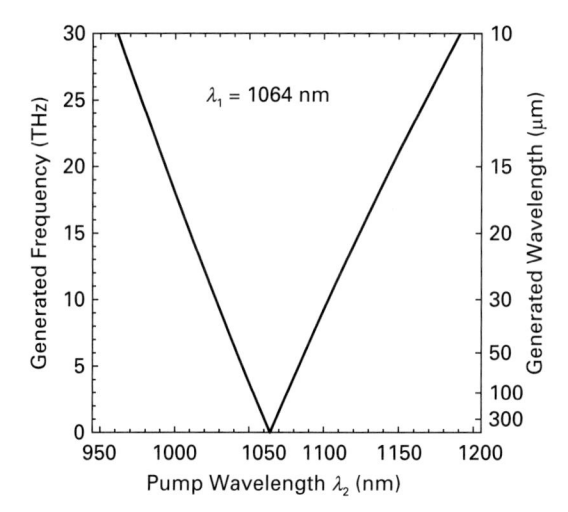

Figure 10.6 The range of the THz and far-IR frequency generated by difference-frequency generation, as a function of λ_2 for $\lambda_1 = 1064$ nm.

Figure 10.6 shows the dependence of the THz frequency on one pump wavelength for the case with the other wavelength fixed to the common 1064 nm Nd:YAG laser line.

For the efficient generation of THz waves by difference-frequency generation, besides a high second-order nonlinear optical susceptibility, the most important parameter is the low refractive index mismatch $\Delta n = n_{\text{THz}} - n_g$ between the generated THz and the pump optical waves. This is where organic materials have a big advantage compared with standard inorganic materials such as LiNbO$_3$. Because of the relatively low contribution of the lattice phonon vibrations to the dielectric constant, the dispersion of the refractive index between the optical and the THz frequency range is low (see also Section 3.2) and therefore the phase-matching condition can easily be satisfied even for the diagonal susceptibility element, while for inorganic materials such as LiNbO$_3$ special phase-matching configurations are needed as in the optical range. Note that a moderate index mismatch of e.g. $\Delta n = 0.1$ in the optical range ($\lambda \sim 1$ μm) leads to a coherence length of only $l_c \sim 5$ μm, while in the THz range, because of the longer wavelength (1 THz corresponds to $\lambda \sim 300$ μm), the condition is more relaxed, leading to $l_c \sim 1.5$ mm, which is already in the bulk thickness range of materials.

For difference-frequency generation in the case of phase-matching and neglecting the pump-light absorption, the visible-to-THz conversion efficiency is given by [9]

$$\eta_{\text{THz}} = \frac{\omega_{\text{THz}}^2 d_{\text{THz}}^2 L^2 I_0}{2\varepsilon_0 c^3 n_0^2 n_{\text{THz}}} \exp(-\alpha_{\text{THz}} L/2) \frac{\sinh^2(-\alpha_{\text{THz}} L/4)}{(-\alpha_{\text{THz}} L/4)^2}, \quad (10.11)$$

where

$$d_{\text{THz}} = \frac{1}{4} n_0^4 r \quad (10.12)$$

Table 10.1 Organic and inorganic nonlinear optical materials that have been investigated for optical-to-THz frequency conversion and their most relevant parameters*. Where possible, the parameters close to the velocity-matched optical wavelengths and THz frequencies are given.

	n_o	n_g	n_{THz}	r (pm/V)	d_{THz}[a] (pm/V)	FOM_{THz}[b] (pm/V)	ν_{phonon} (THz)	α_{THz} (cm^{-1})	Φ_{sat} (mJ/cm^2)	GVM (nm)	λ (nm)
DAST	2.13	2.3[c]	2.26	47	240	5600	22	20	>100	1500	1500
DSTMS	2.13	2.3	2.26	49	250	6100	22	15	>100	1500	1500
OH1	2.16	2.33	2.28[d]	52	280	7400	8	2[f]	60	1350	1350
LAPC[e]	1.6	1.8	1.7	52	85	1700	>17	15			1500
GaAs	3.37	3.61	3.63	1.6	52	66	7.6	0.5		1420	1560
ZnTe	2.83	2.18	3.16	4	64	160	5.3	1.3	0.01	820	840
InP	3.2	3.16	3.54[f]	1.45	38	40	10				
GaP	3.12		3.34	1.0	24	17	10.8	0.2			100
ZnS	2.3		2.88	1.5	10	7	9.08				
CdTe	2.82		3.24	6.8	110	470		4.8			
LiNbO$_3$	2.2	2.18	4.96	28	160	1100		17	>60		

[a] $d_{THz} = \frac{1}{4}n_0^4 r$

[b] $FOM_{THz} = \frac{d_{THz}^2}{n_0^2 n_{THz}} = \frac{n_0^6 r^2}{16 n_{THz}}$

[c] $\nu > 1.5$ THz

[d] $\nu > 1.9$ THz

[e] LAPC guest–host polymer [27]

[f] $\nu \approx 1$ THz

* Refractive index n_0 at the pump optical wavelength λ; group index n_g at λ; refractive index n_{THz} in the THz frequency range; the electro-optic coefficient r; the susceptibility a_{THz} for THz-wave generation; figure-of-merit FOM_{THz} for THz generation by optical rectification; optical phonon frequency of the material ν_{phonon} in the THz range; the absorption in the THz frequency range α_{THz}; pump pulse fluence Φ_{sat} at which saturation has been observed; best wavelength for group-velocity matching, GVM.

is the nonlinear optical susceptibility for THz-wave generation, ω_{THz} the angular frequency of the generated THz wave, L the length of the THz-generation material, I_0 the pump intensity, α_{THz} the absorption constant at the THz frequency, r the electro-optic coefficient, n_0 and n_{THz} the refractive indices at the pump optical and the generated THz frequencies, respectively. Besides phase-matching and minimal THz absorption, the main material figure-of-merit for THz generation (FOM$_{THz}$) according to Eq. (10.11) is

$$FOM_{THz} = \frac{d_{THz}^2}{n_0^2 n_{THz}} = \frac{n_0^6 r^2}{16 n_{THz}}. \qquad (10.13)$$

Table 10.1 shows most of these parameters for a series of inorganic and organic crystals, as well as for an electro-optic polymer. As can be seen in this table, the organic crystals DAST, DSTMS, and OH1 show the largest figure-of-merit and can also be phase matched using pump lasers at telecommunication wavelengths 1.3–1.55 μm. In addition, OH1 shows a very small absorption constant at THz frequencies, thus allowing large interaction lengths to be used [10]. The optical damage threshold of these organic

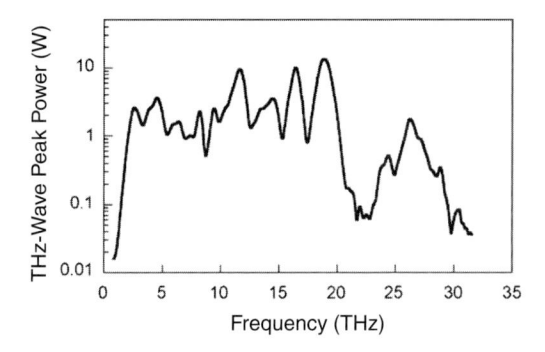

Figure 10.7 THz-wave peak power spectrum generated in a 1-mm-thick DAST crystal using difference-frequency generation (from Takahashi et al. [20], reprinted with permission from Elsevier). A coherent, widely tunable THz wave in the broad range of 2–31.5 THz with a high peak power has been generated [20].

crystals mainly depends on the optical quality, both of the bulk crystal quality and the quality of surface polishing. Very slow cooling growth with good temperature stability of (± 0.002 °C) has to be used for high damage threshold materials reaching $I_{damage} > 150$ GW/cm^2 for 150 fs pulses at 1550 nm [11].

By using the process of difference-frequency generation, frequency-tunable THz-wave generation in DAST crystals under or close to the phase-matched conditions has been demonstrated [12 to 25]. The accessible THz wavelength range is intrinsically limited by the absorption of the nonlinear optical material employed. While with inorganic materials it is difficult to generate waves above 5 THz owing to strong optical phonon resonances, THz waves reaching frequencies above 30 THz with a high peak power can be generated in DAST, as shown in Fig. 10.7 [20], and up to at least 10 THz in OH1 [26].

For the THz range larger than several THz and for the far-IR range, a slightly non-collinear difference-frequency generation geometry may be used to achieve exact phase-matching conditions [21]. Figure 10.8 shows the generated THz spectra from DAST crystals using different non-collinear angles between signal and idler beams. In this and the majority of other literature examples, as-grown plate type organic crystals are used without further processing beside polishing.

It is therefore straightforward to produce narrowband THz waves by difference-frequency generation in organic nonlinear optical crystals, which are phase matched without a need for special crystal cuts, at least for some pump-wavelength ranges and thinner crystals [12 to 23], but also can be specially cut to optimize the phase matching and therefore the efficiency [24,25]. It may be more challenging to develop an appropriate pump-beam pair with a desired separation in the THz frequency range. This is usually done by designing complex OPO systems, usually with a tuning option to tune the generated THz frequency [12 to 21, 23]. These systems are still very bulky and sensitive at present. Much more compact generation systems can be achieved by using optical rectification, as described in the following section.

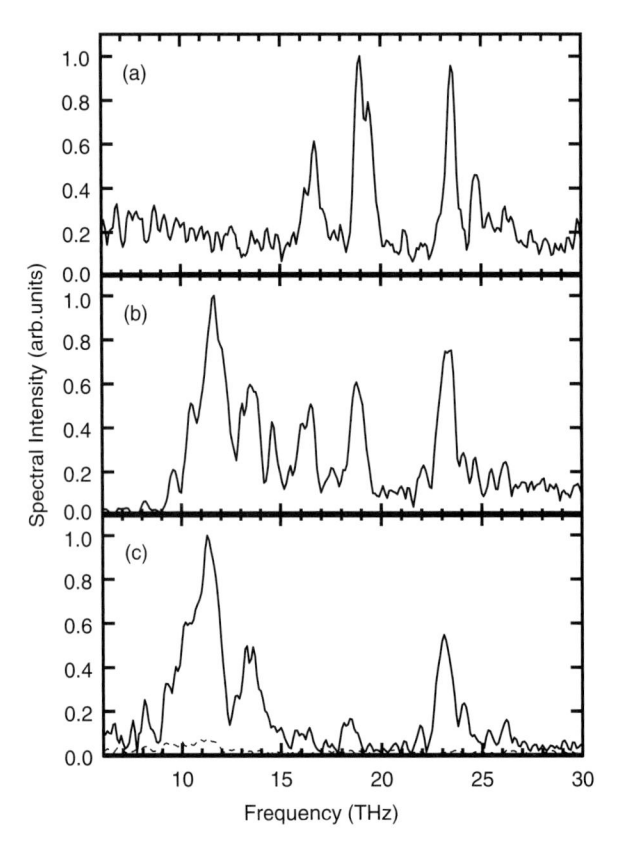

Figure 10.8 Spectra of pulses generated in a 0.5-mm-thick DAST crystal by difference-frequency generation for non-collinear angles between signal and idler of (a) $\theta = 1.4°$, (b) $\theta = 0.7°$, and (c) $\theta = 0.°$ with the center wavelengths of the signal beams at (a) 1514, (b) 1537, and (c) 1548 nm (from Satoh et al. [21], reprinted with permission from OSA).

10.2.2 THz-wave generation by optical rectification

By using the process of optical rectification (OR), broadband THz radiation can be efficiently generated in non-centrosymmetric NLO materials pumped by femtosecond pulses. An ultrashort laser pulse (10–200 fs) induces a quasi-static polarization in such materials through optical rectification, which follows the amplitude of the pump pulse and thus acts as a source for the THz pulse. In other words, a short laser pulse has an intrinsically broad bandwidth, i.e. a laser beam with a pulse length of 10–200 fs has a bandwidth of roughly 2–40 THz, depending on the pulse shape. Different frequency components in such a pulse can mix with each other in a nonlinear crystal by difference-frequency generation, producing a broadband THz wave.

THz generation by optical rectification can be therefore understood as mixing of two frequency components in the frequency spectrum of the ultrashort optical pump pulse in the electro-optic material. A nonlinear polarization is generated at all difference frequencies, which gives rise to a THz pulse, limited by the bandwidth of the optical pump pulse, and the relevant materials parameters of the nonlinear material. The nonlinear polarization propagating along the z-axis far away from electronic resonances is given by [27]

$$P_{OR}(z,\omega_{THz}) = 2\varepsilon_0 d_{THz} \int_{-\infty}^{\infty} E(z,\omega) \cdot E(z,\omega - \omega_{THz}) d\omega = \frac{2d_{THz}}{n_0 c} I(z,\omega_{THz}), \quad (10.14)$$

where $I(z, \omega_{THz})$ is the autocorrelation of the optical electric field or the Fourier transform of the intensity profile of the optical pulse $I(z,t)$. The nonlinear polarizations scales with the second-order NLO susceptibility $\chi^{(2)}(-\omega_{THz}, \omega, \omega_{THz} - \omega) = 2d_{THz}$ of the material, where ω_{THz} is the THz angular frequency and ω the optical angular frequency, and is assumed to be constant over the range of the generated THz frequencies.

Dispersion and absorption of the nonlinear material at THz and optical frequencies make it difficult to solve the nonlinear wave equation

$$\left(\frac{d^2}{dz^2} + \hat{n}_{THz}^2 \frac{\omega_{THz}^2}{c^2}\right) E_{THz}(z,\omega_{THz}) = -\omega_{THz}^2 \mu_0 P_{OR}(z,\omega_{THz}) \quad (10.15)$$

with $P_{OR}(z, \omega_{THz})$ from Eq. (10.14) analytically; $\hat{n}_{THz}$ is the complex refractive index at THz frequency (i.e. including the absorption). However, under certain simplifying assumptions, analytical solutions for the THz fields $E_{THz}(z, \omega_{THz})$ have been obtained [28, 29]. These solutions show that the THz field $E_{THz}(z, \omega_{THz})$ has a maximum for $n_g(\omega) = n_{THz}$, which is, in analogy to phase matching, called velocity matching: the group optical index $n_g(\omega)$ of the pump optical pulse at the angular frequency ω should ideally match the refractive index n_{THz} at the THz frequency. Owing to the low dielectric constant of organic materials, velocity matching between the pump pulse and the generated THz wave is possible, and this, in addition to the high second-order NLO susceptibilities, is essential to achieve high generation efficiencies. In analogy to phase matching, the coherence length for velocity matching is defined as

$$l_{c,VM} = \frac{\pi c}{\omega_{THz}(n(\omega_{THz}) - n_g(\omega))} = \frac{\lambda_{THz}}{2(n_{THz} - n_g)}. \quad (10.16)$$

THz-wave generation by using fs pump pulses is illustrated in Fig. 10.9 for the case of velocity-mismatched and velocity-matched generation, which is clearly indicating the advantage of velocity matching.

Further simplifications of the solution of the wave equation can be made by assuming the velocity matching is fulfilled. We can also assume a Gaussian intensity profile of the pump beam

$$I(\rho,t) = I_c(t) e^{-\frac{\rho^2}{2s^2(z)}}, \quad (10.17)$$

where $I_c(t)$ is the time-dependent intensity at the center of the beam, ρ is the radial distance from the center, and $s(z)$ the beam size parameter. For a crystal length $L \ll s^2(z)/\lambda$, i.e. for crystal lengths shorter than the Rayleigh length of the optical

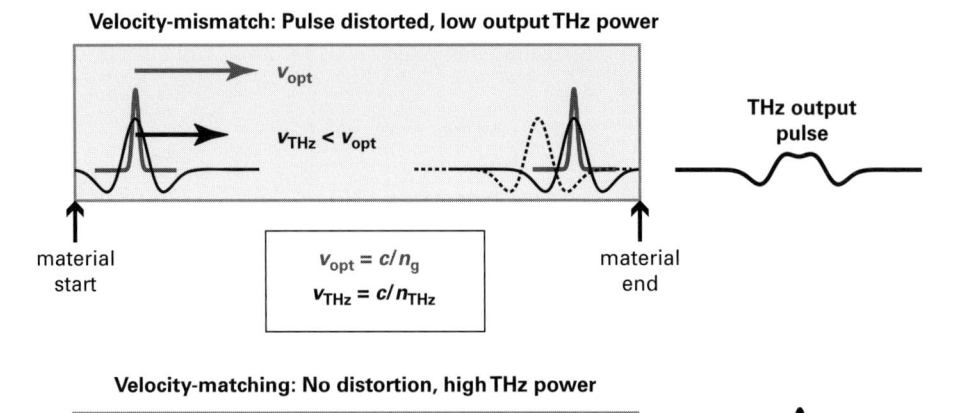

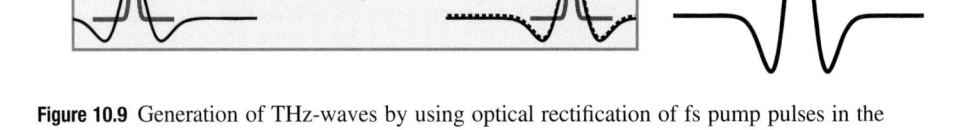

Figure 10.9 Generation of THz-waves by using optical rectification of fs pump pulses in the case of velocity-mismatching and velocity-matching. The longer the propagation distance, the higher the effect of velocity-matching.

beam (e.g. $\lambda = 1\,\mu m$, $L = 0.5\,mm$, $s \ll 23\,\mu m$), neglecting the absorption at optical and THz frequencies, and assuming velocity-matching, we obtain, at the end of the crystal $z = L$ and at the beam axis $\rho = 0$, the following spectrum of the THz pulse [29]

$$|E(\omega_{THz}, L)| = \frac{\mu_0 d_{THz}}{c n_0 n_g} s^2 I_0(\omega_{THz}) \omega_{THz}^2 \operatorname{arsinh}\left(\frac{cL}{s^2 \omega_{THz}}\right) \quad (10.18)$$

and for the intensity of the THz wave, considering further $cL \ll s^2 \omega_{THz}$

$$I(\omega_{THz}, L) = \frac{1}{2}\varepsilon_0 c n_{THz} |E(\omega_{THz}, L)|^2 \quad (10.19)$$

$$\approx \frac{d_{THz}^2 I_0^2 \omega_{THz}^2 L^2}{2\varepsilon_0 c^3 n_0^2 n_{THz}}. \quad (10.20)$$

The conversion efficiency is then given by

$$\eta = \frac{I(\omega_{THz}, L)}{I_0} = \frac{d_{THz}^2 I_0 \omega_{THz}^2 L^2}{2\varepsilon_0 c^3 n_0^2 n_{THz}} = \frac{1}{2\varepsilon_0 c^3} \operatorname{FOM}_{THz} \omega_{THz}^2 I_0 L^2, \quad (10.21)$$

which has the same form as the expression for difference-frequency generation – Eq. (10.11) with $\alpha_{THz} = 0$ – and the figure-of-merit FOM_{THz} defined in Eq. (10.13) and listed in Table 10.1 for a series of nonlinear optical materials. For more general expressions for the optical and THz-wave absorption, as well as velocity mismatching, see e.g. Ref. [29].

Figure 10.10 illustrates how velocity matching can be achieved in DAST. Owing to the small absorption and the corresponding dispersion near 1.1 THz, phase matching

10.2 Terahertz-wave generation

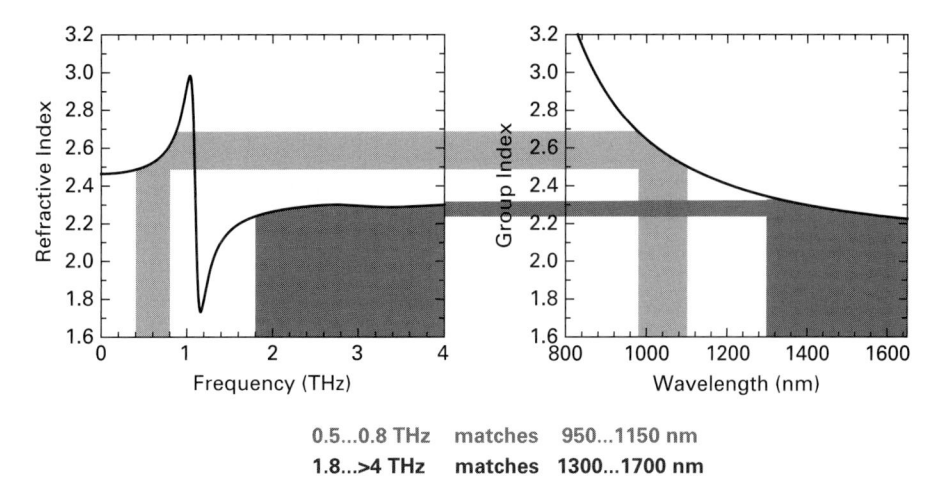

0.5...0.8 THz matches 950...1150 nm
1.8...>4 THz matches 1300...1700 nm

Figure 10.10 Refractive index dispersion in the optical and the group index dispersion in the THz frequency range for DAST, indicating the optimal ranges for velocity matching.

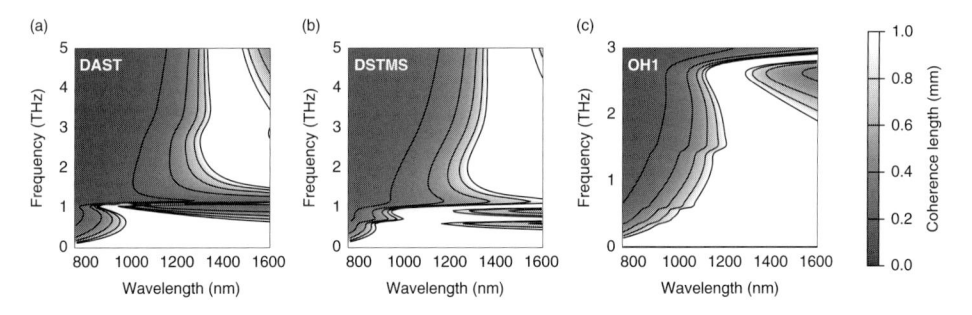

Figure 10.11 Coherence length for velocity matching $l_{c,VM}$ for THz generation by using optical rectification in (a) DAST, (b) DSTMS, and (c) OH1 as a function of the pump optical wavelength and the generated THz frequency. The white area represents the range with best velocity matching leading to coherence lengths larger than 1 mm [29].

can be realized either for pump wavelengths between 830 nm and 1300 nm, yielding THz waves between 0.2 THz and 1 THz, or for pump waves with wavelengths between 1300 nm and 1700 nm for the generation of THz waves above 1.8 THz [29,30].

Best conditions for velocity matching in DAST are also illustrated in Fig. 10.11(a), which shows the coherence length $l_{c,VM}$ as a function of the pump optical wavelength and the generated THz frequency. The generation of THz waves with frequencies around ~1.1 THz is limited in DAST owing to a transverse optical phonon [31]. Other organic materials have different velocity matching conditions and can be used for an available pump wavelength and the desired THz frequency range [32 to 37]. Figures 10.11(b) and (c) show similar plots for some newer materials, DSTMS [38] and OH1 [10]. In DSTMS, the optical phonon absorption near 1.1 THz is suppressed by a heavier counter-anion (see Fig. 6.5) and therefore this material is superior to DAST within this frequency range [38]. OH1 crystal is based on hydrogen bonds and has even

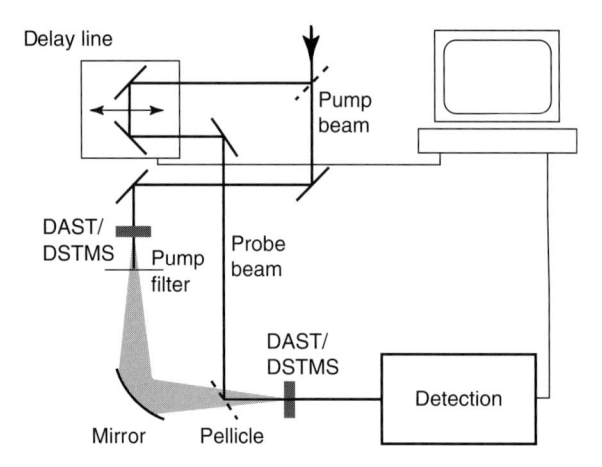

Figure 10.12 Scheme of the set-up for generation and detection of THz pulses.

higher figure-of-merit for THz-wave generation (see Table 10.1) compared with DAST and DSTMS and optimum velocity-matching between 1200 nm and 1460 nm for 0.3–2.5 THz, has no absorption at about 1 THz but at about 3 THz [10]. THz frequencies from 0.5 THz to 10 THz have been generated in OH1 by difference-frequency generation [26].

THz-wave generation by optical rectification in organic materials has been first demonstrated using DAST in 1992 [39] and later on investigated by several groups (see e.g. Refs. [20,24,29,31, 40 to 45]), extended also to other organic crystals [10, 32 to 38, 46 to 48] and polymers [27, 49 to 53]. An example of the experimental set-up used for THz generation by optical rectification is illustrated in Fig. 10.12. The incoming infrared laser pulse is split into a pump beam and a probe beam. The pump beam is used to generate a THz pulse in e.g. a DAST crystal and is then blocked by a material transparent to THz radiation. The THz pulse is focused by an ellipsoidal mirror onto a second DAST crystal, where it induces a temporally and spatially varying change in the refractive index through the linear electro-optic effect. This refractive index profile can be read out with the coinciding probe beam through the technique of THz-induced lensing [54], which is an alternative to the usual electro-optic sampling techniques that work for non-birefringent materials. Time-resolution – and therefore coherent detection – is achieved by varying the path length of the probe beam with a computer-controlled delay line.

THz generation efficiency of optical rectification in DAST is under similar conditions in the velocity-matched regime more than one order of magnitude higher compared with that of the commonly employed inorganic semiconductor ZnTe [29]. Using 160 fs pump pulses with a pulse energy of 25 µJ at a wavelength of 1500 nm, broadband THz electric field spectrum up to 8 THz, limited only by the length of the pump pulse, in a 0.6 mm thick DAST crystal with very high electric field strengths of up to 50 kV/cm was generated [30], as shown in Fig. 10.13.

As mentioned above, the extension of the THz spectrum and the electric-field amplitude of the generated THz wave depends, beside the nonlinear optical susceptibilities and velocity matching of the generation and detection crystals (see Table 10.1),

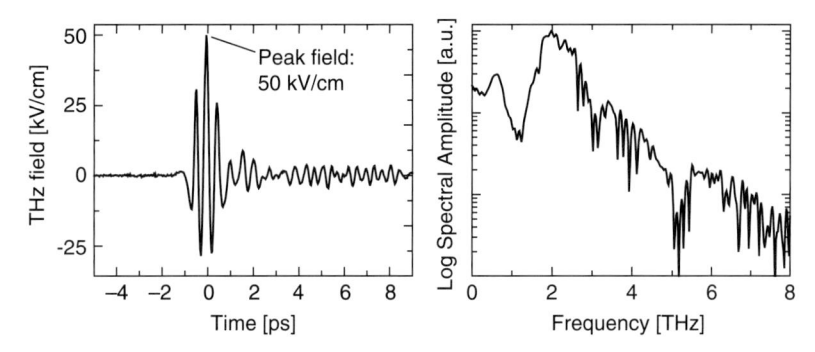

Figure 10.13 THz transient from DAST at a wavelength of 1500 nm (left) and its Fourier transform (right). Peak electric field strength: 50 kV/cm. The oscillations for $t > 0.5$ ps are due to ambient water vapor absorption. Reprinted with permission from Ref. [30].

on the characteristics of the pump source. The shorter the pump pulses, the broader the THz spectrum. For example, using 17-fs pump pulses, it was possible to generate a spectrum extending from 0.1 to beyond 25 THz in DAST crystals by optical rectification [44], while using 5-fs short pulses, the spectrum extended up to 200 THz [55]. By using DAST pumped with the signal wavelength of a powerful optical parametric amplifier, high-power single-cycle THz pulses with electric field strengths as high as MV/cm and magnetic field strengths beyond 0.3 T were generated [45].

Beside organic crystals, poled polymers have also been exploited for the generation of THz waves [27, 49 to 53]. Like organic crystals, poled polymers exhibit parameters close to velocity matching, resulting in coherence lengths of up to several 100 μm, depending on the pump wavelength and the generated THz frequencies. Polymers are attractive for broadband THz generation, since they allow for certain parameters tuning to optimize the phase matching and they do not exhibit gaps in spectra due to phonon-absorption bands as seen in inorganic and organic crystals. On the other hand, it is more challenging to reach larger thicknesses for efficient THz generation in the velocity-matched regime, as well as their damage threshold and long-term stability are considerably lower as for crystals. Nevertheless, poled polymers could represent excellent THz sources for broadband THz spectroscopy provided that thicker polymers with a high effective nonlinearity and low losses can be produced. Figure 10.14 shows terahertz spectra generated by a thin, 160-μm-thick DAST emitter in comparison with a 15-μm-thick poled polymer AJTB203/APC (AJTB203 chromophore – see Fig. 10.14 – in amorphous polycarbonate APC) emitter pumped with 50-fs pulses at 1300 nm [52]. Both spectra extend over a similar THz range and while the DAST crystal provides a higher efficiency owing to its larger thickness, the polymer film produces a gap-free spectrum over the whole bandwidth, limited by the width of the pump pulse.

10.2.3 THz-wave detection

Organic crystals can also be used for the coherent detection of THz waves. The most commonly used technique for detecting THz waves is electro-optic sampling based on a

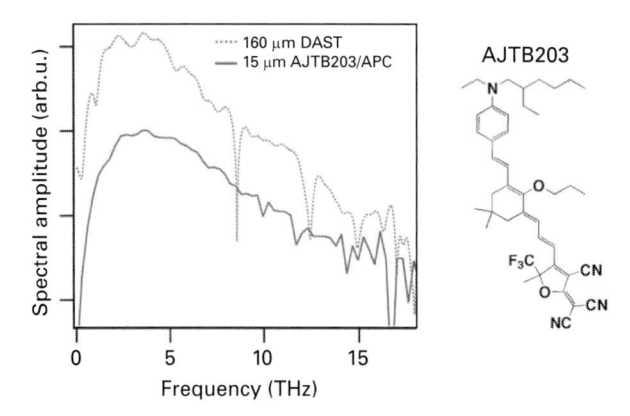

Figure 10.14 THz spectra generated by a DAST crystalline emitter and an AJTB203/APC poled-polymer emitter, pumped with 50-fs pulses at 1300 nm. Both DAST and the poled polymer exhibited an effective electro-optic coefficient of 53 pm/V in this experiment. Reprinted with permission from Ref. [52].

change of polarization state of the probe light in the detecting EO material due to the THz field. This is, however, limited to materials with a relatively small birefringence [56]. For highly birefringent materials such as DAST, an alternative technique was developed, which is based on a lensing effect in the detecting EO material due to changes of refractive index induced by the THz waves [54]. This relatively simple and reliable technique depends on the spatial profile of the refractive index change, induced by the spatial profile of the THz wave given by

$$\Delta n(x, y, t) = -\frac{1}{2} n^3 r E_{\text{THz}}(x, y, t). \quad (10.22)$$

This time-varying refractive index profile acts as a time-varying optical lens and leads to a focusing/defocusing of the probe beam, which is detected by a small-aperture detector or a fiber as a function of the delay of the probe beam with respect to the THz pulse (Fig. 10.15) [54] or by exploiting the effect of the two-photon absorption in a photodiode [57]. A calibration relates the detected intensity changes to the THz field. An example of such a measurement is shown in Fig. 10.16.

10.2.4 Applications of THz waves

Some of the applications of THz waves are related to the unique ability of these waves to excite molecular vibrations and lattice vibrations in the "Reststrahlen" range. In addition, the THz waves show low absorption and are transmitted through most non-conductive homogeneous plastics, paper, cardboard, most clothes, etc., and can therefore detect hidden hazardous substances. Therefore, besides THz spectroscopy of materials, these waves are potentially useful for security applications, but also for the identification of defects in non-conductive materials. For conductive and partially conductive materials, THz spectroscopy can give useful insights into the mechanisms

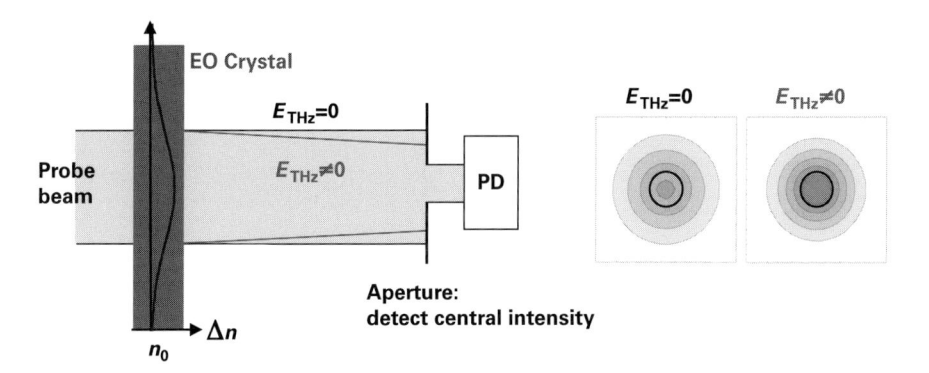

Figure 10.15 Principle of THz-induced lensing: via the electro-optic effect, the THz field induces a refractive index profile in the material, which depends on the profile of the THz beam. If the refractive index increases/decreases in the center, the probe beam becomes focused/defocused, and this is observed by monitoring the power through a small aperture.

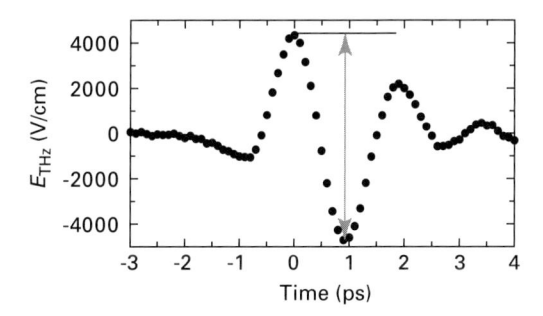

Figure 10.16 An example of the calibrated THz-field temporal waveform using pump pulses at 1125 nm and pulse energy of 48 µJ, detected by THz-induced lensing in an organic crystal.

of charge transport in these materials. Many overviews of possible THz-wave applications can be found in the literature, e.g. in Refs. [8, 58 to 65]. In this section we give some examples of THz spectra and materials testing, using organic nonlinear optical materials for generation and detection of THz waves.

Figure 10.17(a) shows the THz spectra of several explosives as measured with an arrangement similar to that shown in Fig. 10.12. Figure 10.17(b) shows a Semtex explosive sample hidden behind two Teflon plates as seen by optical waves (left) and THz waves (right). The Semtex is revealed as a mid-gray rectangle; the dark arc-shaped region on the upper right corresponds to the paper sticker shown in the left picture). Figure 10.17(c) shows an optical picture and THz image of a sample of *Bacillus cereus* spores (anthrax) hidden in an envelope.

Examples of materials testing are shown in Figs. 10.17(d) and (e). Figure 10.17(d) shows the optical and THz pictures of a pile of overhead transparencies, where the label "ETH" has been cut out in one of the transparencies (not seen in visible light), and its THz transmission image giving a full contrast image owing to the phase shift of the THz wave in the cut-out area (the "defect"). The second picture shows how metallic defects

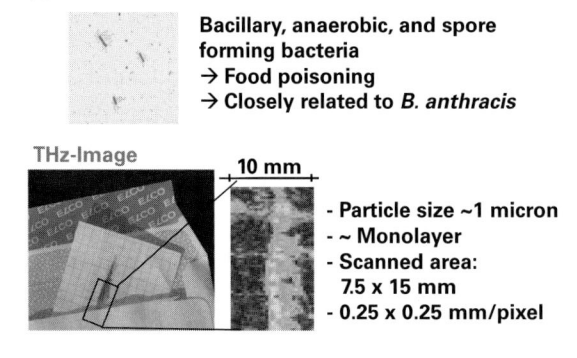

Figure 10.17 Examples of applications demonstrated by using organic nonlinear optical materials for THz generation and detection. For descriptions, see text.

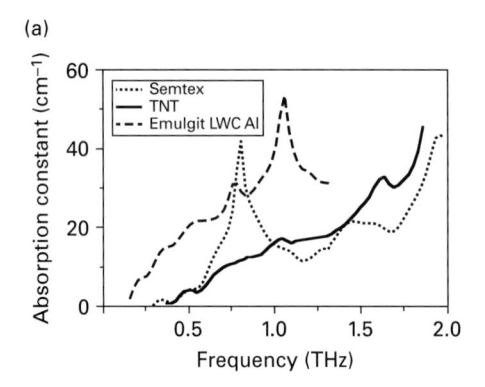

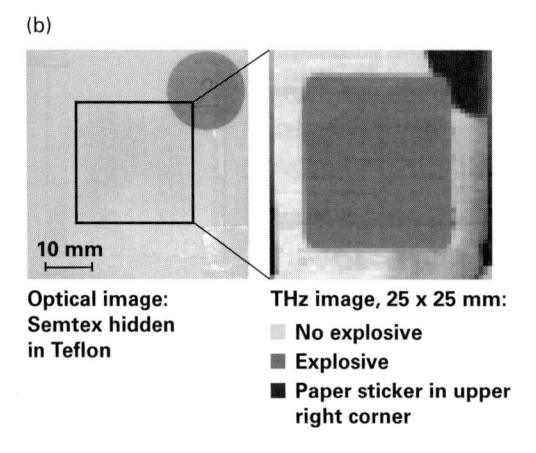

or inclusions (a metal wire with the symbol "NLO") embedded in a plastic can be made visible by THz waves. The last picture in Figure 10.17(d) shows the identification of a void in a piece of plastic. Figure 10.17(e) shows the impressed credit card number from a credit card hidden in an envelope.

10.2 Terahertz-wave generation 247

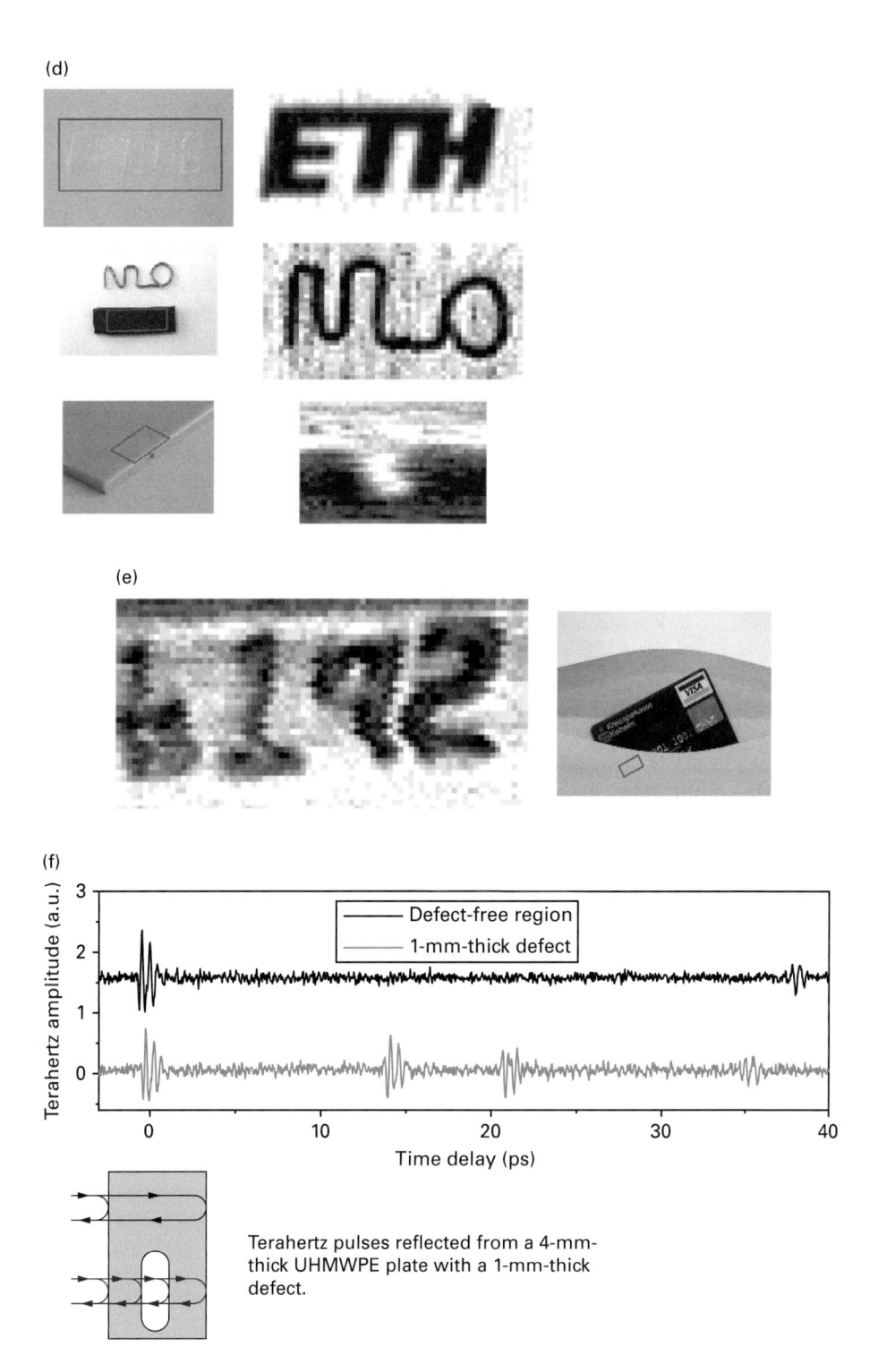

(d)

(e)

(f)

Terahertz pulses reflected from a 4-mm-thick UHMWPE plate with a 1-mm-thick defect.

Figure 10.17 (*cont.*)

Figure 10.17(f) shows examples of THz reflection images of polyethylene samples with and without defects. The voids can be made visible in 3D with a resolution of less than 10 μm. This sub-wavelength (longitudinal) resolution is due to the fact that the phase shift and temporal resolution of the reflected wave can be determined very precisely.

Organic nonlinear optical materials show great potential for such applications, since both the generation and detection efficiencies are superior to those obtained with inorganic materials. Table 10.1 shows a comparison of the most relevant parameters of various materials that have been exploited for nonlinear optical THz frequency conversion.

References

[1] P. Gunter, P. M. Asbeck, and S. K. Kurtz, *Appl. Phys. Lett.* **35**, 461–463 (1979).
[2] C. Bosshard, M. Bösch, I. Liakatas, M. Jäger, and P. Günter, in *Nonlinear Optical Effects and Materials*, P. Günter, Ed., Berlin Heidelberg New York, Springer Series in Optical Science, Vol. **72** (2000), p. 163.
[3] B. Wyncke and F. Brehat, *J. Phys. B* **22**, 363 (1989).
[4] U. Meier, M. Bosch, C. Bosshard, and P. Gunter, *Synth. Met.* **109**, 19–22 (2000).
[5] J. L. Oudar and D. S. Chemla, *J. Chem. Phys.* **66**, 2664 (1977).
[6] L. Mutter, F. D. J. Brunner, Z. Yang, M. Jazbinsek, and P. Gunter, *J. Opt. Soc. Am. B* **24**, 2556–2561 (2007).
[7] D. H. Auston, K. P. Cheung, and P. R. Smith, *Appl. Phys. Lett.* **45**, 284–286 (1984).
[8] M. Tonouchi, *Nat. Photon.* **1**, 97–105 (2007).
[9] R. L. Sutherland, *Handbook of Nonlinear Optics*, New York, Dekker, (2003).
[10] F. D. J. Brunner, O. P. Kwon, S. J. Kwon, et al., *Opt. Express* **16**, 16496–16508 (2008).
[11] Rainbow Photonics AG, www.rainbowphotonics.com
[12] K. Kawase, M. Mizuno, S. Sohma, et al., *Opt. Lett.* **24**, 1065 (1999).
[13] K. Kawase, T. Hatanaka, H. Takahashi, et al., *Opt. Lett.* **25**, 1714–1716 (2000).
[14] T. Taniuchi, J. Shikata, and H. Ito, *Electron. Lett.* **36**, 1414–1416 (2000).
[15] K. Kawase, J. Shikata, and H. Ito, *Solid-State Mid-Infrared Laser Sources* **89**, 397–423 (2003).
[16] T. Taniuchi, S. Okada, and H. Nakanishi, *J. Appl. Phys.* **95**, 5984 (2004).
[17] I. Taniuchi, H. Adachi, S. Okada, T. Sasaki, and H. Nakanishi, *Electron. Lett.* **40**, 549 (2004).
[18] I. Taniuchi, H. Adachi, S. Okada, T. Sasaki, and H. Nakanishi, *Electron. Lett.* **40**, 549–551 (2004).
[19] T. Taniuchi, S. Ikeda, S. Okada, and H. Nakanishi, *Jpn. J. Appl. Phys. 2* **44**, L652 (2005).
[20] Y. Takahashi, H. Adachi, T. Taniuchi, et al., *J. Photochem. Photobiol. A* **183**, 247 (2006).
[21] T. Satoh, Y. Toya, S. Yamamoto, et al., *J. Opt. Soc. Am. B* **27**, 2507–2511 (2010).
[22] M. Tang, H. Minamide, Y. Wang, et al., *Opt. Express* **19**, 779–786 (2011).
[23] M. Koichi, K. Miyamoto, S. Ujita, et al., *Opt. Express* **19**, 18523–18528 (2011).
[24] J. Liu and F. Merkt, *Appl. Phys. Lett.* **93**, 131105 (2008).
[25] J. Liu, H. Schmutz, and F. Merkt, *J. Mol. Spectrosc.* **256**, 61–63 (2009).
[26] H. Uchida, T. Sugiyama, K. Suizu, T. Osumi, and K. Kawase, *Terahertz Sci. Technol.* **4**, 132–136 (2011).

10.2 Terahertz-wave generation

[27] X. M. Zheng, C. V. McLaughlin, P. Cunningham, and L. M. Hayden, *J. Nanoelectron. Optoelectron.* **2**, 58–76 (2007).
[28] J. Faure, J. Van Tilborg, R. A. Kaindl, and W. P. Leemans, *Opt. Quantum Electron.* **36**, 681–697 (2004).
[29] A. Schneider, M. Neis, M. Stillhart, et al., *J. Opt. Soc. Am. B* **23**, 1822 (2006).
[30] A. Schneider, M. Stillhart, and P. Gunter, *Opt. Express* **14**, 5376–5384 (2006).
[31] M. Walther, K. Jensby, S. R. Keiding, H. Takahashi, and H. Ito, *Opt. Lett.* **25**, 911–913 (2000).
[32] O. P. Kwon, S. J. Kwon, M. Stillhart, et al., *Cryst. Growth Des.* **7**, 2517–2521 (2007).
[33] F. D. J. Brunner, A. Schneider, and P. Gunter, *Appl. Phys. Lett.* **94**, 061119 (2009).
[34] K. Miyamoto, S. Ohno, M. Fujiwara, et al., *Opt. Express* **17**, 14832–14838 (2009).
[35] P. J. Kim, J. H. Jeong, M. Jazbinsek, et al., *Cryst. Eng. Comm.* **13**, 444–451 (2011).
[36] J. Y. Seo, S. B. Choi, M. Jazbinsek et al., *Cryst. Growth Des.* **9**, 5003–5005 (2009).
[37] P. J. Kim, J. H. Jeong, M. Jazbinsek et al., *Adv. Funct. Mater.* **22**, 200–209 (2012).
[38] M. Stillhart, A. Schneider, and P. Gunter, *J. Opt. Soc. Am. B* **25**, 1914–1919 (2008).
[39] X. C. Zhang, X. F. Ma, Y. Jin, et al., *Appl. Phys. Lett.* **61**, 3080–3082 (1992).
[40] P. Y. Han, M. Tani, F. Pan, and X. C. Zhang, *Opt. Lett.* **25**, 675 (2000).
[41] J. J. Carey, R. T. Bailey, D. Pugh, et al., *Appl. Phys. Lett.* **81**, 4335–4337 (2002).
[42] K. Kuroyanagi, K. Yanagi, A. Sugita et al., *J. Appl. Phys.* **100**, 043117 (2006).
[43] E. Kwon, S. Okada, and H. Nakanishi, *Jpn. J. Appl. Phys. Part 2* **46**, L46–L48 (2007).
[44] J. Takayanagi, S. Kanamori, K. Suizu, et al., *Opt. Express* **16**, 12859–12865 (2008).
[45] C. P. Hauri, C. Ruchert, C. Vicario, and F. Ardana, *Appl. Phys. Lett.* **99**, 161116 (2011).
[46] K. Akiyama, S. Okada, Y. Goto, and H. Nakanishi, *J. Cryst. Growth* **311**, 953–955 (2009).
[47] T. Matsukawa, Y. Mineno, T. Odani, et al., *J. Cryst. Growth* **299**, 344–348 (2007).
[48] K. Kuroyanagi, M. Fujiwara, H. Hashimoto, et al., *Jpn. J. Appl. Phys.* Part 1 **45**, 4068–4073 (2006).
[49] A. Nahata, D. H. Auston, C. J. Wu, and J. T. Yardley, *Appl. Phys. Lett.* **67**, 1358–1360 (1995).
[50] A. M. Sinyukov and L. M. Hayden, *Opt. Lett.* **27**, 55–57 (2002).
[51] X. M. Zheng, A. Sinyukov, and L. M. Hayden, *Appl. Phys. Lett.* **87**, 081115 (2005).
[52] C. V. McLaughlin, L. M. Hayden, B. Polishak, et al., *Appl. Phys. Lett.* **92**, 151107 (2008).
[53] P. D. Cunningham and L. M. Hayden, *Opt. Express* **18**, 23620–23625 (2010).
[54] A. Schneider, I. Biaggio, and P. Gunter, *Appl. Phys. Lett.* **84**, 2229 (2004).
[55] I. Katayama, R. Akai, M. Bito, et al., *Appl. Phys. Lett.* **97**, 021105 (2010).
[56] Q. Wu and X. C. Zhang, *Appl. Phys. Lett.* **68**, 1604–1606 (1996).
[57] A. Schneider and P. Gunter, *Appl. Phys. Lett.* **90**, 121125 (2007).
[58] E. Knoesel, M. Bonn, J. Shan, and T. F. Heinz, *Phys. Rev. Lett.* **86**, 340–343 (2001).
[59] S. P. Mickan and X. C. Zhang, *Int. J. High Speed Electron. Systems* **13**, 601–676 (2003).
[60] F. Wang, J. Shan, M. A. Islam, et al., *Nat. Mater.* **5**, 861–864 (2006).
[61] W. L. Chan, J. Deibel, and D. M. Mittleman, *Rep. Prog. Phys.* **70**, 1325–1379 (2007).
[62] W. Withayachumnankul, G. M. Png, X. Yin, et al., *Proc. IEEE* **95**, 1528–1558 (2007).
[63] J. B. Baxter and G. W. Guglietta, *Analyt. Chem.* **83**, 4342–4368 (2011).
[64] M. C. Hoffmann and J. A. Fueloep, *J. Phys. D: Appl. Phys.* **44**, 083001 (2011).
[65] R. Ulbricht, E. Hendry, J. Shan, T. F. Heinz, and M. Bonn, *Rev. Mod. Phys.* **83**, 543–586 (2011).

11 Photorefractive effect and materials

While literally the word photorefraction may describe all kinds of photo-induced changes of the refractive index of a material and therefore any photo-induced phase grating would belong to this category, it has become customary in the literature to consider only a smaller class of materials as being photorefractive. These materials possess two important properties: they are photoconductive and exhibit an electro-optic effect. Photoconductivity ensures charge transport, resulting in the creation of a space-charge distribution under inhomogeneous illumination. The electro-optic effect translates the internal electric fields induced by the inhomogeneous space-charges into a modulation of the material refractive index. This is the main mechanism for photorefraction in inorganic and organic crystals [1]. In polymers and liquid crystals, the concept of photorefraction has been expanded to include refractive index changes governed by a field-assisted molecular reorientation of the chromophores. The photorefractive effect is also distinguished from many other mechanisms leading to optically induced refractive index gratings by the fact that it is an intrinsically non-local effect, in the sense that the maximum refractive index change does not need to occur at the spatial locations where the light intensity is largest.

In the strict sense mentioned above, the photorefractive effect was first observed in the mid-1960s by Ashkin and co-workers [2]. They found that intense laser radiation focused on ferroelectric $LiNbO_3$ and $LiTaO_3$ crystals induced semi-permanent index changes. This phenomenon was unwanted for their purposes, and therefore they referred to it as "optical damage". However, the potential of this new effect for use in high-density optical storage of data was soon realized by Chen and co-workers [3]. The effect later became known as the photorefractive effect, and it is understood as a modulated refractive index change. Although the photorefractive effect was first discovered in inorganic-ferroelectric materials, in the past few years highly polarizable and photoconductive organic crystals and polymeric materials with extended π-electron systems have presented themselves as possible candidates for photorefractive applications.

The photorefractive process, which culminates with the formation of the phase grating, is described by the mechanisms shown in the schematic diagram of Fig. 11.1. The three most important properties that a material must fulfill are depicted in the figure in the shaded boxes: optical absorption, charge transport, and electro-optic effect or field-assisted molecular reorientation. Optical absorption and charge transport together give rise to photoconductivity, while the electro-optic effect or molecular reorientation translates the internal electric field into refractive index changes. The mechanisms in the top loop indicated by dashed arrows are also necessary in most materials under low-intensity continuous illumination.

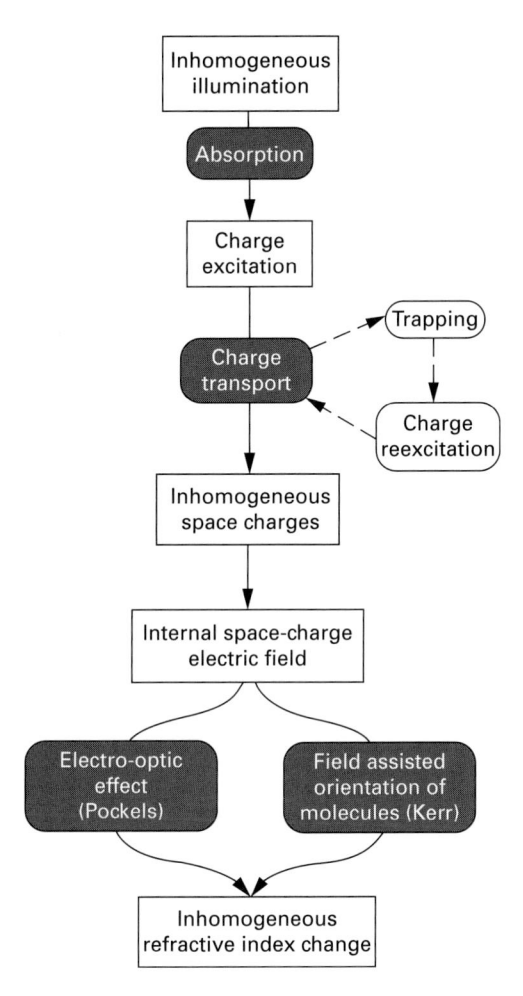

Figure 11.1 The important mechanisms involved in the photorefractive effect and in field assisted chromophore orientation in polymers.

Under these conditions, a large number of trapping sites permits creation of a considerable space-charge modulation amplitude, even though the number of mobile charges is small at any moment in time. Large photorefractive effects may be observed without the necessity of charge trapping by studying the initial response to intense short-pulsed light [4], or, in some conditions, by using continuous light with a wavelength short enough to produce interband photoexcitations [5] and thus create a large number of mobile electrons and holes.

11.1 Theoretical models of the photorefractive effect

11.1.1 Band transport model

Charge photoexcitation from band-gap impurity levels is likely to generate a free electron in the conduction band or a free hole in the valence band (see also

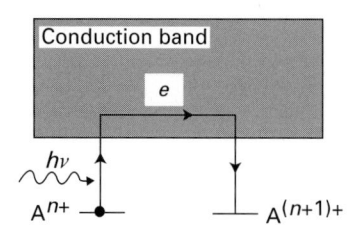

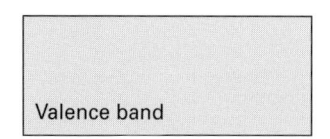

Figure 11.2 Simple band scheme for a photorefractive crystal. A^{n+} indicates an impurity ion (i.e. Fe^{2+}, Cu^+, Cr^{3+}, ...).

Fig. 11.2). These charges quickly relax to the bottom of the conduction band or the top of the valence band and move by a band transport mechanism or as large or small polarons. The direction of collective movement is driven either by drift in an electric field, by random diffusion, or by the photogalvanic effect [6,7].

Crystalline photorefractive materials can be considered as wide band-gap semiconductors containing mid-gap impurity levels. The dynamics of charge redistribution can therefore be described by rate equations similar to those that are common in semiconductor physics. In the general case, there can be a number of deep and shallow defect levels active in the photorefractive process, and numerous theoretical models have been analyzed in the literature; for a review, see for instance [8]. Here we consider only the simplest model with the band scheme shown in Fig. 11.2. It is assumed that there is a single active impurity level between the two bands. The valence of these impurity centers can change between the states A^{n+} and $A^{(n+1)+}$ by excitation/retrapping of electrons to and from the conduction band. Charge transport occurs only in the conduction band by means of drift, diffusion, or the photogalvanic effect. The equations governing charge redistribution and the creation of a space-charge electric field are [9]:

$$\frac{\partial N_D^+(\mathbf{r})}{\partial t} = (sI(\mathbf{r}) + \beta)(N_D - N_D^+(\mathbf{r})) - \gamma n(\mathbf{r})N_D^+(\mathbf{r}) \tag{11.1}$$

$$\frac{\partial n(\mathbf{r})}{\partial t} = \frac{\partial N_D^+(\mathbf{r})}{\partial t} + \frac{1}{e}\nabla \mathbf{J}(\mathbf{r}) \tag{11.2}$$

$$\mathbf{J}(\mathbf{r}) = e\mu n(\mathbf{r})\mathbf{E}(\mathbf{r}) + \mu k_B T \nabla n(\mathbf{r}) + \chi w(\mathbf{r})(N_D - N_D^+(\mathbf{r}))e\mathbf{L}_{ph} \tag{11.3}$$

$$\nabla \mathbf{E}(\mathbf{r}) = \frac{e}{\varepsilon_{eff}\varepsilon_0}\left(N_D^+(\mathbf{r}) - n(\mathbf{r}) - N_A\right), \tag{11.4}$$

where n is the free electron concentration in the conduction band, s is a photoexcitation constant, $I(\mathbf{r})$ is the light intensity distribution, N_D is the total donor concentration, N_D^+ is the concentration of ionized donors, N_A is the concentration of ionized donors in the dark, **J** is the current density vector, **E** is the electric field vector, **L**$_{ph}$ is the photogalvanic drift-vector connected to photovoltaic charge transport [7], $\chi \cong [2\hbar(N_D - N_A)]^{-1}$ is a normalization constant, β is the dark generation rate, γ is the recombination constant, μ is the electronic mobility, ε_0 is the permittivity of vacuum, ε_{eff} is the effective dielectric constant for the given photorefractive configuration, e is the absolute value of the elementary charge, k_B is the Boltzmann constant, T is the absolute temperature, and **r** is the position vector. Finally, the driving quantity in the above equations is the spatially modulated optical intensity $I(\mathbf{r})$, which is locally dissipated for the generation of mobile charge carriers [10].

Equations (11.1 to 11.4) describe the formation of a non-homogeneous charge distribution, and thus of a non-homogeneous electric field as a result of a non-uniform photo-excitation of charge carriers. For the treatment in this section we consider specifically the interference of two coherent plane waves with the absorbed energy given by

$$I(\mathbf{r}) = I_0[1 + m\cos(\mathbf{Kr})] \quad (11.5)$$

where **K** is the wave vector of the interference grating, with |**K**| = $2\pi/\Lambda$, where Λ is the fringe spacing of the interference pattern of the two plane waves leading to Eq. (11.5), and m is the intensity modulation index. The set of Eqs. (11.1 to 11.5) permits calculation of the photo-induced space-charge field **E**(**r**) for any intensity distribution, e.g. for a sinusoidal elementary grating given by Eq. (11.5). The results will be discussed in Section 11.2.

11.1.2 Hopping transport

The band transport mechanism is inadequate for a complete description of charge transport in molecular crystals or polymers. The molecules of organic crystals are usually held together by Van der Waals forces or hydrogen bonds. These are weak interactions, and therefore the electronic states of a molecule in a solid do not strongly differ from the states of the isolated molecule. This results in very narrow conduction bands with widths comparable to the thermal energy k_BT at room temperature. In very pure molecular crystals, the room temperature carrier transport falls into an intermediate category between band-like movement and charge hopping (Fig. 11.3) between localized states [11]. With increasing concentration of impurities, necessary for good photorefractive performance, hopping transport becomes dominant and the transport mobilities μ decrease considerably to values much lower than $\mu \approx 1$ cm^2/(Vs), which may be reached in pure crystalline samples at room temperature.

Disordered unsaturated polymers do not possess any true conduction or valence bands to allow band-like charge transport. The lack of any long-range order precludes the formation of extended electron states. In these materials charge movement is clearly of the hopping type [12]. Charges travel mainly by hopping through the side-chains or guest molecules. In general, most polymers are hole-conducting with carrier mobilities of the order of 10^{-9} to 10^{-6} cm^2/(Vs), much lower than in organic or inorganic crystals. The hopping mobility increases with electric field, and hole mobilities approaching

Figure 11.3 Hopping charge transport.

10^{-3} cm^2/(Vs) [13,14] and electron mobilities of the order of 10^{-5} cm^2/(Vs) [15] have been measured in polymers for fields exceeding 50 V/μm.

As shown by Feinberg et al. [12], for a photorefractive material, charge hopping may be modeled by considering equidistant localized sites at different electric potentials ϕ_i and different local values of the light intensity I_i. Light-induced charge motion is expressed in terms of changes in the probability W_n that site n is occupied as a result of hopping to and from sites $m \neq n$

$$\frac{dW_n}{dt} = -\sum_m D_{mn} W_n I_n \exp\left(\frac{q(\phi_m - \phi_n)}{2k_B T}\right) + \sum_m D_{nm} W_m I_m \exp\left(\frac{q(\phi_n - \phi_m)}{2k_B T}\right), \quad (11.6)$$

where q is the carrier charge and D_{mn} is the transition probability per second and unit time that a carrier in site m hops to site n. It has been shown [12,16,17] that upon appropriate choice of the transition probabilities D_{mn}, a photorefractive model in which charge motion is described by Eq. (11.6) gives results equivalent to the conventional band model presented in Section 11.1.1.

11.2 Steady-state space-charge field

The set of coupled Eqs. (11.1 to 11.4) and (11.5) was solved by Kukhtarev et al. [9] in some limiting cases. Here we concentrate only on the analytic solutions that are obtained in the case of small modulation ($m \ll 1$) and weak coupling. First, we assume no contributions from photogalvanic currents ($\mathbf{L}_{ph}$=0). In this case the steady-state field distribution can be shown to be [18]

$$\mathbf{E}(\mathbf{r}) - \mathbf{E}_0 = -m\hat{\mathbf{k}}\left[\frac{E_q^2 E_0}{(E_q + E_D)^2 + E_0^2}\cos(\mathbf{K}\cdot\mathbf{r}) + \frac{E_q E_D^2 + E_q E_0^2 + E_q^2 E_D}{(E_q + E_D)^2 + E_0^2}\sin(\mathbf{K}\cdot\mathbf{r})\right]$$
(11.7)

In Eq. (11.7), $\mathbf{E}_0$ is an externally applied electric field and $E_0 = \mathbf{E}_0 \cdot \hat{\mathbf{k}}$ is its projection in the direction of the unit vector along the grating $\hat{\mathbf{k}} = \mathbf{K}/|\mathbf{K}|$. The scalar quantities E_D and E_q have the dimension of an electric field and are called the diffusion and the trap-limiting field, respectively. They are defined as

$$E_D = \frac{|\mathbf{K}|k_B T}{e}, \quad \text{and} \quad (11.8)$$

$$E_q = \frac{e}{\varepsilon_{eff}\varepsilon_0 |\mathbf{K}|}\frac{N_{D0}^+(N_D - N_{D0}^+)}{N_D} \equiv \frac{e}{\varepsilon_{eff}\varepsilon_0 |\mathbf{K}|}N_{eff}, \quad (11.9)$$

11.2 Steady-state space-charge field

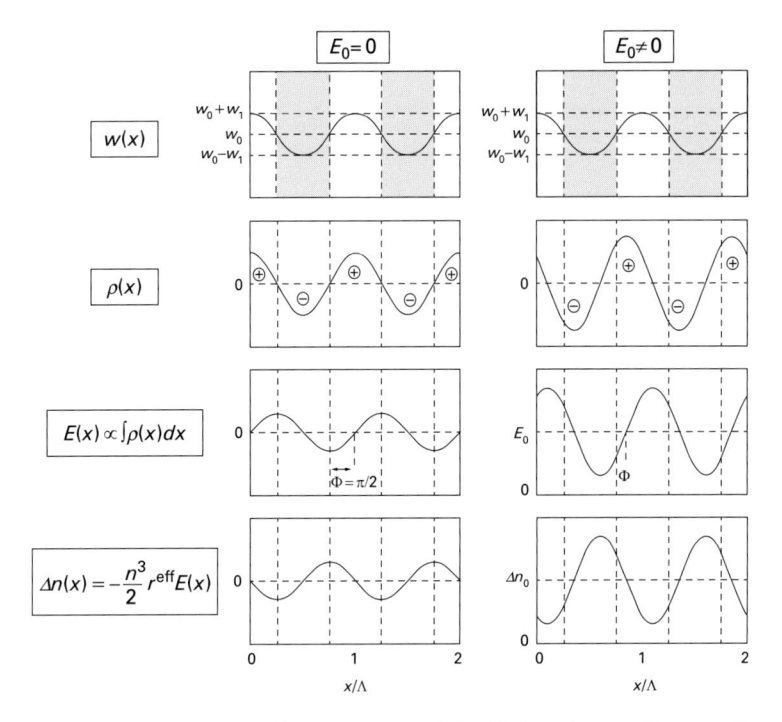

Figure 11.4 Phase relationship between "usefully dissipated energy", space-charge, space-charge field and refractive index distributions.

where N_{D0}^+ is the number of ionized donors in the dark and N_{eff} is an effective trap concentration which become maximal if $N_{D0}^+ = N_D/2$. For the case of no applied field ($\mathbf{E}_0 = 0$), Eq. (11.7) simplifies to

$$\mathbf{E}(\mathbf{r}) = -m\hat{\mathbf{k}} \frac{E_q E_D}{E_q + E_D} \sin(\mathbf{K} \cdot \mathbf{r}). \tag{11.10}$$

For this purely diffusive transport mechanism, the electric-field grating is phase-shifted by $\pi/2$ with respect to the fringes defined by the rate of photo-excitation, as is seen by comparing Eqs. (11.10) and (11.5). Figure 11.4 shows schematically the phase relationship between $I(\mathbf{r})$, the total space-charge ϱ, the space-charge field and the refractive index change for the diffusion-only ($\mathbf{E}_0 = 0$) and drift-assisted cases ($\mathbf{E}_0 \neq 0$).

Figure 11.5 shows the space-charge field amplitude as a function of the grating spacing $\Lambda = 2\pi/|\mathbf{K}|$, as obtained from Eq. (11.10) or (11.8). The maximum for the solid curve ($\mathbf{E}_0 = 0$) is at the Debye grating spacing

$$\Lambda_0 = 2\pi \sqrt{\frac{\varepsilon_{\text{eff}} \varepsilon_0 k_B T}{e^2 N_{\text{eff}}}}, \tag{11.11}$$

for which $E_q = E_D$. In typical inorganic crystals, the Debye grating spacing ranges between approximately 0.3 μm and 1.5 μm.

In Fig. 11.6, we plot again the predictions of Eq. (11.7), this time with the external electric field E_0 as a variable. The unshifted ($\propto \cos \mathbf{Kr}$) and $\pi/2$-phase-shifted ($\propto \sin \mathbf{Kr}$)

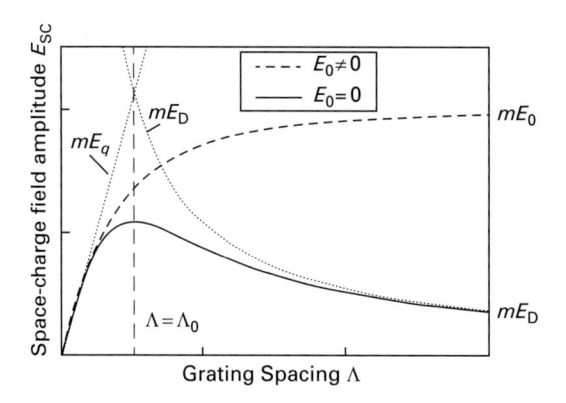

Figure 11.5 Dependence of the space-charge field amplitude on grating spacing.

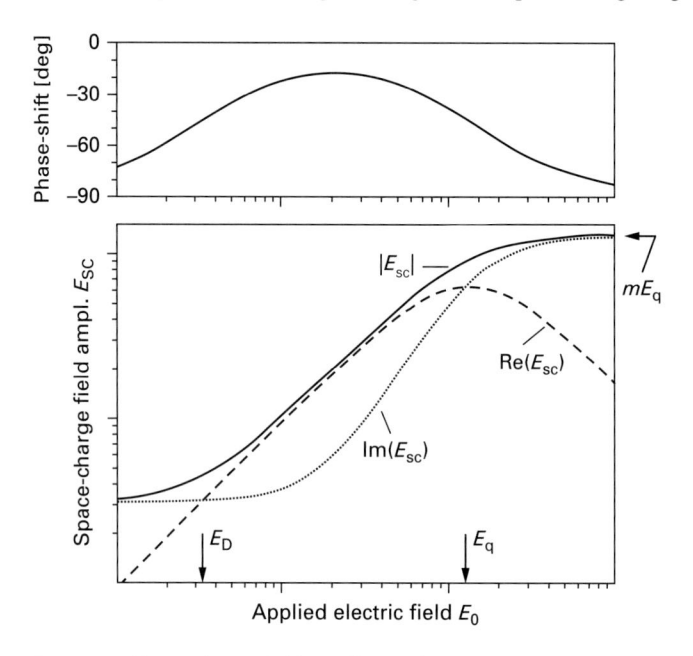

Figure 11.6 Space-charge field amplitude (lower graph) and refractive index grating phase-shift (upper graph) as a function of the applied electric field ($\Lambda = 5$ μm).

components of the space-charge field amplitude are plotted separately together with the total amplitude E_{sc}. They are denoted in the graph as $\mathrm{Re}(E_{sc})$ and $\mathrm{Im}(E_{sc})$. For the grating spacing considered here (5 μm), the unshifted component dominates at intermediate electric fields, but $\mathrm{Im}(E_{sc})$ becomes dominant again when E_0 exceeds the value of the field E_q.

11.3 Space-charge field dynamics

The response time of the photorefractive material can also be calculated from the set of equations (11.1–11.5). The amplitude E_{sc} of the space-charge field evolves exponentially in time as

$$E_{\text{sc}}(t) = E_{\text{sc,sat}}[1 - \exp(-t/\tau)] \quad (11.12)$$

during the build-up of the grating, and as

$$E_{\text{sc}}(t) = E_{\text{sc}}(t=0)\exp(-t/\tau) \quad (11.13)$$

during the grating decay, either under homogeneous illumination, or in the dark.

In the absence of applied fields and photovoltaic currents, the time constant τ is given by:

$$\tau = \tau_{\text{die}}\left[1 + \frac{K^2}{K_e^2}\middle/ 1 + \frac{K^2}{K_0^2}\right], \quad (11.14)$$

where $K = |\mathbf{K}|$, $K_0 = 2\pi/\Lambda_0$,

$$K_e = \sqrt{\frac{e\gamma N_A}{\mu k_B T}} \quad (11.15)$$

is the inverse diffusion length, and

$$K_D = \frac{eE_0}{k_B T}. \quad (11.16)$$

The dielectric relaxation time τ_{die} is expressed by

$$\tau_{\text{die}} = \frac{\varepsilon_{\text{eff}}\varepsilon_0}{e\mu n_0}, \quad (11.17)$$

with the concentration of free electrons n_0 which depends on the average light intensity.

11.4 Model for photo-induced refractive index changes in crystals and polymers

11.4.1 Organic crystals

The presence of an electric field can modify the optical properties of a material. In non-centrosymmetric, piezoelectric materials, the induced change in the refractive index is linear in the electric field. This linear electro-optic effect (commonly known as the Pockels effect) may be expressed mathematically by a third-rank tensor r_{ijk} relating the change of the tensor of the (real) optical indicatrix $(1/\varepsilon)_{ij}$ to the electric field vector $\mathbf{E} = (E_1, E_2, E_3)$ as

$$\Delta\left(\frac{1}{\varepsilon}\right)_{ij} = \Delta\left(\frac{1}{n^2}\right)_{ij} = r_{ijk}E_k, \quad (11.18)$$

where the Einstein summation rule over equal indices is used. The linear electro-optic tensor r_{ijk} is symmetric in the first two indices. A reduced form r_{mk} (with $ij \to m$) is often used for easier tabulation. The symmetry-allowed tensor elements for every crystallographic point group to which inorganic or organic crystals may belong are found in many textbooks [19]. A photo-induced space-charge field therefore changes the indices of refraction according to Eq. (11.18). The photo-induced refractive index change in crystals can be derived from Eqs. (11.18) and (11.7) and in scalar notation is given by:

$$\Delta n = n^3/2 r_{\text{eff}} E(\mathbf{r}), \qquad (11.19)$$

where r_{eff} is an effective electro-optic coefficient which depends on the grating configuration and the relevant tensor components. In Eqs. (11.18) and (11.19), it is assumed that no piezoelectric local strain is induced by the space-charge fields, an assumption which is not valid in general (see Refs. [20 to 24] for a detailed description).

11.4.2 Polymers

The mechanisms leading to photo-induced space-charge fields in electro-optic polymers are similar to the one present in inorganic materials; however, some major differences do exist [25 to 28]. We start by indicating the important constituents of a photorefractive polymer. Figure 11.7 corresponds to the diagram presented in Ref. [25], which shows these constituents and their energy location. During the photorefractive process a hole is created by photo-excitation of a sensitizer molecule. This hole is transported by hopping between charge transport agent (CTA) molecules, until it falls into a trap. The nonlinear optical chromophore (NLO) finally provides the necessary electro-optic response to the generated space-charges, but does not participate in the charge transport. We notice in

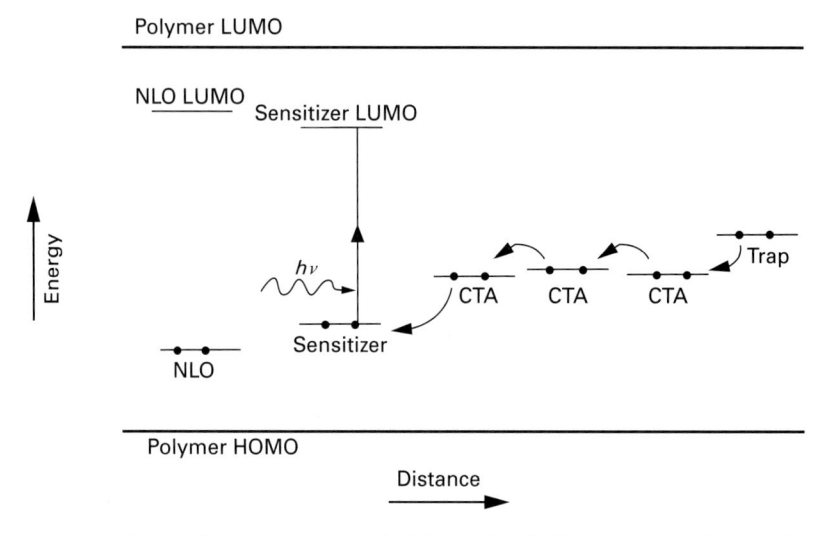

Figure 11.7 The various components of a "photorefractive" polymer and the ideal location of the energy levels which ensures optimum hole transport and low overall absorption. The HOMO–LUMO optical transition energy of the sensitizer agent has to be smaller than those of the nonlinear optical chromophores (NLO) and the host polymer. The charge-transport agents (CTA) are not photo-excited; they form a network over which photo-generated holes can hop into trapping sites. Note that, in this diagram, holes hop to the right [25].

11.4 Modeling refractive index changes

Fig. 11.7 that there is an ideal ordering of the optical absorption energies for the various building blocks discussed above. The absorption energy of the charge generator should be smaller than that of the polymer host, which corresponds to the energy difference between the lowest unoccupied molecular orbital (LUMO) and the highest occupied molecular orbital (HOMO) of the polymer. The charge generator absorption energy also has to be smaller than that of the NLO molecules. This prevents unwanted absorption in channels not participating in the photorefractive effect. To allow the hopping transport of holes in the CTAs, it is important that their energy levels are higher than the levels for the sensitizer and the NLO molecule. This means that the electrochemical oxidation potential E_{ox} has to be smaller for the CTA than for the other molecules [25]. Examples of the most commonly used polymer hosts, sensitizer, CTA, and NLO molecules are given in Section 11.8.2. Here we discuss instead a model for the calculation of the space-charge field in a polymer by means of rate equations, which has been developed in two papers by Schildkraut and co-workers [29,30]. The treatment follows closely the analysis of semiconductor-like crystals described in Section 11.1.1, and we discuss only the key differences from that analysis. Unlike in the original work, here we use dimensioned expressions for all physical quantities.

As mentioned in Section 11.1.1 and in contrast to inorganic materials, the quantum efficiency for charge generation and charge carrier mobility for polymers are highly dependent on the electric field [31]. The quantum efficiency ϕ (see (Section 11.1.2)) is low at small fields owing to geminate recombination, which is encouraged by the small dielectric constants of organics [32]. In Schildkraut's work, an empirical dependence of the form

$$\phi(E) \propto s(E) \propto E^p; \quad s(E) = s_0(E/E_0)^p \qquad (11.20)$$

is assumed for the hole-conducting polymers. E is the local electric field, p is a numeric exponent that has to be determined experimentally, and E_0 is the value of the projection of the applied field in the direction of the grating vector. The functional dependence of Eq. (11.20) is in accordance with the behavior of some organics in the range of electric fields between 10 and 100 V/µm. The hole mobility μ is modeled as

$$\mu(E) = \mu_0 \exp\left[C\left(\left(\frac{E}{E_0}\right)^{1/2} - 1\right)\right] \qquad (11.21)$$

where C is an experimentally determined constant and μ_0 is the value of the mobility at field E_0. There are theoretical and experimental foundations for this kind of dependence. Schildkraut et al. use the Einstein relation $D = \mu k_B T/e$ to relate the diffusion coefficient and the charge mobility. There is evidence that this relationship may not hold for some polymers for high electric fields [33]. However, carrier diffusion is only important in regions of small electric field, where the above relation is always fulfilled.

For polymers it is assumed that there exist some charge-generator molecules G (the sensitizer in Fig. 11.7) which can be photo-ionized to generate a hole. The constant s governing this process is proportional to the quantum efficiency as described above. The inverse process of hole recombination is mediated by a recombination constant γ_R

$$G \underset{\gamma_R}{\overset{sI}{\rightleftharpoons}} G^- + \text{hole} \tag{11.22}$$

where I is the light intensity. In addition to the generator centers, there are some trapping centers T for holes. Trapping is mediated by a trapping constant γ_T, while thermal detrapping can occur through a detrapping rate r

$$T + \text{hole} \underset{r}{\overset{\gamma_T}{\rightleftharpoons}} T^+. \tag{11.23}$$

For recombination and trapping processes, following Langevin theory [34] the constants γ_R and γ_T can be expressed in terms of the carrier mobilities [30], e.g.

$$\gamma_R(E) = \gamma_T(E) = \frac{e\mu(E)}{\varepsilon_{\text{eff}}\varepsilon_0}. \tag{11.24}$$

The Langevin model is appropriate to the case where a mobile carrier and the generation site are both charged, so that a strong Coulomb attraction between them controls the recombination probability when they are within a sphere with radius equal to the Coulomb radius $r_C = e^2/4\pi\varepsilon_{\text{eff}}\varepsilon_0 k_B T$. Since the generator molecule G is usually initially uncharged, this assumption is reasonable for the recombination process [35]. However, Langevin trapping may not be appropriate if the trapping centers T are initially uncharged in the process [36].

Equations (11.20 to 11.24) are used to build a set of rate equations similar to the one valid for crystalline materials; see Eqs. (11.1 to 11.4). The space-charge field distribution under a sinusoidal light illumination of the form shown in Eq. (11.5) was found first using a numerical approach [30]. In a second paper [29], the equations were linearized to give the first Fourier harmonic of the space-charge field grating for some limiting situations. Under the assumption of very deep hole traps, untrapping is negligible ($r = 0$) and in the steady state all traps have been filled. The steady-state space-charge field is then

$$E_{\text{sc}} = mAE_g \frac{B_2 E_0 + B_1 E_D}{B_1^2 + B_2^2} \cos(\mathbf{Kr}) + mAE_q \frac{B_2 E_D - B_1 E_0}{B_1^2 + B_2^2} \sin(\mathbf{Kr}) \tag{11.25}$$

which may be directly compared with the amplitude of the space-charge field for band transport (11.7). Under the further assumption that the average hole density p_0 is small compared with the total concentrations of generator (N_G) and trapping molecules (N_T), the quantities appearing in Eq. (11.25) are expressed as follows.

$$A = \frac{s_0 I_0 + \gamma_T N_T}{\gamma_T N_T} \tag{11.26}$$

is a dimensionless factor, E_D is the diffusion field defined in (11.8), and

$$E_q = \frac{e}{\varepsilon_{\text{eff}}\varepsilon_0 K} \frac{\gamma_T N_T (N_G - N_T)}{s_0 I_0 + \frac{e\mu_0 N_G}{\varepsilon_{\text{eff}}\varepsilon_0}} \equiv \frac{e}{\varepsilon_{\text{eff}}\varepsilon_0 K} N_{\text{eff}} \tag{11.27}$$

11.4 Modeling refractive index changes

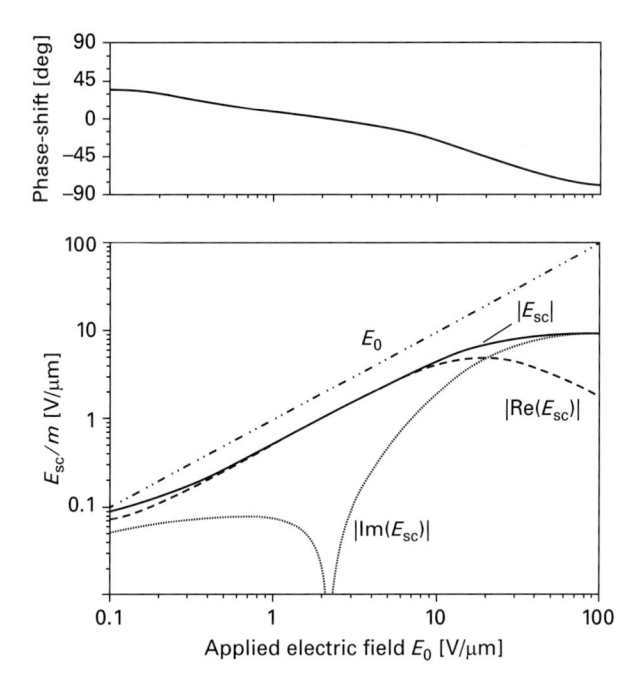

Figure 11.8 Space-charge field amplitude (lower graph) and refractive index grating phase-shift (upper graph) as a function of the applied electric field as obtained using the photorefractive model for polymers described above. Note that the space-charge field never exceeds the value of the applied electric field E_0. The parameters used are listed in the text.

is a limiting field similar to Eq. (11.9), but with the effective number of photorefractive centers N_{eff} which depends on light intensity and on the mobility (owing to Langevin recombination). The quantities B_1 and B_2 are expressed as

$$B_1 = E_D + (1 + Ap)E_q + \left(1 + \frac{1}{2}C\left(\frac{E_0}{E_{\text{ref}}}\right)^{1/2}\right)E_I, \qquad (11.28)$$

and

$$B_2 = E_0 + \frac{E_D E_q}{E_0}\left(\frac{1}{2}C\left(\frac{E_0}{E_{\text{ref}}}\right)^{1/2} - Ap\right), \qquad (11.29)$$

where E_{ref} is a reference value of the applied field for which the constant C in Eq. (11.21) is determined. Finally, E_I is a field similar to E_q

$$E_I = \frac{s_0 I_0}{s_0 I_0 + \gamma_T N_T} \frac{e}{\varepsilon_{\text{eff}}\varepsilon_0 K}(N_G - N_T). \qquad (11.30)$$

If the traps are initially charged, one can assume a Langevin trapping process, thus substituting the constant γ_T with $e\mu_0/\varepsilon_{\text{eff}}\varepsilon_0$ in Eqs. (11.26, 11.27 and 11.29).

In analogy with Fig. 11.6, we plot in Fig. 11.8 the dependence of the real and imaginary components of the space-charge field on electric field. The curves are obtained using Eqs. (11.20 to 11.30) with the parameters $\Lambda = 1$ μm, $E_{ref} = 1$ V/μm, $s(E_{ref}) = 1$ cm^2/Ws, $p = 1$, $I_0 = 1$ W/cm^2, $\mu(E_{ref}) = 10^{-7}$ cm^2/Vs, $C = 3$, $\varepsilon_{eff} = 3$, $N_G = 5 \times 10^{15}$ cm^{-3}, $N_T = 10^{16}$ cm^{-3}, and $T = 295$ K. Some of the details seen in Fig. 11.8 are due to the particular assumptions that have been made and do not all necessarily need to represent real physical phenomena. However, some qualitative conclusions may be drawn. The most important is that, as is the case for conventional photorefractivity in inorganic materials, the 90° phase-shifted component of the space-charge field E_{sc} can become dominant if the applied field is large enough. With few exceptions, photorefractive polymers are always used in a very high-field regime, with E_0 ranging between 100 and 500 V/μm. Experimentally, in polymer composites the phase shift between the light fringes and the refractive index grating can approach 90° for large applied fields (see for instance [35]). Like the one discussed above, any model describing the formation photorefractive gratings in polymers may need a large number of parameters and may be valid only in limiting situations. It is therefore not surprising that most researchers in polymeric photorefractive materials use the conventional band model presented in Section 11.1.1 to explain their results [27]. This might be considered as justified because the most relevant general features are common to the two situations, at least in first approximation, as can be seen for instance by comparing Fig. 11.6 and Fig. 11.8 in the high field limit. However, it is evident that more theoretical and experimental work is needed in order to establish a generally recognized theory for the charge redistribution in polymers.

In materials containing individual polar molecular entities, such as liquid crystals or dipolar nonlinear optical polymers, the polar molecules may reorient under the action of an electric field. If these molecules are strongly birefringent and are present in large quantities, this reorientation results in a large change in the refractive index of the material.

Reorientational effects upon application of an electric field are the base of applications of liquid crystals in display technology [37]. The electric field vector **E** of the optical wave itself can induce a reorientation proportional to **E**^{2}, which may be used to create optical gates [38] or to record phase gratings using the two-wave mixing configuration shown in Fig. 11.9 [39]. The two incident light waves are not symmetric with respect to the alignment of molecules, so that there exist components of the resultant light electric field vector both parallel and perpendicular to the average molecular dipole moment. In this optimum situation, there is no threshold value of the electric field which would have to be exceeded in order to reorient the molecules. The configuration of Fig. 11.9 with oblique beams is always employed for the study of photorefraction in polymers, where considerations of the symmetry of the electro-optic tensor impose this geometry with the grating wave vector **K** not normal to the poling direction.

Electro-optic polymers with an elevated glass transition temperature may be designed for optimum stability, so that upon removal of the poling field their electro-optic and nonlinear optic response will remain stable for years [40]. There is evidence that photorefractive polymers profit from exactly the opposite property. Low glass transition

11.4 Modeling refractive index changes

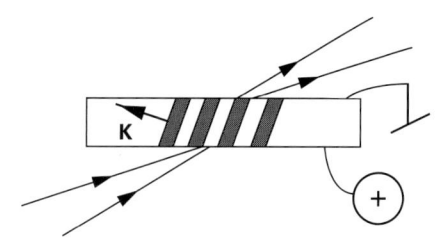

Figure 11.9 Oblique geometry usually used to record phase gratings in nematic liquid crystal cells and polymers. The grating wave vector **K** of the light fringes is not perpendicular to the applied electric field.

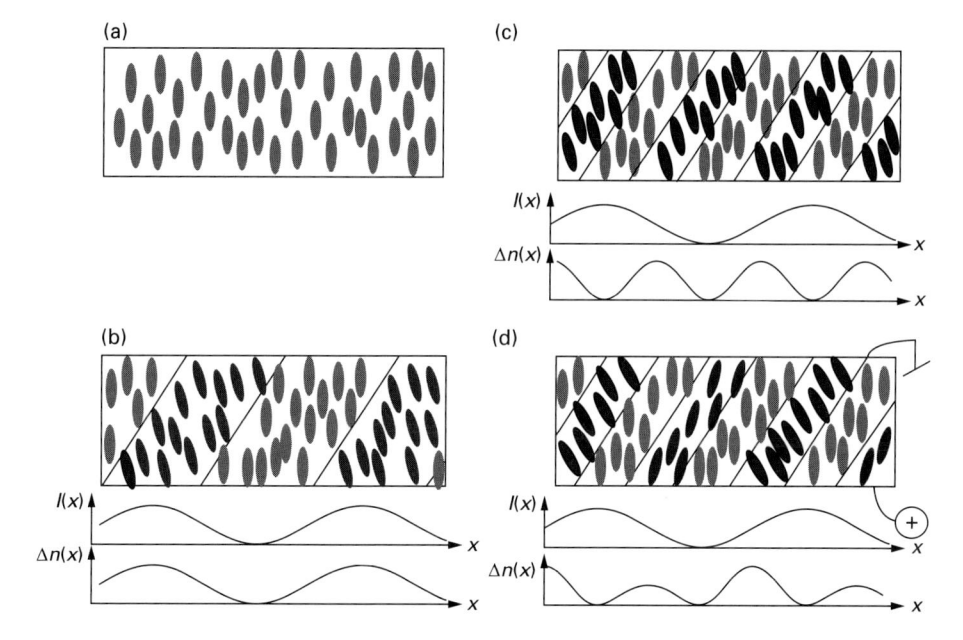

Figure 11.10 Orientational gratings in a homeotropic nematic liquid crystal cell. (a) Cell configuration; (b) reorientation by the electric field vector of two optical waves creating oblique interference fringes; (c) reorientation by a static light-induced space-charge electric field grating, no external electric field applied; (d) like (c) but with externally applied field along the cell normal. The graphs for light intensity $I(x)$ and refractive index change $\Delta n(x)$ are given at the entrance face of the sample. The reorientation of nonlinear optical chromophores in polymers having low T_g is analogous to case (d).

temperature (T_g) polymers with nonlinear chromophore guest molecules show very large photo-induced refractive index changes [41,42]. These changes are much larger than those obtained by the linear Pockels effect and are due to the field-induced rotation of the flexible chromophores. This leads to a refractive index change analogous to the one in nematic liquid crystals, as depicted in Fig. 11.10(d). In analogy with Fig. 11.4, the photorefractive process in such low T_g polymers may be represented by the quantities plotted in Fig. 11.11.

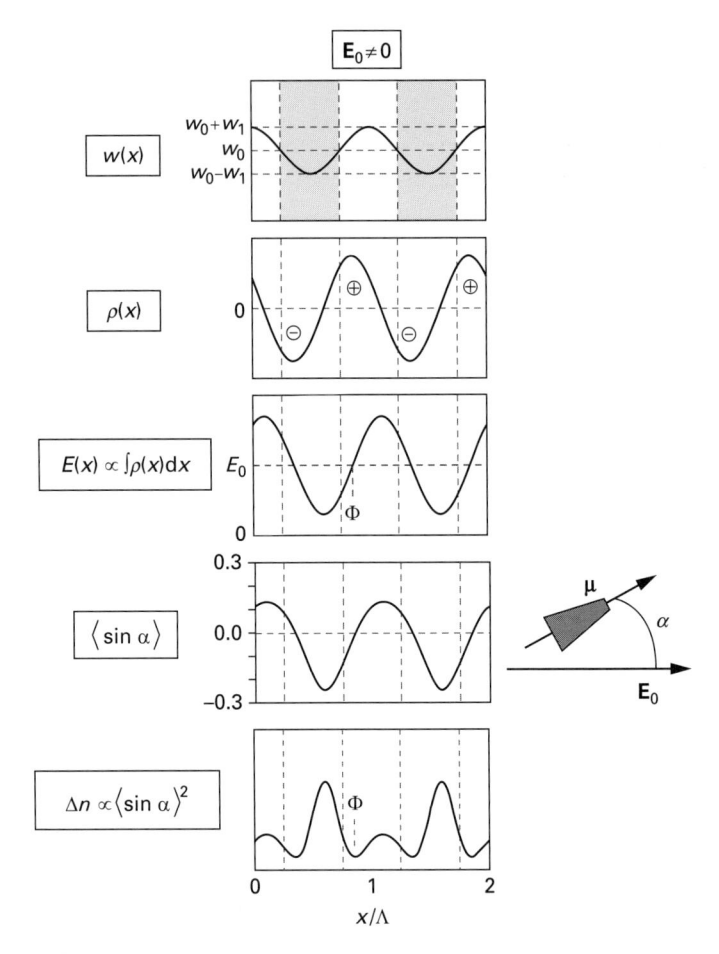

Figure 11.11 Schematic diagram of "usefully dissipated energy", space-charge distribution, space-charge electric field, orientation direction, and refractive index distribution along the direction x parallel to the grating vector $\mathbf{K}$ of photo-assisted orientational gratings in polymers. E_0 is the projection of the external field along $\mathbf{K}$ and α is the angle between the field $\mathbf{E}_0$ and the chromophore dipole moment.

11.5 Measurement of photo-induced refractive index changes

This section describes two principal measuring techniques used to probe the strength of photorefractive gratings. Two-wave mixing probes only the component of the space-charge field which is 90° out-of-phase with respect to the fringes defined by the distribution of dissipated energy. Bragg diffraction, in contrast, probes the total amplitude of the space-charge field, with no information on its phase.

11.5.1 Two-wave mixing

Light amplification by two-wave mixing is one of the key phenomena occurring in photorefractive materials. Figure 11.9 shows a general two-wave mixing configuration.

The signal (S) and reference wave (R) interfere in the material. The nonlinear interaction gives rise to a transfer of energy from one wave to the other for the 90° phase-shifted (with respect to the intensity grating) component of the photo-induced refractive index grating. For the case of a weak signal ($S \ll R$), the amplification of the signal S in the direction of the coordinate s parallel to its direction of propagation (Poynting vector) is given by

$$\frac{\partial S(s)}{\partial s} = \frac{\Gamma}{2} S(s). \tag{11.31}$$

For its intensity one has

$$\frac{\partial I_S(s)}{\partial s} = \Gamma I_S(s), \tag{11.32}$$

and thus

$$I_S(s) = I_S(s=0)\exp(\Gamma s). \tag{11.33}$$

The exponential gain Γ can be calculated as follows

$$\Gamma = \frac{2\pi}{\lambda} n^3 r_{\text{eff}} \text{Im}(E_{\text{sc}}), \tag{11.34}$$

where n is the refractive index and r_{eff} is the effective electro-optic coefficient. The quantity $\text{Im}(E_{\text{sc}})$ is the $\pi/2$-phase-shifted component of the space-charge field amplitude (second term in the bracket of Eq. (11.7)); its electric field dependence has been shown in Fig. 11.6.

A measurement of the intensity changes of the signal beam I_s by adding the reference beam I_R allows one to determine Γ, which according to Eq. (11.34) is a measure of the refractive index change of the $\pi/2$-phase-shifted component of $\mathbf{E}(\mathbf{r})$.

Besides its utility for a number of applications (optical image amplification, optical phase conjugation etc. [1]), the energy transfer by two-wave mixing is also often used to characterize the important parameter N_{eff}, the effective number of trapping centers contained in a material, e.g. by measuring $\Gamma(\Lambda)$ and analysing this dependence with the model equations (11.34) and (11.7).

11.5.2 Bragg diffraction

In a photorefractive Bragg diffraction experiment, a grating is recorded in a similar way to the case of two-wave mixing by interference of two laser beams. A third beam (the readout beam) is then used to probe the grating. Under the appropriate conditions, this beam generates a fourth scattered beam. In most cases, a weak readout beam at a longer wavelength than the two recording beams is used. This approach ensures that the readout process does not influence the grating.

The diffraction efficiency for a weakly absorbing transmission grating, defined as the diffracted wave power divided by the incident wave power inside an isotropic material, is calculated as [1]:

$$\eta = \sin^2\left(\frac{\pi}{2\lambda}\frac{(n_{\text{inc}}n_{\text{diff}})^{3/2}(g_{\text{inc}}g_{\text{diff}})^{1/2}}{(\cos\theta_{\text{inc}}\cos\theta_{\text{diff}})^{1/2}}r_{\text{eff}}E_{\text{sc}}d\right), \tag{11.35}$$

where λ is the vacuum wavelength of the readout light, n is the refractive index, θ is the angle between the sample normal and the Poynting vectors of the incident and probe wave, and d is the sample thickness.

Measuring the diffraction efficiency of the photo-induced grating allows one to determine the amplitude of the refractive index change and the photo-induced space-charge field.

11.6 Applications

Applications of the photorefractive effect have been proposed in a number of areas. They include optical data storage, real-time holography, real-time image processing, edge enhancement, image amplification, phase conjugation, spatial light modulation, and all-optical associative memories [1]. A complete list is beyond the scope of this work; the reader is referred to the large number of references given in [1]. Here we limit ourselves to some considerations on the potential of organic photorefractive materials for optical storage.

11.7 Materials requirements and figures-of-merit

For the different applications, a series of materials properties have to be considered. Among those requirements, the diffraction efficiencies, photo-sensitivities, recording and dark storage times, and other special figures-of-merit are most relevant.

One first important criterion is the dynamic range, i.e. the maximum photo-induced refractive index change Δn_{max} achievable in a material. This quantity is of great importance for holographic storage and for many of the thick grating applications mentioned above. In many cases rather than Δn_{max} it is the grating strength ($\Delta n_{\text{max}}d/\lambda$) which is of practical importance (where d is sample thickness). The total grating strength gives a better measure to judge the achievable diffraction efficiencies and the Γd (two-wave mixing gain × length) products.

For many real-time dynamic applications, it is important not only to achieve a large saturation refractive index change, but also to achieve it fast and at low costs in optical energy. In this case it is desirable to use a material with high sensitivity, i.e. large refractive index change per unit incident or absorbed energy. Four differently defined sensitivities have been used in the literature [11.1a]

$$S_{n1} = \frac{1}{\alpha}\frac{\partial(\Delta n)}{\partial W_0}, \tag{11.36}$$

$$S_{n2} = \frac{\partial(\Delta n)}{\partial W_0}, \tag{11.37}$$

$$S_{\eta 1} = \frac{1}{\alpha d}\frac{\partial \eta^{1/2}}{\partial W_0}, \tag{11.38}$$

$$S_{\eta 2} = \frac{1}{d} \frac{\partial \eta^{1/2}}{\partial W_0}, \qquad (11.39)$$

where W_0 is the incident fluence of optical energy per unit area and Δn is the refractive index change induced by the photorefractive process. The derivatives in the above expressions are taken at the beginning of the recording process.

In order to predict the theoretical performance and the potential for applications of a given material starting from its physical parameters, one can introduce some appropriate figures-of-merit. It has to be kept in mind that different experimental regimes may require a different figure-of-merit, so that materials may not be absolutely compared only in terms of a unique quantity. We discuss here a couple of figures-of-merit which are important in particular situations.

11.7.1 Figures-of-merit for refractive index change

A set of figures-of-merit describes the maximum refractive index change obtainable in a given experimental situation. We recall from Section 11.2 that the space-charge field amplitude in (11.10) is proportional to the field E_q in two different regimes, at small fringe spacings Λ or at large fringe spacing in the case where the applied field E_0 exceeds E_q. In these cases the induced refractive index change is $\Delta n \approx 0.5(n^3 r_{\text{eff}} E_q)$ and one can write a material figure-of-merit

$$Q_1 \equiv \frac{n^3 r_{\text{eff}}}{\varepsilon_{\text{eff}}} N_{\text{eff}}. \qquad (11.40)$$

The effective number of traps N_{eff} depends strongly on the individual crystal sample and often is not considered within the expression for the figure-of-merit, giving the very common expression

$$Q_2 \equiv \frac{n^3 r_{\text{eff}}}{\varepsilon_{\text{eff}}}. \qquad (11.41)$$

In many inorganic crystals the electro-optic coefficients are so high that the space-charge field does not need to approach the value of the field E_q to obtain the desired large nonlinearities. The experiments are then performed at intermediate fringe spacings with only moderate applied fields. The space-charge field is proportional either to the applied field E_0 or to the diffusion field E_D in the case of no applied field. Both are independent of material parameters and the figure-of-merit becomes

$$Q_3 \equiv n^3 r_{\text{eff}}. \qquad (11.42)$$

Polymers are often operated in a regime where the figures-of-merit Q_1 and Q_2 are the limiting factors; inorganic crystals often rely on a regime where Q_3 is important.

11.7.2 Figures-of-merit for the photosensitivity

In analogy with the above figures-of-merit, one can calculate expressions giving the combinations of material parameters for optimum sensitivity. Here we limit ourselves to expressions proportional to the photorefractive sensitivity S_{n1} defined in Eq. (11.36) above. In the limits $\Lambda \ll \Lambda_0$, K_e^{-1} or $E_0 \gg E_q$ the sensitivity is proportional to $(\Delta n/\tau_{\text{die}})$ $(K_e/K_0)^2$ (see Section 11.3, Eqs. (11.7,11.10)) and a material figure-of-merit can be defined as

$$Q_4 \equiv \frac{n^3 r_{\text{eff}}}{\varepsilon_{\text{eff}}} \phi \lambda, \qquad (11.43)$$

where ϕ is the quantum efficiency and λ is the light wavelength. In contrast, in the regime of validity of Eq. (11.36) one has $S_{n1} \propto \Delta n/\tau_{\text{die}}$ and the figure-of-merit is

$$Q_5 \equiv \frac{n^3 r_{\text{eff}}}{\varepsilon_{\text{eff}}} \phi \mu \tau_R \lambda, \qquad (11.44)$$

where μ is the carrier mobility and τ_R is the carrier lifetime.

11.8 Photorefractive materials and their properties

This section summarizes some of the properties of important organic photorefractive materials and their properties which have been investigated in the past. These materials are partially compared with some inorganic crystals in terms of some of the above-defined figures-of-merit for certain applications.

11.8.1 Organic crystals

A potential advantage of organic crystals for photorefractive applications is the possibility of altering the molecular structure for optimizing the electro-optic or nonlinear optical properties. Furthermore, organic materials have smaller dielectric constants than inorganic crystals. Therefore large photorefractive space-charge fields can be created at short grating spacings (large field E_q in (11.9)) even with a crystal sample containing a relatively small number of traps. Some organic crystals show large electro-optic effects, as described in Chapters 3 and 6, and good photoconductivity, and are therefore potentially interesting for photorefraction. Therefore the first observation of photorefraction was made in TCNQ-doped COANP crystals [43].

To our knowledge, the photorefractive effect has been observed only in three different organic single crystals, known by the short names COANP:TCNQ, MNBA, and DAST [43 to 46]. As mentioned above, the first observation of the photorefractive effect in an organic material was reported in 1990 [43,45] in single crystals of 2-cyclooctylamino–5–nitropyridine (COANP) doped with 7,7,8,8–tetracyanoquinodimethane (TCNQ). The material in this study was chosen because of its large nonlinear optical and electro-optic effects. As a result of the doping with TCNQ, the charge transport properties as well as the photorefractive effects were enhanced. The amplitudes of both absorption and phase gratings as well as their phase shifts with respect to the fringe pattern were investigated using an interferometric technique, which confirmed that the observed effect is based on a light-induced charge separation and a refractive index change mediated by the linear electro-optic effect. Photoabsorption or photochemical refractive index changes were excluded as mechanisms for the observed effects. The phase grating build-up in COANP:TCNQ was slow, with build-up times ranging between 30 and 50 min.

In an effort to find an organic material with better photorefractive response, single crystals of 4′-nitrobenzylidene-3-acetamino-4-methoxyaniline (MNBA) were investigated [44]. MNBA belongs to the point group m and has nearly optimized structure for electro-optic applications, with all molecular charge transfer axes lying nearly parallel along the crystallographic a-axis. The largest electro-optic coefficient is $r_{111} = 50$ pm/V at a wavelength of 532 nm [46]. High-quality single crystals can be grown from supersaturated solutions, and (010) faces can be used in optical experiments without prior polishing. Two-wave mixing experiments gave beam coupling gains as high as 2.2 cm^{-1} and response times between 5 and 300 s for writing intensities of 130 mW/cm^2 at visible wavelengths. It was found that the dependence of the response time and two-beam coupling gain on fringe spacing is consistent with the predictions of the single carrier band model.

More recently the crystal DAST [47] was established as the first organic crystal showing net photorefractive gain ($\Gamma = 2.2$ cm^{-1}, $\alpha = 1.8$ cm^{-1} at $\lambda = 750$ nm). The response of this material was found to be relatively fast ($\tau = 500$ ms at 0.3 W/cm^2 intensity).

Potentially, organic crystals combine many of the advantages of polymers and inorganic crystals. Like polymers, they have small dielectric constants and can be designed for optimum properties. Like inorganic crystals, they can be grown in relatively large sizes and can show large electro-optic effects. It might be expected that future optimized organic crystals may compete with the best inorganics for photorefractive applications.

11.8.2 Polymers

The mechanisms leading to photorefractivity in these classes of materials are similar to that in inorganic materials, but a few major differences do exist, as discussed in Section 11.4. There are two potential advantages of polymers with respect to inorganic crystals. First, the ingredients necessary for charge generation, charge transport, trapping and for obtaining an electro-optic response can be introduced into the material quite easily by mixing appropriate agents in the polymer solution. In this way, one can design a wide spectrum of materials with different properties. Second, but no less important, polymer films can be produced at low cost. A third advantage is given by the low dielectric constant, which allows the formation of quite large electric space-charge fields with a relatively small effective number of traps. This advantage, however, is important only in certain experimental regimes, and the large dielectric constants are not an absolute limitation for inorganics. Potential drawbacks of polymers are their small maximum thickness, limited to typically 100–300 μm, the relatively slow time response, and some stability issues in low-T_g compounds related to crystallization [34].

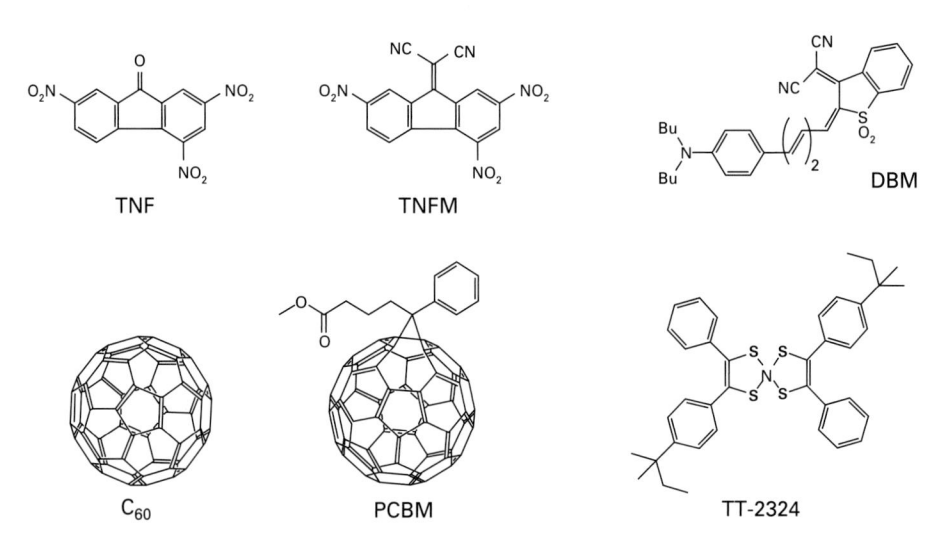

Figure 11.12 Sensitizers used in organic photorefractive materials.

Since the first observation of a photorefractive effect in a not yet optimized polymer in 1991 [48], many papers have been published on this subject [49] and the performance of these materials has greatly increased, as listed in many review publications [25, 28, 50 to 53]. The increase in diffraction efficiencies and two-beam coupling gains has been made possible in large extent by using softer materials with a lower T_g [41,42]. In this regime a quadratic electro-optic effect related to reorientation of the nonlinear chromophores constituents dominates the material response to the space-charge field grating.

Essentially there are four classes of polymers used for photorefractive studies: (i) electro-optic polymers doped with sensitizer and charge transport molecules; (ii) photoconducting polymers doped with electro-optically active chromophores; (iii) inert host polymers used as a binder for all active components; and (iv) fully functionalized polymers having all the necessary components attached to the polymer backbone. The role of all the key constituents of a photorefractive polymer is shown in the energy diagram of Fig. 11.7. The chemical structures of some of the most commonly used molecules and polymers in a photorefractive polymeric material are shown in Figs. 11.12 and 11.13.

Trapping centers represent the first necessary constituents for a photorefractive polymer, and they are usually present in sufficient amount in the material compound. The selection of trapping centers and the control of their concentration have received little attention until now [54,55]. In contrast, the other constituents need to be carefully chosen for material properties optimization.

The second building block is a sensitizer which creates movable charge carriers upon photon absorption. In inorganics, the charge generation process is guaranteed by the presence of impurity dopants, such as Fe, Cu, Co, or Rh. In polymers the charge generator can be introduced by adding appropriate molecules to the polymer solution. TNF (2,4,7-trinitro-9-fluorenone) and C_{60}-fullerenes are often used to this purpose [42,56]. Usually the sensitizing molecules are not attached directly to the

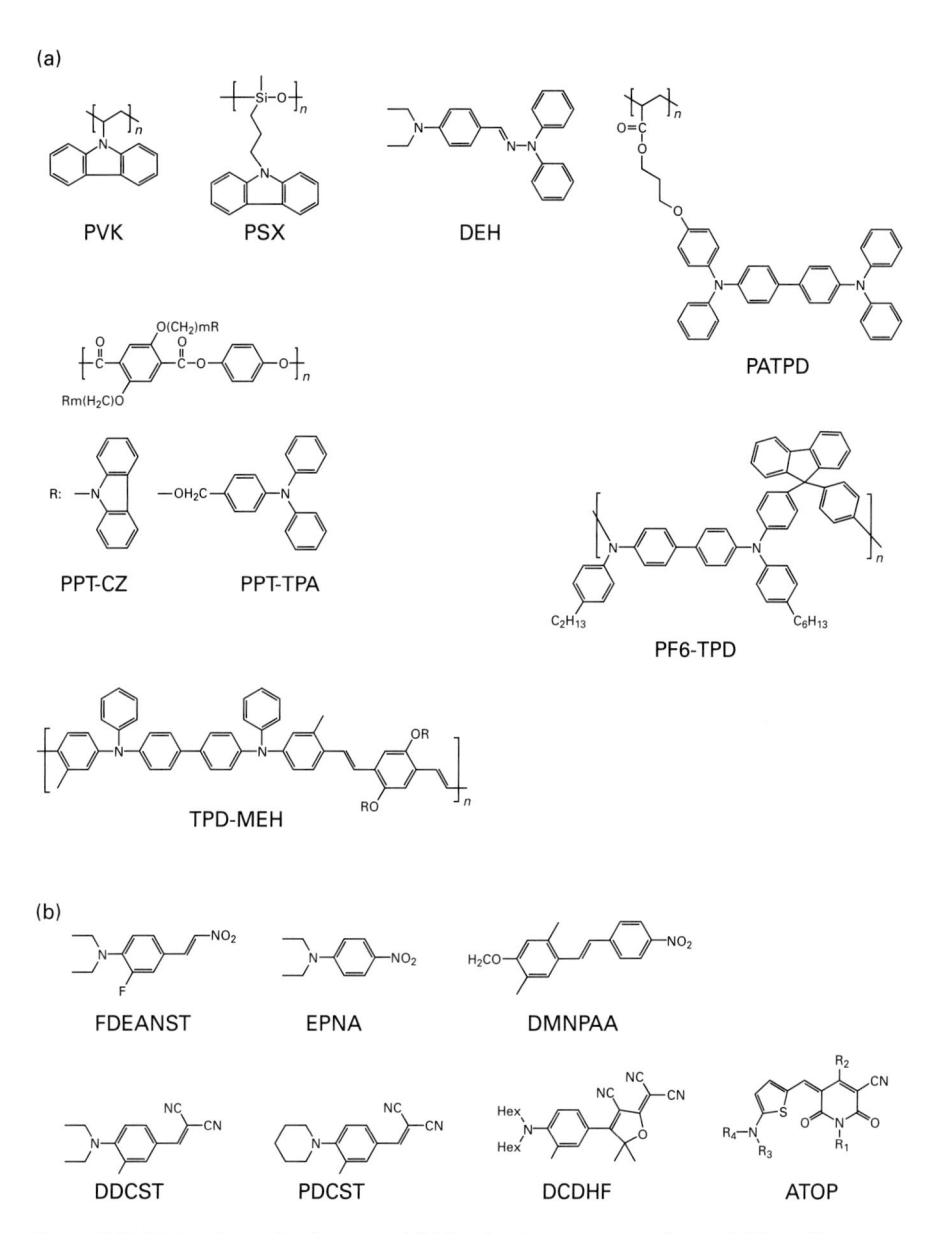

Figure 11.13 Molecules and polymers exhibiting (a) charge-transporting and (b) nonlinear optical properties for photorefractive materials.

polymer backbone. Their concentration is usually in the less than 1.0 wt% region. Recently, for near infrared (NIR) and infrared (IR) sensitivities, new sensitizers were developed. For example, (2,4,7-trinitro-9-fluorenylidene)malononitrile (TNFM) derivatives, which are TNF analogs, and phenyl-C_{61}-butyric acid methyl ester (PCBM) derivatives, which are C_{60} derivatives, were used in the NIR region near 830 nm [57 to 59]. Over the wavelength of 1 μm, the sensitizers Ni-dithiolene complex (TT-2324) or PCBM with hole-transporting PF6-TPD copolymers, where fluorine group and tetraphenylbiphenyl-4,4′-diamine (TPD) group are in main chains, were demonstrated at the wavelength of 1064 nm [60]. The photorefractive polymeric composites based on two photon absorption sensitizer DBM (2-(2-(5-(4-di-n-butylamino)phenyl)-2,4-pentadienylidene)-1,1-dioxido-1-benzothien-3(2H)-ylidene)malononitrile), exhibit high photorefractive performances at wavelengths of 975 and 1550 nm, the latter of which is a telecommunications wavelength [61, 62]. In addition, for use at telecommunication wavelength, quantum dots PbSe and PbS, which exhibit strong absorption from the NIR to IR region, were used as sensitizers at 1340 nm and 1550 nm [63, 64].

The charge created by photoionization of the sensitizer moves by hopping through charge-transport agents (CTA). Very often the molecule (diethylamino)benzaldehyde-diphenylhydrazone (DEH) is used for this purpose [25]. Large concentrations of the order of 30–40 wt% are necessary in order to increase the hopping probability. The alternative to doping with a CTA is to use a polymer host, in which charge transporting groups are in the side-chains or main chain. The most commonly used photoconducting host is polyvinylcarbazole (PVK) [55], which constitutes the host in most of the best-performing photorefractive polymers to date. Polysiloxane (PSX) has also proven to be a very attractive photoconducting polymer for photorefraction, and very high two-wave mixing gains have already been obtained [65]. More recently, charge-transporting triphenylamine, carbazole, and fluorene groups have been incorporated into polymers as side-chains and main chains [50 to 53].

Finally, the nonlinear element that produces a change in the composite refractive index in response to an internal electric field is provided by doping with nonlinear optical chromophores. In some cases the NLO component can also provide the necessary photosensitivity, so that doping with an additional sensitizer is unnecessary. The NLO chromophore may be attached directly to the polymer chain, like in the early experiment by Ducharme *et al.* [49] that used 4-nitro-1,2-phenylenediamine (NPDA) chromophores connected to a bisphenol-A-diglycidylether (bisA) chain. However, the first high photorefractive performances have been achieved using NLOs decoupled from the main polymer chain. In materials with low T_g, these chromophores keep sufficient rotational mobility to enhance the electro-optic response. Chromophores studied until now include 2,5-dimethyl-4-(p-nitrophenylazo)anisole (DMNPAA) [42], 3-fluoro-4-(diethylamino)-(E)-β-nitrostyrene (FDEANST) [41], N,N,diethyl-substituted p-nitroaniline (EPNA) [65], 4-piperidinobenzylidene-malononitrile (PDCST) [28], thiophene-based barbituric acid derivatives (ATOP) [66], tricyanohydrofuran derivatives (DCDHF) [59,67], and many others [50 to 53, 66]. In general, high-concentration doping with NLO chromophores is necessary to achieve optimum performance, but high doping often has the consequence of reduced stability of the compound.

The first high-performance photorefractive polymer was based on PVK:DMNPAA:ECZ:TNF in the ratio 33:50:16:1 wt% [42]. N-ethylcarbazole (ECZ) is a plasticizer added in order to decrease the glass temperature T_g and facilitate the reorientation of the DMNPAA chromophores. This compound exhibits Bragg diffraction efficiencies of nearly 100% and net two-beam coupling gain of 207 cm^{-1} at external electric fields of 90 V/µm in a 105-µm-thick sample. The performance of this compound has been approached or equaled later by other compounds based on PVK or PSX polymers [38, 68 to 72]. Later on, a compound based on the inert polymer poly(methyl methacrylate-co-tricyclodecylmethycrylate-co-*N*-cyclohexylmaleimide-co-benzyl methacrylate) (PTCB) and the NLO dye 2,*N*,*N*-dihexylamino-7-dicyanomethylidenyl-3,4,4,6,10-pentahydro-naphtalene (DHADC-MNP) showed an impressive gain of $\Gamma = 225$ cm^{-1} for an applied field of 50 V/μm, and the first maximum of diffraction efficiency at a field of 28 V/μm in 105-μm-thick samples [71]. This compound has excellent thermal stability, and the chromophores play the triple function of charge generator, charge-transport agent, and electro-optic dye.

Many photorefractive polymers still show a relatively slow time response; however, quick progress has been achieved in this respect too. An example of a polymer compound with the fastest response is based on PVK:7-DCST:BBP:C$_{60}$ in the ratio 49.5:35:15:0.5 wt% [73], where 7-DCST is the nonlinear optical chromophore and BBP is a liquid plasticizer that shows better resistance against crystallization than ECZ. The response time for this material is 4 ms at the intensity of 1 W/cm^2 ($\lambda = 647$ nm) and with an applied field of 100 V/μm.

11.8.3 Organic amorphous glass

Beside the classes of photorefractive polymers mentioned above, there are two additional types of material that have shown very encouraging results and merit consideration. These are (1) organic glasses formed by glassy chromophores, and (2) organic oligomers that permit combination of all functionalities in a single component material. The potential of organic glasses for photorefraction has been proven by Lundquist *et al.* [72], demonstrating a gain coefficient of 69 cm^{-1} at the comparatively low field of 40 V/μm in a compound containing nearly 90 wt% of the glassy chromophore 2BNCM. The response time of these material is rather slow and could be decreased at least to the level of ~80 s by adding about 10 wt% of PMMA polymer to the glassy chromophore matrix. Organic oligomers [72 to 76] are very interesting because of their multifunctionality. In contrast to guest–host systems, phase separation and crystallization issues are not major problems. In Ref. [74] a maximum gain coefficient of 80 cm^{-1} and diffraction efficiency of 20% were demonstrated for a field of 31 V/cm in a 130-μm-thick sample of a photoconductive electro-optic carbazole trimer doped with 0.06 wt% TNF as charge generator.

11.8.4 Selected examples of organic photorefractive materials

This section describes a few successful examples of overcoming major material problems (such as the need for a large external electric field for obtaining high performance, and insufficient long-term stability in low-T_g materials) and also describes demonstrations with a reflection grating geometry and large-area three-dimensional (3D) holographic display.

(1) **Under zero electric field.** While the large majority of experiments using polymers are performed under the conditions of large external electric field applied longitudinal

to the thin samples, a few investigations have been reported under zero-field conditions [77,78]. Yu *et al.* reported on a novel kind of conjugated photorefractive polymer where the backbone also plays the role of charge generator and charge transporter [78]. The synthesis of such fully functionalized materials is much more challenging than for guest–host systems but promises better long-term stability. In a compound with the nonlinear chromophores covalently bonded to the backbone, a gain coefficient of 5.7 cm^{-1} was measured at zero field in a 1.4-µm thin sample for a grating spacing of ~1.25 µm [78]. While the grating phase shift and the spatial dispersion of the gain coefficient are consistent with the conventional photorefractive band model of Section 11.1.1, the value of the calculated refractive index modulation amplitude ($\Delta n = 3 \times 10^{-5}$) calculated assuming negligible dichroism is too large to be consistent with the measured electro-optic coefficient of 11.3 pm/V and a diffusion-limited process. Some kind of enhancement of the electro-optic response might be active in this case. An even larger gain coefficient of about 200 cm^{-1} at zero applied field has been observed in another polymer compound that contains an ionic tri(bispyridyl) ruthenium complex as the charge-generating species [79]. There are indications that some kind of internal electric field may enhance the zero-external field charge transport in this kind of polymers [80]. More investigations are needed to clarify the origin of the refractive index gratings in these compounds.

(2) **Mesophase photorefractive polymers.** Since the orientation enhancement effect was observed [42], low-T_g materials (with T_g near or less than room temperature) have been widely developed. There are two types of such materials: one is low-T_g polymeric composites, usually a guest–host system, and the other is amorphous glass materials, usually a fully functionalized system. Low-T_g polymeric composites show not only high steady-state photorefractive performances such as two-beam coupling gain and diffraction efficiency, but also fast dynamic photorefractive performance (i.e., response time). However, the high mobility of NLO chromophores at room temperature, because of high content of plasticizer, often produces phase separation, leading to low temporal stability. In contrast to low-T_g polymeric composites, amorphous glass materials show high temporal stability without phase separation, because of fully functionalized groups, but exhibited slow photorefractive response time [50 to 52]. Kwon *et al.* reported on a novel kind of mesophase photorefractive polymeric composite [81 to 87]. The composites, based on mesophase conductive polymers consisting of poly(*p*-phenyleneterephthalate) (PPT) backbone with pendant hole-transporting carbazole (CZ) or triphenylamine (TPA) groups, exhibit a layered mesophase for PPT-CZ and nematic-like mesophase for PPT-TPA, respectively. The hole-transporting groups and rigid PPT backbone are linked by flexible alkyl chains. The rigid PPT backbone contributes to high temporal stability, and the flexible side-chains contribute to high mobility of NLO chromophores owing to low T_g. The mesophase photorefractive composites consisted of charge transporting PPT-CZ or PPT-TPA, nonlinear optical DDCST(diethylaminodicyanostyrene) or PDCST (piperidinodicyanostyrene) and sensitizer C$_{60}$. Despite having no plasticizers, PPT-CZ and PPT-TPA composites exhibit very low T_g of −7 °C and 15 °C, respectively, and high steady-state and fast dynamic photorefractive performances, up to a gain of

$\Gamma = 250$ cm^{-1} for an applied field of 60 V/μm in PPT-CZ:DDCST:C$_{60}$ and photorefractive sensitivity S_{n2} of 2 cm^2/kJ in PPT-TPA:DDCST:C$_{60}$ [81 to 83].

(3) **Reflection grating geometry**. Compared with tilted transmission geometry, reflection grating geometry, in which beams are incident on opposite surfaces, is better for applications such as switchable and/or tunable narrowband reflection filters and phase conjugation mirrors. However, owing to the short grating spacing (~0.2 μm), which leads to a small space-charge field in reflection grating geometry, most organic photorefractive materials including polymeric systems and glasses have been used in a tilted transmission geometry [84 to 89]. The mesophase photorefractive composites PPT-CZ and PPT-TPA exhibit a two-beam coupling gain of up to $\Gamma = 104$ cm^{-1} for an applied field of 60 V/μm with a grating spacing of 0.205 μm in reflection grating geometry, which is attributed to an increased effective number of trapping sites [84 to 86]. In addition, the high photorefractive performance of layered PPT-CZ composites in reflection grating geometry showed, for the first time, backward beam fanning due to self-diffraction [87].

(4) **Dynamic holography.** Mecher et al. [90] reported on fast organic photorefractive materials with video rate in the NIR region. The polymeric composites based on TPD-PPV polymers, where charge-transporting TPD and PPV (phenylenevinylene) groups are in the main chain, exhibited enhanced response time on pre-illumination with a shorter wavelength, a so-called 'gating' procedure, compared with the same composites without pre-illumination. The composite AZO:TPD-PPV:DPP:PCBM with pre-illumination of 633 nm exhibits a response time one order of magnitude faster than without pre-illumination, and a photorefractive sensitivity one order of magnitude higher than conventional organic photorefractive materials [90].

Peyghambarian, Tay and Blanche et al. reported on novel refreshable prototype 3D holographic displays by using photorefractive polymeric composites in order to overcome the limitation of small sizes of inorganic crystals [53,91,92]. The photorefractive polymeric composites were based on hole-transporting polyacrylic PATPD/CANN copolymer, where PATPD is poly(acrylic tetraphenyldiaminobiphenol) [93] and CANN is a pendant carbaldehyde aniline group attached by an alkyl spacer to the polyacrylic backbone [91,92]. The meltable polymeric composite PATPD/CANN:FDCST:ECZ:PCBM was used for preparing a large-area photorefractive device with a size of 12 inch ×12 inch and a thickness of 100 μm [91]. The prototype organic photorefractive device exhibited quasi-real-time response, with a refresh rate of 2 seconds per image, and demonstrated multicolor 3D display and, for the first time, 3D telepresence through the Ethernet by using angular and spatial multiflexing [91].

11.8.5 Materials properties

There are many ways by which the performance of different photorefractive materials may be compared. One approach is to consider the achievable diffraction efficiencies (or exponential gain) and the time needed to achieve them. The higher the diffraction efficiency and the shorter the response time, the better the material. The diagram of Fig. 11.14 shows the steady-state diffraction efficiency and the inverse response time

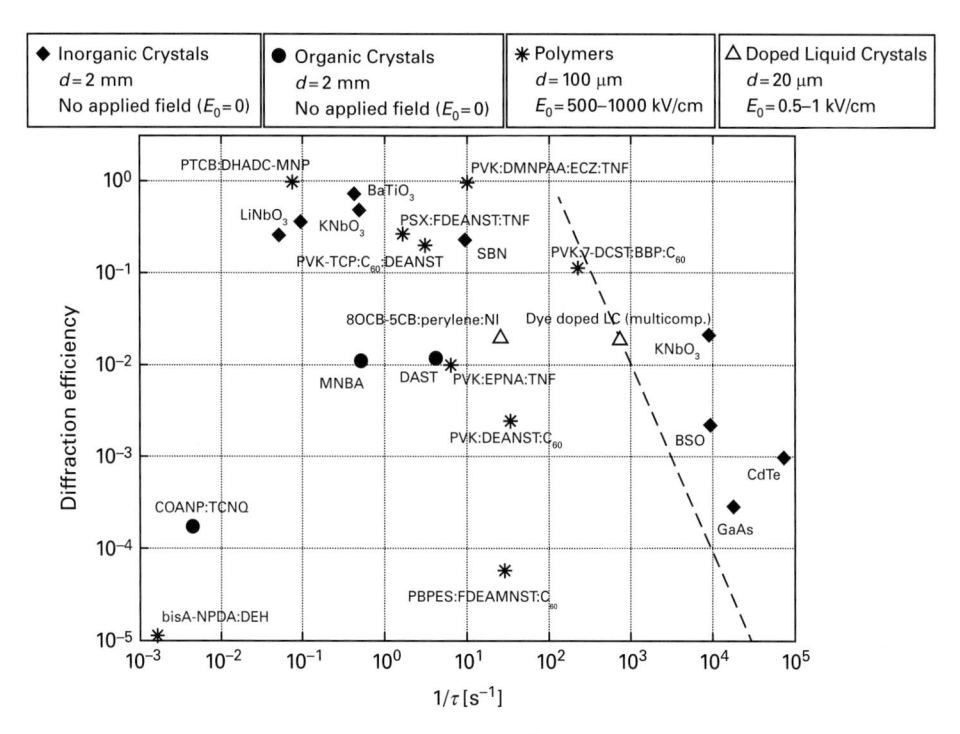

Figure 11.14 Diffraction efficiency versus growth rate τ^{-1} for a selection of photorefractive crystals, polymers, and doped liquid crystals. The values are scaled to 1 W/cm² light intensity and to a typical thickness of 2 mm for crystals, 100 μm for polymer films, and 20 μm for liquid crystal cells. The dashed line connects points of equal sensitivity.

obtained in a selection of inorganic and organic crystals, photorefractive polymers, and dye doped liquid crystals. The diffraction efficiencies are scaled to a typical thickness d of actual samples, i.e. $d = 100$ μm for polymers, $d = 20$ μm for liquid crystalline cells, and $d = 2$ mm for inorganic and organic crystals. The response time is scaled to a light intensity of 1 W/cm². The data for polymers are taken under an applied electric field of typically 500–1000 kV/cm, those for liquid crystals under a field of 0.5–1 kV/cm, and those for inorganic and organic crystals under zero-field conditions. The dashed line in the diagram connects points of constant sensitivity. However, the same sensitivity for the different categories of materials is not on the same line because of the different values assumed for the sample thickness. The region of the best materials (the upper right corner in the diagram) is still occupied mainly by inorganic crystals, but the best polymers and liquid crystals are approaching this region very fast. The best organic crystals have made it to the mid-field region.

Another way of comparing the performance of different photorefractive materials is through the photorefractive sensitivities defined in Section 11.6. In Table 11.1 the sensitivities S_{n1} and S_{n2} are compared for a selection of experiments in organic and inorganic photorefractive materials. The refractive index change Δn_{max} achieved in the referenced works is also reported in Table 11.1. In many inorganic materials, values of

11.8 Photorefractive materials and properties

Table 11.1 Absorption constants, photorefractive sensitivities, refractive index changes, and grating strengths for selected experiments reported in inorganic and organic photorefractive materials.

Material	α (cm^{-1})	S_{n1} (cm^3/kJ)	S_{n2} (cm^2/kJ)	Δn_{max} [a]	$\frac{\Delta n_{max} d}{\lambda}$ [b]	E_0 (kV/cm)
LiNbO$_3$	0.1–100	0.005–0.05	0.17 [c,d]	10^{-5}–10^{-3}	0.04–4	0
KNbO$_3$	3.8	12	50	~10^{-5}	0.04	7
	5500 [c]	0.4 [c]	30	~10^{-5}	0.04	0
			2000 [c]	2×10^{-5}	0.08	2.2
BaTiO$_3$	2.2	1.3 [e]	7.6 [f]	5×10^{-5}	0.2	10
		3.4 [f]		1.5×10^{-4} [f]	0.6	0
SBN	1.0	0.4	0.4	1.5×10^{-4}	0.6	0
Bi$_{12}$SiO$_{20}$	2.3	72	167	~10^{-6}	0.004	6
	~0.8	10	8			0
GaAs	~1.2 [g]	35 [g]	42 [g]	3×10^{-6}	0.006	0
CdTe	1.8 [h]	230 [h]	410 [h]	5×10^{-6}	0.007	0
GaAs/AlGaAs MQW	8000	2000 [i,j]	16×106 [i,j]	10^{-2} [i]	0.02 [i]	75
CdZnTe/ZnTe MQW		<700 [i,k]	7×106 [i,k]	10^{-2} [i]	0.03 [i]	320
PLZT ceramic (9/65/35)	2–4	~0.02	0.06	~10^{-3}	~1	
COANP:TCNQ	5 [l]	7×10^{-7} [l]	4×10^{-6} [l]	2×10^{-6} [l]	0.006	0
MNBA	70	2×10^{-5}	0.001	10^{-5}	0.04	0
bisA-NPDA:DEH	10 [k]	1.4×10-5 [k]	1.4×10^{-4} [k]	2×10^{-6}	0.0003	114
PVK:DMNPAA:ECZ:TNF	13 [p]	0.8 [n,p]	10 [n,p]	5×10^{-3} [n,p]	0.7 [n]	900
PSX:FDEANST:TNF	32 [l]	0.07 [n,l]	2.1 [n,l]	1.4×10^{-3} [n,l]	0.29 [n]	770
PVK-7-DCST:BBP:C60	26 [m]	0.8 [m,n]	20 [m,n]	8×10^{-4} [m,n]	0.1 [m,n]	1000
Anthraquinone dye doped liquid crystal	462 [q]	16 [q]	7400 [q]	1.9×10^{-3} [q]	0.05 [q]	10

Notes: Data for $\lambda = 488$ nm or $\lambda = 514$ nm unless otherwise indicated. [a] Representative for the cited experiments; higher values have been observed in most of the inorganic crystals in different experimental situations; [b] calculated for a typical thickness of 2 mm for crystals, 100 μm for photorefractive polymers, 500 μm for ceramics, and 2 μm for MQW devices; [c] $\lambda = 351$ nm; [d] $E_0 = 50$ kV/cm; [e] $\lambda = 458$ nm; [f] 45°-cut; [g] $\lambda = 1064$ nm; [h] $\lambda = 1523$ nm; [i] quantum confined Stark effect; [j] $\lambda = 830$ nm; [k] $\lambda = 596$ nm; [l] $\lambda = 676$ nm; [m] $\lambda = 647$ nm; [n] reorientationally enhanced; [p] $\lambda = 675$ nm; [q] $\lambda = 633$ nm.

Δn_{max} considerably larger than those listed here have been observed. However, large refractive index changes are often obtained at the expense of a slower response, and the sensitivities are not increased. It can be easily shown that the maximum achievable sensitivity in a material is ultimately limited by the speed of charge production [94], explaining the fact that samples with moderate refractive index changes often show sensitivities comparable to or better than crystal samples with stronger space-charge gratings at saturation. In many cases, rather than Δn_{max}, it is the grating strength ($\Delta n_{max} d/\lambda$) which is of practical importance. The total grating strength gives a better measure to judge the achievable diffraction efficiencies and two-wave mixing gain × length product. This quantity is also given in Table 11.1 for typical sample thicknesses.

Table 11.2 Refractive index, electro-optic coefficients, dielectric constant, mobility, and figures-of-merit for some inorganic and organic photorefractive materials. The expressions for the quantities $Q_1 \ldots Q_5$ are defined in the text.

Material	λ (nm)	n	r_{eff} (pm/V)	ε_{eff}	μ (cm^2/Vs)	$\varphi\mu\tau_R$ (10^{-16} m^2/V)	N_{eff} (10^{22} m^{-3})	Q_1 (V^{-1} µm^{-2})	Q_2 (pm/V)	Q_3 (pm/V)	Q_4 (nm^2/V)	Q_5 (10^{-32} m^4/V^2)
LiNbO$_3$	633	$n_e = 2.20$	$r_{333}= 31$	29	0.8	0.03–0.6	20	2.3	11.4	331		0.002–0.04
KNbO$_3$	514	$n_c = 2.21$	$r_{33\text{eff}}= 59$	35	0.5	10–10^5	5	0.9	18.2	634	3.3	0.9–9000
		$n_b = 2.40$	$r_{23\text{eff}}= 460$	989				0.3	6.4	6359	1.2	0.3–3000
BaTiO$_3$	514	$n_e = 2.42$	$r_{33\text{eff}}= 90$	83	0.5	10–1000	5	0.8	15.4	1275	2.6	0.8–80
		$n_o = 2.49$	$r_{23\text{eff}}=1326$	4350				0.24	4.7	20470	0.8	0.24–24
Bi$_{12}$SiO$_{20}$	633	2.54 [a]	$r_{231}= 5.0$	55	3.4	10^5	2	0.03	1.5	82		940
GaAs	1064	3.50 [b]	$r_{231}= 1.2$	13.2	$\mu_e = 8500$	10^5	0.2	0.008	3.9	51		6900
					$\mu_h = 400$							
CdTe	1550	2.74	$r_{231}= 6.8$ [c]	9.4	$\mu_e = 1050$	$\mu\tau \approx 10^7$	0.5	0.07	14.9	140		$<10^6$
					$\mu_h = 5$							
PLZT ceramic (10/65/35) [d]	514	~2.5	200–800	4500		10	50	~0.9	~1.7	~8000		~0.1
COANP:TCNQ	633	$n_c = 1.65$	$r_{333}= 15$	2.6		0.05–0.2	0.1	0.03	25.9	67	2×10^{-4}	0.008–0.032
MNBA	514	$n_1 = 2.23$	$r_{111}= 62$	4.0		0.013	0.01	0.02	172	688	10^{-5}	0.01
bisA-NPDA: DEH [e]	647	1.63	0.33	2.9	small		0.2	0.001	0.5	1.4		
PVK: FDEANST: TNF [f]	647	~1.7	~1.8 [g]	~3	small				~3 [g]	~9 [g]		
PVK: DMNPAA: ECZ:TNF [h]	675	1.75	~18 [g]	~3.1	small				~32 [g]	~100 [g]	< 0.2	

Notes: [a] $\lambda = 620$ nm; [b] $\lambda = 1020$ nm; [c] $\lambda = 3.39$ µm; [d] $E_0 = 5$–10 kV/cm; [e] $E_0 = 120$ kV/cm; [f] $E_0 = 400$ kV/cm; [g] estimated including the quadratic effect of chromophore reorientation; [h] $E_0 = 900$ kV/cm.

The performance of a material in view of a certain application may be related to material parameters by means of figures-of-merit, such as the quantities Q_1 to Q_5 defined in Section 11.6. Large values of Q_1, Q_2, and Q_3 indicate materials where large refractive index changes can be obtained. Large values of Q_4 and Q_5 indicate materials with good holographic sensitivity. Table 11.2 shows the typical quantities $Q_1\ldots Q_5$, as well as a few other relevant parameters, for some photorefractive materials.

Because of the small dielectric constants, some organic materials have a larger figure-of-merit Q_2 than the best inorganics, but the latter are still better in terms of Q_1, and especially Q_3. The faster response of most inorganic crystals is also reflected in higher values for the sensitivity figures-of-merit Q_4 and Q_5.

In the past, the photorefractive effect has proved invaluable for demonstrating new concepts in industrial inspection, locking of laser diodes, image processing, and optical data storage. At present, inorganic crystals are still the materials of choice for most applications, but further improvements of the materials properties are required. Ultimately it is not expected that a single class of materials will excel in all classes of applications. The selection of the materials that will reach the best commercial maturity will also be based not only on pure performance characteristics, but also on other criteria, such as reliability, stability, and cost. The research efforts into known and new materials have to continue in order to optimize all these factors.

References

[1] (a) P. Günter, *Phys. Rep.* **93**, 199 (1982).
   (b) P. Günter and J. P. Huignard, Eds., *Photorefractive Materials and their Applications Vols 1 & 2*, Berlin, Springer Verlag (1988).
   (c) P. Günter and J. P. Huignard, Eds. *Photorefractive Materials and their Applications Vols 1–3*, Berlin, Springer Verlag (2007).
   (d) H. J. Eichler, P. Günter, and D. W. Pohl, *Laser-induced Dynamic Gratings*, Berlin, Springer Verlag (1986).
[2] A. Ashkin, G. D. Boyd, J. M. Dziedzic, *et al.*, *Appl. Phys. Lett.*, **25**, 233 (1966).
[3] F. S. Chen, J. T. LaMacchia, and D. B. Fraser, *Appl. Phys. Lett.*, **13**, 223 (1968).
[4] I. Biaggio, M. Zgonik, and P. Günter, *J. Opt. Soc. Am. B*, **9**, 1480 (1992).
[5] G. Montemezzani, P. Rogin, M. Zgonik, and P. Günter, *Phys. Rev. B*, **49**, 2484 (1994).
[6] A. M. Glass, D. VonderLinde, and T. J. Negran, *Appl. Phys. Lett.*, **25**, 233 (1974).
[7] W. Kraut and R. von Baltz, *Phys. Rev. B*, **19**, 1548 (1979).
[8] K. Buse, *Appl. Phys. B Lasers Opt.*, **64**, 273 (1997).
[9] N. V. Kukhtarev, V. B. Markov, S. G. Odulov, M. S. Soskin, and V. L. Vinetskii, *Ferroelectrics*, **22**, 949 (1979).
[10] G. Montemezzani, C. Medrano, P. Gunter, and M. Zgonik, *Phys. Rev. Lett.*, **79**, 3403 (1997).
[11] M. Pope and C. E. Swenberg, *Electronic Processes in Organic Crystals*, Oxford, Clarendon Press (1982).
[12] J. Feinberg, J. D. Heimen, A. R. Tanguay, and R. W. Hellwarth, *J. Appl. Phys.*, **51**, 1297 (1980).
[13] P. M. Borsenberger, E. H. Magin, and J. J. Fitzgerald, *J. Phys. Chem.*, **97**, 9213 (1993).

[14] S. M. Silence, J. C. Scott, F. Hache, et al., *J. Opt. Soc. Am. B*, **10**, 2306 (1993).
[15] H. Tokuhisa, M. Era, T. Tsutsui, and S. Saito, *Appl. Phys. Lett.*, **66**, 3433 (1995).
[16] R. A. Mullen, In: *Photorefractive Materials and their Applications I: Fundamental Phenomena*, Vol. **1**, P. Günter and J. P. Huignard, Eds., Springer Verlag, Berlin (1988), pp. 167–194.
[17] R. A. Mullen, R. W. Hellwarth, *J. Appl. Phys.*, **58**, 40 (1985).
[18] G. Montemezzani, C. Medrano, M. Zgonik and P. Günter, In: *Nonlinear-Optical Effects and Materials*, P. Günter, Ed., New York, Springer Verlag (2000).
[19] J. F. Nye, *Physical Properties of Crystals*, Oxford, Clarendon Press (1985).
[20] P. Günter and M. Zgonik, *Opt. Lett.*, **16**, 1826 (1991).
[21] M. Zgonik, R. Schlesser, I. Biaggio, et al., *J. Appl. Phys.*, **74**, 1287 (1993).
[22] M. Zgonik, P. Bernasconi, M. Duelli, et al., *Phys. Rev. B*, **50**, 5941 (1994).
[23] M. Zgonik, K. Nakagawa, and P. Günter, *J. Opt. Soc. Am. B*, **12**, 1416 (1995).
[24] G. Montemezzani, A. A. Zozulya, L. Czaia, et al., *Phys. Rev. A*, **52**, 1791 (1995).
[25] W. E. Moerner and S. M. Silence, *Chem. Rev.*, **94**, 127 (1994).
[26] S. M. Silence, D. M. Burland, and W. E. Moerner, In: *Photorefractive Effects and Materials*, D. D. Nolte, Ed., Boston, Kluwer Academic (1995), pp. 265–309.
[27] S. Ducharme, *Proc. SPIE*, **2526**, 144 (1995).
[28] W. E. Moerner, A. Grunnetjepsen, and C. L. Thompson, *Annu. Rev. Mater. Science*, **27**, 585 (1997).
[29] J. S. Schildkraut and Y. Cui, *J. Appl. Phys.*, **72**, 5055 (1992).
[30] J. S. Schildkraut and A. V. Buettner, *J. Appl. Phys.*, **72**, 1888 (1992).
[31] W. D. Gill, In: *Photoconductivity and Related Phenomena*, J. Mort and D. M. Pai, Eds., Amsterdam, Elsevier (1976), pp. 303–334.
[32] A. Twarowski, *J. Appl. Phys.*, **65**, 2833 (1989).
[33] L. Pautmeier, R. Richert, H. Bässler: *Philos. Mag. B* **63**, 587 (1991).
[34] K. Meerholz, R. Bittner, Y. Denardin, et al., *Adv. Mater.*, **9**, 1043 (1997).
[35] M. Liphardt and S. Ducharme, *J. Opt. Soc. Am. B*, **15**, 2154 (1998).
[36] J. K. Mack, L. B. Schein, and A. Peled, *Phys. Rev. B*, **39**, 7500 (1989).
[37] P. G. deGennes and J. Prost, In: *The Physics of Liquid Crystals*, edn 2, J. Birman, S. F. Edwards, C. H. L. Smith, and M. Rees, Eds., International Series of Monographs on Physics, vol. **83**, Oxford, Clarendon (1993).
[38] F. Wang, Z. Chen, Q. Gong, Y. Chen, and H. Chen, *Solid State Commun.*, **106**, 299 (1998).
[39] I. C. Khoo, *IEEE J. Quantum Electron.*, **22**, 1268 (1986).
[40] P. Prêtre, P. Kaatz, A. Bohren, et al., *Macromolecules*, **27**, 5476 (1994).
[41] M. C. J. M. Donckers, S. M. Silence, et al., *Opt. Lett.*, **18**, 1044 (1993).
[42] K. Meerholz, B. L. Volodin, Sandalphon, B. Kippelen, and N. Peyghambarian, *Nature*, **371**, 497 (1994).
[43] K. Sutter, J. Hulliger, and P. Günter, *Solid State. Commun.*, **74**, 867 (1990).
[44] K. Sutter, J. Hulliger, R. Schlesser, and P. Günter, *Opt. Lett.*, **18**, 778 (1993).
[45] K. Sutter and P. Günter, *J. Opt. Soc. Am. B*, **7**, 2274 (1990).
[46] G. Knöpfle, C. Bosshard, R. Schlesser, and P. Günter, *IEEE J. Quantum Electron.*, **30**, 1303 (1994).
[47] S. Follonier, C. Bosshard, F. Pan, and P. Günter, *Opt. Lett.*, **21**, 1655 (1996).
[48] S. Ducharme, J. C. Scott, R. J. Twieg, and W. E. Moerner, *Phys. Rev. Lett.*, **66**, 1846 (1991).
[49] B. Kippelen, B. Sandalphon, N. Peyghambarian, S. R. Lyon, A. B. Padias, H. K. Hall, *Electron. Lett.*, **29**, 1873 (1993).
[50] O. Ostroverkhova and W. E. Moerner, *Chem. Rev.*, **104**, 3267 (2004).
[51] B. Kippelen and N. Peyghambarian, *Adv. Polym. Sci.*, **161**, 87 (2003).

[52] S. Köber, M. Salvador, and K. Meerholz, *Adv. Mater.* **23**, 4725 (2011).
[53] J. Thomas, C. W. Christenson, P.-A. Blanche, *et al.*, *Chem. Mater.*, **23**, 416 (2011).
[54] G. G. Malliaras, V. V. Krasnikov, H. J. Bolink, and G. Hadziioannou, *Appl. Phys. Lett.*, **66**, 1038 (1995).
[55] A. GrunnetJepsen, D. Wright, B. Smith, *et al.*, *Chem. Phys. Lett.*, **291**, 553 (1998).
[56] Y. Zhang, Y. Cui, and P. N. Prasad, *Phys. Rev. B*, **46**, 9900 (1992).
[57] S. Köber, F. Gallego-Gómez, M. Salvador, *et al.*, *J. Mater. Chem.*, **20**, 6170 (2010).
[58] E. Mecher, F. Gallego-Gómez, K. Meerholz, *et al.*, *Chem. Phys. Chem.*, **5**, 277 (2004).
[59] O. Ostroverkhova, D. Wright, U. Gubler, *et al.*, *Adv. Funct. Mater.*, **12**, 621 (2002).
[60] S. Köber, J. Prauzner, M. Salvador, *et al.*, *Adv. Mater.*, **22**, 1383 (2010).
[61] M. Eralp, J. Thomas, S. Tay, *et al.*, *Appl. Phys. Lett.*, **85**, 1095 (2004).
[62] S. Tay, J. Thomas, M. Eralp, *et al.*, *Appl. Phys. Lett.*, **87**, 171105 (2005).
[63] K. Roy Choudhury, Y. Sahoo, S. Jang, and P. N. Prasad, *Adv. Funct. Mater.*, **15**, 751 (2005).
[64] K. Roy Choudhury, Y. Sahoo, and P. N. Prasad, *Adv. Mater.*, **17**, 2877 (2005).
[65] G. G. Malliaras, V. V. Krasnikov, H. J. Bolink, and G. Hadziioannou, *Appl. Phys. Lett.*, **65**, 262 (1994).
[66] F. Würthner, R. Wortmann, and K. Meerholz, *Chem. Phys. Chem.*, **3**, 17 (2002).
[67] D. Wright, U. Gubler, Y. Roh, *et al.*, *Appl. Phys. Lett.*, **79**, 4274 (2001).
[68] O. Zobel, M. Eckl, P. Strohriegl, and D. Haarer, *Adv. Mater.*, **7**, 911 (1995).
[69] A. M. Cox, R. D. Blackburn, D. P. West, *et al.*, *Appl. Phys. Lett.*, **68**, 2801 (1996).
[70] A. GrunnetJepsen, C. L. Thompson, R. J. Twieg, and W. E. Moerner, *Appl. Phys. Lett.*, **70**, 1515 (1997).
[71] E. Hendrickx, J. Herlocker, J. L. Maldonado, *et al.*, *Appl. Phys. Lett.*, **72**, 1679 (1998).
[72] P. M. Lundquist, R. Wortmann, C. Geletneky, *et al.*, *Science*, **274**, 1182 (1996).
[73] D. Wright, M. A. Díaz-García, J. D. Casperson, *et al.*, *Appl. Phys. Lett.*, **73**, 1490 (1998).
[74] L. Wang, Y. Zhang, T. Wada, and H. Sasabe, *Appl. Phys. Lett.*, **69**, 728 (1996).
[75] Y. D. Zhang, L. M. Wang, T. Wada, and H. Sasabe, *Appl. Phys. Lett.*, **70**, 2949 (1997).
[76] T. Wada, Y. Zhang, and H. Sasabe, *RIKEN Rev.* no. 15, p. 334 (1997).
[77] Y. M. Chen, Z. H. Peng, W. K. Chan, and L. P. Yu, *Appl. Phys. Lett.*, **64**, 1195 (1994).
[78] L. Yu, Y. Chen, W. K. Chan, and Z. Peng, *Appl. Phys. Lett.*, **64**, 2489 (1994).
[79] Z. H. Peng, A. R. Gharavi, and L. P. Yu, *Appl. Phys. Lett.*, **69**, 4002 (1996).
[80] L. Yu, Y. M. Chen, and W. K. Chan, *J. Phys. Chem.*, **99**, 2797 (1995).
[81] O. P. Kwon, S. H. Lee, G. Montemezzani, and P. Günter, *Adv. Funct. Mater.*, **13**, 434 (2003).
[82] O. P. Kwon, S. H. Lee, G. Montemezzani, and P. Günter, *J. Opt. Soc. Am. B*, **20**, 2307 (2003).
[83] O. P. Kwon, S. J. Kwon, M. Jazbinsek, S. H. Lee, and P. Günter, *Polymer*, **46**, 10301 (2005).
[84] O. P. Kwon, G. Montemezzani, P. Günter, and S. H. Lee, *Appl. Phys. Lett.*, **84**, 43 (2004).
[85] O. P. Kwon, S. J. Kwon, M. Jazbinsek, and P. Günter, *J. Chem. Phys.*, **124**, 104705 (2006).
[86] O. P. Kwon, S. J. Kwon, M. Jazbinsek, P. Günter, and S. H. Lee, *Appl. Phys. Lett.*, **87**, 121910 (2005).
[87] O. P. Kwon, M. Jazbinsek, and P. Günter, *Appl. Phys. Lett.*, **89**, 021905 (2006).
[88] K. Meerholz, E. Mecher, R. Bittner, and Y. D. Nardin, *J. Opt. Soc. Am. B*, **15**, 2114 (1998).
[89] T. Sassa, T. Muto, and T. Wada, *J. Opt. Soc. Am.*, **21**, 1255 (2004).
[90] E. Mecher, F. Gallego-Gomez, H. Tillmann, *et al.*, *Nature*, **418**, 959 (2002).
[91] P. A. Blanche, A. Bablumian, R. Voorakaranam, *et al.*, *Nature*, **468**, 80 (2010).
[92] S. Tay, P.-A. Blanche, R. Voorakaranam, *et al.*, *Nature*, **451**, 694 (2008).
[93] J. Thomas, C. Fuentes-Hernandez, M. Yamamoto, *et al.*, *Adv. Mater.*, **16**, 2032 (2004).
[94] P. Yeh, *Appl. Opt.*, **26**, 602 (1987).

12 Conclusions and future prospects

12.1 General conclusions

In the preceding chapters, an introduction has been provided into the fundamentals and state-of-the-art of organic nonlinear optical materials and devices with particular emphasis on electro-optic, second-harmonic generation, difference-frequency generation, optical rectification, and photorefractive materials, devices, and applications related to organic second-order nonlinear optical materials. Organic electro-optic materials exhibit extremely attractive features with respect to temporal response to time-varying electric fields and with respect to the magnitude of second-order optical nonlinearity, which can translate into energy-efficient devices. Organic materials also are attractive in providing a wide variety of processing options including crystal growth (from solution, melt, and vapor phase), sequential synthesis/self-assembly (both Langmuir–Blodgett and Merrifield methods), electric field poling of macromolecular materials near their glass transition temperature, and even laser-assisted electric field poling. Organic materials are amendable to nano-imprint lithography, leading to stamping out complex circuitry, and they are amenable to lift-off techniques for production of conformal and flexible devices. Organic second-order nonlinear optical materials are compatible with a wide variety of materials including semiconductors, metal oxides, and metals; this feature greatly facilitates the production of hybrid devices including those based on silicon photonics, plasmonics, photonic crystals, and metamaterials. They have been integrated into stripline and resonant device structures, cascaded prism device structures, and even structures for slow wave propagation. The low dielectric constants of organic electro-optic materials can also be an advantage for certain applications such as sensing.

Of course, there is a great diversity of organic nonlinear optical materials with a corresponding diversity of properties. However, organic second-order nonlinear optical materials do not typically possess the extremely low optical loss and high optical damage threshold of crystalline inorganic materials. Nevertheless, they frequently exhibit lower optical loss than inorganic electro-absorptive materials and much higher dielectric breakdown properties than semiconductor materials.

The performance properties of organic nonlinear optical materials continue to evolve at a significant rate. However, the commercialization of organic nonlinear optical materials is very limited and immature relative to inorganic materials, and thus the cost of devices based on organic materials does not benefit from large-scale production.

Moreover, device engineering utilizing organic nonlinear optical materials is relatively immature compared with the device engineering focused on inorganic materials. As is noted in the next section, cost can be a deciding factor in the adoption of organic nonlinear optical materials for specific applications.

There are many potential applications of organic nonlinear optical materials, and particularly electro-optic materials, ranging from telecommunications, high-performance computing, RF photonics, sensing and nonlinear optical materials for terahertz wave generation and detection, and other wave mixing applications. They can conceivably play a significant role in the emerging technology related to chipscale integration of electronics and photonics. However, their future prospects will depend upon competing technologies including the rapid evolution of technologies discussed in the next section.

12.2 Future prospects: competing technologies for electrical-to-optical signal transduction

Organic electro-optic materials must compete not only with inorganic electro-optic and electro-absorptive materials but also with other technologies for controlling light and effecting electronic/photonic integration. Which technology will be the technology of choice for a particular application will depend very much on the application in question and upon factors such as cost and commercial availability. For example, vertical-cavity surface emitting lasers (VCSELs) have become inexpensive and are very competitive for point-to-point optical interconnection when electrical-to-optical signal transduction is required. In the past, VCSEL technology has been limited by bandwidth and power requirements, but these aspects are continuing to be improved. Single-mode operation with high phase stability continues to be a concern. However, it is clear that the low cost of VCSELs makes them the choice for many applications, including for some applications related to chipscale integration of electronics and photonics. VCSELs by their very nature are not suited for applications such as EM field sensing as in phased-array radar receivers, or complex information management such as active wavelength-division-multiplexing (WDM).

Like VCSELs, inorganic light-emitting diode (LED) technology is becoming increasingly attractive for single point-to-point applications. This technology also has the potential for very low cost, and the bandwidth is increasing at a significant rate and may become cost- and performance-competitive with VCSELs.

Another very serious competitor, particularly for chipscale integration, will be silicon modulators. Again, the advantages will be cost and leveraging off the mature technology of CMOS electronics. Homogeneous integration of electronics and photonics continues to be seductive, although it is still difficult to imagine silicon yielding competitive sources and detectors as well as modulators. In the past, the most serious limitations of silicon modulators have been in power requirements and bandwidth, but the technology continues to improve and is certainly amenable to large-scale integration. As with VCSELs, there are applications for which silicon modulators will not be suited but, again like VCSELs, silicon modulators will be attractive for applications where cost and CMOS manufacturing are central issues. Virtual foundry operations such as OPSIS (http://spie.org/X85621.xml) may impact the adoption of silicon modulator technology.

Research on the improvement of the performance characteristics of inorganic electro-optic materials has been limited; however, the device performance (including drive voltage and bandwidth performance) has been greatly improved by clever engineering. Some continued improvement is likely to occur. The high dielectric permittivity of lithium niobate (and other inorganic materials) will be an issue for applications such as EM sensing. Lithium niobate is a well-entrenched technology, effectively exploiting the low optical loss and high thermal stability properties of lithium niobate and carefully engineering around liabilities such as optical/radiofrequency velocity mismatch. Lithium niobate will likely continue to be a commercially competitive technology for the production of single modulator devices despite the high cost of crystal growth and difficulties associated with integration with diverse material technologies. Chipscale integration will be a challenge for lithium niobate, and it is difficult to see it competing with other technologies such as those noted above.

Inorganic electro-absorptive modulator technologies offer advantages of low drive voltages and increasingly high bandwidth. Power consumption and CHIRP (associated with absorption and phase change occurring at the same time) continue to be serious limitations.

Liquid crystalline materials afford very large electro-optic coefficients and dominate applications such as beam steering. However, because switching requires movement of both electron and nuclear mass, liquid crystals are relatively slow (response times greater than a microsecond) in their response to time-varying electrical fields. If the electro-optic coefficients of "electronic response" organic π-electron materials can be increased to significantly above 1000 pm/V, then such materials may start to compete with current liquid crystalline materials for beam steering and spatial light modulation applications. This will particularly be true for applications that can tolerate cascaded prism device architectures. Based on counting electrons as discussed by Kuzyk (see Chapter 4), the maximum achievable electro-optic activity is approximately 3000 pm/V for CLD-class chromophores.

12.3 Future prospects: fundamental issues facing the development and utilization of organic electro-optic materials

Two fundamental approaches to the generation of organic electro-optic materials have come to dominate the production of materials for devices: (1) growth of single crystals and crystalline thin films of molecules with suitable side groups to allow self-assembly by ionic or hydrogen bonding; and (2) preparation of macromolecular electro-optic materials by electric field poling or laser-assisted electric field poling. The fundamental limitation related to the preparation of single crystals is chromophore size. It is increasingly difficult to prepare high-quality crystals as the size and internal degrees of freedom of chromophores increase. Dipole moments typically increase with chromophore size, and chromophore–chromophore dipolar interactions strongly inhibit acentric crystal growth of prolate ellipsoidal-shaped chromophores as has been noted in previous chapters. Stronger ionic and hydrogen-bonding interactions are required to overcome chromophore dipolar interactions, and the competition of these interactions will ultimately define the maximum electro-optic activity that can be achieved by this approach.

The fundamental limitation of electrically poled macromolecular materials is the maximum poling field that can be achieved at the chromophores, which is defined by electrical conductivity and dielectric breakdown of the organic electro-optic material. Competing with the poling ordering force is chromophore–chromophore dipolar interactions. As noted in the preceding chapters, the introduction of additional interactions such as dipolar (coumarin) or quadrupolar (arene/perfluoroarene) can be used to enhance poling efficiency. Further improvement of the acentric order of macromolecular electro-optic materials will require utilization of stronger interactions. Both single crystal growth and electric field poling of macromolecular materials depend upon finding the right interactions to overcome chromophore dipolar interactions.

A major change in the organic electro-optic landscape over the past decade is the changing focus from consideration of all-organic device structures to the integration of organic electro-optic materials into silicon photonic, plasmonic, and metamaterial device architectures. For example, voltage–length parameters of less than 0.5 V-mm together with digital signal processing bandwidths of greater than 100 Gbit/s and power efficiencies of less than 1 fJ/bit have already been realized for silicon–organic hybrid (SOH) devices. Issues associated with each of these types of hybrid device generation can dominate the choice of materials utilized. The fact that organic electro-optic materials can be integrated into these device architectures both by solution and vapor phase processing can be an important advantage, permitting utilization of the bandwidth and low drive-voltage potential of organic electro-optic materials while exploiting the unique characteristics of silicon photonics, plasmonics, and metamaterials. Under the right circumstances, silicon photonics, plasmonics, and metamaterials device architectures can reduce the sizes and drive-voltage requirements of electro-optic devices. As with organic electro-optic materials, silicon photonics, plasmonics, and metamaterials are evolving technologies, and it is difficult to anticipate the long-term role that these technologies will play in chipscale integration of electronics and photonics and other electro-optic applications. Unanticipated developments in materials and device concepts of hybrid device technologies could well have a huge impact on commercial applications. Of course, these technologies must compete against the technologies discussed in the preceding section.

Organic electro-optic materials can be used to fabricate conformal and flexible devices, and techniques such as nano-imprint lithography can be used to produce complex photonic circuitry incorporating organic electro-optic devices. However, this technology has not yet been translated to practical application. The same thing can be said for integration of organic electro-optic materials with fiber sensor technology.

12.4 Future prospects: optical sum and difference-frequency generation, optical rectification, and THz generation

In the 1980s, there was considerable interest in second-order nonlinear optical materials for the generation of visible wavelength lasers; however, with the production of blue GaN type lasers diodes, this interest has been greatly reduced.

Optical difference-frequency generation and optical parametric oscillation, however, are still attractive techniques for producing infrared and far-infrared radiation where no other efficient and economic sources are available.

A difference-frequency generation phenomenon where organic second-order nonlinear optical materials may have a significant impact is terahertz generation and detection, particularly at frequencies above 1 THz (currently to 20 THz and more) where electronic sources become ineffective. Organic second-order nonlinear optical materials offer significant advantages over their crystalline inorganic counterparts by affording larger nonlinear optical coefficients and, in the case of electrically poled macromolecular materials and some organic crystals, of avoiding phonon resonances that limit bandwidth. The small contribution of the lattice or molecular polarizability of organic materials allows interesting phase-matching configurations, particularly for the generation of THz waves in some crystals. By optical rectification and difference-frequency generation in organic materials, wide-bandwidth (0.2–15 THz) and single-frequency THz waves tunable from 1 THz to 20 THz have been generated with excellent conversion efficiencies. Organic materials therefore have shown good potential for wide-bandwidth THz spectroscopy and for measurement of nanoscale dynamics. THz waves generated in organic crystals (DSTMS, OH1, and others) have shown high potential for the identification of defects in high-performance non-conductive polymers and ceramics used, for example, in biomedical applications.

Incorporating organic second-order nonlinear optical materials into silicon photonic devices that concentrate optical fields has permitted observation of optical rectification at microwatt power levels; however, such sensitivity is not competitive with the best photodetectors.

12.5 Future prospects: final comments

Both organic nonlinear optical materials and their related technological applications are rapidly evolving, as are the technologies with which they must compete. Organic nonlinear optical materials differ from their inorganic counterparts in that significant improvement in performance may be achieved by the development of new materials. However, device engineering and large-scale manufacturing are improving the performance of competitive inorganic material technologies and reducing cost. The integration of organic nonlinear optical materials into silicon photonics, plasmonics, and meta-material device architectures to generate a hybrid device technology may be adding a new dimension. The impact of such hybrid technology remains to be seen but certainly has demonstrated the potential for transformative performance.

Index

2-(3-cyano-4,5,5-trimethyl-5H-furan-2-ylidene)-malononitrile (TCF) acceptor, 65

3 dB$_e$ bandwidth, 169, 176

π- and σ-components, 123
π-conjugated bridge, 65
π-electron backbone of a chromophore, 123
π-electron molecular systems, 2

absorption loss, 146
acentric (noncentrosymmetric) order parameter, 118
acentric order, 134
acentric order parameter, 43, 121, 131
acoustic lattice vibrations, 31
acoustic spectrum analyzers, 178
activation enthalpy, ΔH^*, 132
activation entropy, ΔS^*, 132
addition reactions, 145
adiabatic approximation, 47
adiabatic volume adjustment (AVA) method, 169
advanced modulation formats, 162
all-optical modulation, 2
all-organic devices, 155
amorphous polycarbonate (APC), 119
amphiphilic molecules, 70
amplitude (intensity) modulation, 36
analytic derivative methods, 51
angle tuning, 22
applied electric poling field, 121
arene–perfluoroarene modified materials, 138
array waveguide gratings (AWGs), 162, 170
arylamine donors, 65
attenuated total reflection (ATR), 149

B3LYP functional, 56
band transport model, 251
bandwidth, 1, 176
bandwidth performance, 2
bandwidth-sensitivity factor, 179
basis sets, 48
beam steering devices, 181
bending losses, 181

benzocyclobutene, 157
Bessel lattice, 131
bidirectional polarization-independent modulation, 163
binary chromophore (BC) materials, 119, 133–134, 138
birefringence, 168
birefringence (BR) or phase modulator, 177
bit error rates, 163, 169
Bloch theorem, 204
bond length alternation (BLA), 47
bowtie antenna structure, 162
Bragg diffraction, 265
Brillouin zone, 206
buried channel electro-optic waveguide, 219

cascaded prism device, 181, 284
centric (even) and acentric (odd) order parameters, 118
centrosymmetric order parameter, 131, 151
centrosymmetric ordering, 127
channel waveguides, 212
charge blocking layers, 217
charge injection blocking layer, 157
charge redistribution, 252
charge transport, 250
charge transport mobilities, 253
charge trapping centers, 260
chemical modification, 126
chipscale integration, 1, 168, 182–183
    of electronics and photonics, 1, 7, 283, 285
CHIRP, 284
chromophore aggregation, 124
chromophore dipolar interactions, 6, 121, 285
chromophore dipole moment, 121
chromophore dipole–dipole interaction energy, 122
chromophore guest, 119
chromophore loading, 121, 127
chromophore number density, 118, 122, 124
chromophore shape, 123–124, 126
chromophore size, 284
chromophore/polymer composites, 119, 122
circuit miniaturization, 160

cladding layers, 120, 156
Claisen–Schmidt condensation, 66
Claisen–Schmidt olefination, 67
Clausius–Mossotti (CM) equation, 56
CMOS silicon platform, 160
COANP crystal, 215
coarse-grained (pseudo-atomistic) statistical mechanical calculations, 7, 169
co-crystals, 91
coherence length, 18
comb generation, 163
complex refractive index, 151
condensation reactions, 145
conductance, 157
conductivity, 217
configuration interaction (CI), 53
configurationally locked polyene crystals, 104
conformal and flexible devices, 155–156, 216, 285
cooperative processes, 132
co-planar electrode poling, 218
corona needle, 120
corona poling, 219
Coulomb interactions, 90
Coulomb radius, 260
coumarin pendant functionalized chromophore C1, 130
coumarin-containing dendrimers, 145
coumarin–coumarin interactions, 131
coupled cluster (CC), 53
   method, 53
   theory, 48
coupled multiple ring resonators, 179
coupling between waveguides, 195
coupling loss, 147, 162, 218
covalent bond potentials, 128
covalent incorporation, 127
   of chromophores, 126
critical coupling, 201
crystal growth, 88, 108
   from solution, 109
crystalline materials, 4, 88
   electro-optic materials, 119
   organic electro-optic materials, 144

damage threshold, 236
DAST, 91, 94, 212, 228, 230, 241
Debye grating spacing, 255
decomposition of electro-optic chromophores, 145
dendrimers, 126–127
dendronized polymers, 127, 165
density functional theory (DFT), 46, 53–54
deposition of cladding layers, 147
device length, 176
device parameters, 175
dielectric breakdown, 285
dielectric constant, 31

dielectric environment, 121
dielectric permittivity, 45, 118, 127, 168
dielectric relaxation, 142
dielectric screening, 138
Diels–Alder "click" chemistry, 165
   "click" dendronized polymers, 166
Diels–Alder/retro-Diels–Alder crosslinking chemistry, 155
Diels–Alder/retro-Diels–Alder cycloaddition reactions, 145
diene and dienophile reactants, 145
difference frequency generation (DFG), 2, 4, 15, 19, 154, 163, 185, 234
difference-frequency mixing, 234
differential scanning calorimetry (DSC), 132, 142
diffraction efficiency, 265
digital power efficiencies, 169
digital signal processing bandwidths, 2, 285
dimensionality M, 131
dipolar chromophores, 3
dipolar molecules, 118
dipole moment, 40, 52, 119
dipole–dipole interactions, 122
directional coupler, 175, 177, 197
Disperse Red 1 (DR1), 121
dispersion, 13, 31, 34
dispersion-free hyperpolarizability, 42
DLD164 chromophore, 133, 162
donor and bridge functionalization, 167
DR-1-co-PMMA, 135
drive voltage, 176
drive voltage requirements, 160
DSTMS, 94, 100, 241

efficiency of second-harmonic generation, 18
electric field poling, 118, 124, 147, 284
electric field sensing, 178
electric field-induced second harmonic (EFISH), 45, 62
electrical conductivity, 217, 285
electrically poled materials, 119
   macromolecular thin films, 216
   organic materials, 118
   polymer and dendrimer materials, 139
electrical-to-optical signal transduction, 176
electro-absorptive modulator, 183, 284
electromagnetic (EM) sensing, 284
electron acceptor, 65
electron-beam etching, 155
electron-beam induced waveguides, 212
electron-beam structuring, 208
electron density distributions, 126
electron donor, 65
electronic and nuclear intermolecular electrostatic interactions, 124
electronic/photonic integration, 168
electrons, 1

Index

electro-optic (EO) material, 1
electro-optic activity, 3, 53, 124, 126–127, 136, 168–169, 285
electro-optic coefficient, 29, 31
electro-optic effect, 2, 15, 28
electro-optic etalon, 179
electro-optic figure-of-merit, 30
electro-optic modulation, 35
electro-optic tensor, 28
electro-optic waveguide, 176
electrophoretic migration, 126
electrostatic self-assembly, 77
end-fire coupling, 195
engineering strategies, for inducing non-centrosymmetric packing of nonlinear chromophores, 89
epitaxial growth, 209
esterification and/or amidization chemistries, 164
etalons, 175
evanescent optical fields, 147
evaporation-induced local supersaturation growth, 112
exchange-correlation functional, 53
exchange-correlation (xc) potential, 54
excitonic bands, 146
experimental methods, for evaluating r_{33}, 149

Fabry–Perot interferometry, 149
femtosecond laser ablation, 208
ferroelectric materials, 250
fiber coupling, 195
    loss, 176
field-assisted molecular reorientation, 250
figure-of-merit, 29, 37, 177, 228
    for frequency doubling, 229
    for second-harmonic generation, 18
    for THz generation, 236
finite field, 51
finite field (FF) methods, 47, 51
first hyperpolarizability, 47
first-order hyperpolarizability, 40
flexible modulators, 155
flexible spacers, 127
fluorinated and protonated aromatic dendrons, 138
fluorovinyl ether crosslinking, 166
formylation, 67
Fourier transform, 55
four-wave mixing, 25
free radical reactions, 145
free spectral range (FSR), 200
frequency dependence, 31
    of susceptibility tensors, 14
frequency-dependent polarizabilities, 53
frequency-dependent methods, 52
frequency doubling, 4
fully atomistic methods, 126

generalized gradient approximation (GGA) functionals, 54
glass transition temperature, 120, 144–145
gradient bridge chromophores, 57
graphoepitaxial melt growth, 213
grating coupling, 196
grayscale mask, 218
group additivity methods, 56
group velocity, 207
growth anisotropy, 138
guest and host materials, 121

half-wave voltage, 37, 189
hard geometric object approximation, 122
hard object (united atom) approach, 123
Hartree–Fock (HF) theory, 46, 53
heavily crosslinked materials, 119
Hellmann–Feynman theorem, 46
high bandwidth operation, 163
high-density optical storage, 250
high dipole moment chromophores, 119
high extinction, 169
high-frequency A/D converters, 178
highest occupied molecular orbital (HOMO), 59
high-frequency electro-optic modulators, 34
high-frequency modulation, 37
hole-conducting polymers, 259
hopping transport, 253
horizontal slot modulator, 161
horizontal slot silicon photonic devices, 163
Horner–Wadsworth–Emmons (HWE), 67
hybrid functionals, 56
hydrogen bonded crystals, 101
hydrophilicities, 72
hyperpolarizability, 2, 40, 52
hyper-Rayleigh scattering (HRS), 45, 53, 62, 129

imaginary (dichroism) component of the refractive index, k, 151
independent particle approximation, 121, 131
independent particle or oriented-gas model, 42
index of refraction, 127, 168
    mismatch, 147, 176
indium tin oxide, (ITO), 120
INDO (intermediate neglect of differential overlap) method, 47
induced polarization, 11
information technology, 1
inorganic light-emitting diode (LED), 283
insertion loss, 121, 147
integrated phase and amplitude electro-optic modulators, 194
interband electronic transition, 146
interband photoexcitations, 251
interference grating, 253
intermolecular electrostatic interactions, 118

Index

intrinsic bandwidth, of organic electro-optic materials, 177
intrinsic friction analysis (IFA), 131
ion implantation, 208, 211

Jones matrix analysis method, 152

Kerr effect, 28
Kleinman symmetry, 14
Knoevenagel condensation, 67
Kohn–Sham equation, 51
Koopmans' theorem, 58
Kramers–Kronig consistent oscillator layer model, 152

Langevin lattice, 131
Langevin limit, 121
Langevin trapping, 261
Langmuir–Blodgett films, 70
laser-assisted electric field poling / laser-assisted poling (LAEFP or LAP), 119, 137, 139, 284
lateral force microscopy, 132
lattice dimensionality, 118–119, 129, 132–134
lattice hardening by cross-linking, 145
level of detail (LOD), 169
lift-off techniques, 154
$LiNbO_3$ (lithium niobate), 2, 126, 177, 250, 284
line broadening, 146
linear electro-optic effect, 42, 149
linear polarizability, 40
liquid crystalline materials, 284
$LiTaO_3$, 250
local field corrections, 43
local field factor, 121
long-range corrected (LRC) functionals, 54
long-range corrected (LRC) DFT methods, 53
long-range surface plasmon polaritons (LRSPP), 221
long wavelength (zero frequency) limit, 45
long wavelength limit (dipole approximation), 46
Lorentz approximation, 43
Lorentz–Onsager factor, 118
low T_g polymeric composites, 274
lowest unoccupied molecular orbital (LUMO), 59

M05–2X functional, 54
Mach–Zehnder interferometers, 175
Mach–Zehnder modulators, 120, 138, 155, 176, 220
  amplitude (intensity) modulator, 194–195
  interferometer, 36
macromolecular electro-optic materials, 167
macroscopic electro-optic activity, 3
material decomposition temperature, 144
material dispersion, 12
materials requirements, 175
matrix-assisted-poling (MAP) materials, 119, 128, 134
melt capillary grown waveguides, 214

merocyanine dyes, 47
metamaterial device architectures, 1
metamaterials, 285
methylnitroaniline (MNA), 5
microdomain formation, 138
microresonator filter, 201
microring resonators, 198
mode dispersion in optical waveguides, 23
mode size and shape mismatch, 147, 176
mode size transformer, 147
mode transformers, 218
modeling of nonlinear optical effects, 7
modulation in microring resonators, 201
modulators, 188
molecular asymmetry, 88
molecular cooperativity, 131
molecular first hyperpolarizability, 3, 43, 59, 118
molecular quantum mechanics, 46
Møller–Plesset (MP), 53
  and density functional theory (DFT) quantum computational approaches, 7
  perturbation theory, 48
Monte Carlo (MC) and molecular dynamics (MD) theoretical methods, 3, 7, 118, 123
Monte Carlo simulation, 128, 169
multi-chromophore dendrimers, 127, 164

nanoimprint lithography, 145, 216, 220
N-benyl-2-methyl-4-nitroaniline (BNA), 139
neat chromophore materials, 126
neutral ground state/zwitterionic excited state mixing, 59
non-centrosymmetric (acentric) symmetry, 118
non-centrosymmetric organization of chromophores, 3
non-critical phase matching, 22
nonlinear optical (NLO) effects, 4, 11
  frequency conversion, 228
  polarization, 230
  processes, 15
  susceptibilities, 11, 239
non-rod-shaped π-conjugated cores, 90
normal incidence method (NIM), 151
nuclear electrostatic (repulsive or steric) interactions, 122
numerical computations of molecular hyperpolarizabilities, 44

oblate chromophores, 124
oblate ellipsoid, 124
OH1, 95, 106, 209, 228, 241
oligomeric aggregates, 124
operational optical frequency, 118
optic phase modulator, 194
optical absorption spectra, 135
optical couplers/splitters, 170
optical damage, 250

Index

optical data storage, 266
optical frequency, 118
optical frequency doubling, 17
optical gyroscopes, 178
optical indicatrix, 28
optical insulators, 206
optical Kerr effect, 15, 24
optical lattice vibrations, 31
optical loss, 3, 119, 136, 176
   of cladding layers, 156
   insertion loss, 176
optical parametric generation, 19
optical phase conjugation, 25
optical rectification (OR), 4, 13, 15, 19, 53, 185, 228, 238, 286
optical sum and difference frequency generation, 286
optical switch (OS), 170
optical waveguides, 23, 189
optically assisted poling, 136–138
optically induced ordering, 136
optical-quality glasses, 121
optimum poling efficiency, 132
order parameter, 96, 99, 107, 121
order–disorder processes, 3
   transitions, 131
organic crystals, 268
organic electro-optic (OEO) materials, 118
organic molecular beam deposition, 78
orientation-selective melting, 138
overtone vibrational transitions of hydrogen, 146

parallel plate poling, 120, 218
parametric generation, 231
partition function, 123
pendant-functionalized chromophore, 130
periodical poling, 22
perturbation Hamiltonian, 46
perturbation series convention, 47
phase grating, 250
phase matching, 20, 34, 44, 229
   condition, 18
   curves, 230
   mismatch, 18
   techniques, 21
phase modulators, 35, 176
phase relaxation time, 2, 177
phase separation, 126–127
phased array radar, 178, 185
phonons, 31
   resonances, 185, 286
photobleaching, 208, 217
photochemical figure-of-merit, B/σ, 148
photochemical stability, 147
   measurements, 147
photochromic trans–cis–trans isomerization of azobenzene chromophores, 137

photo-excitation, 253
photo-induced crosslinking, 145
photo-induced refractive index changes, 264
   in crystals, 257
photolithography, 208–209
photonic bandgap, 206
photonic crystal, 203
   devices, 220
photonic integration, 1
photonic/electronic integration, 1
photons, 1
photorefractive effect, 4, 250
photorefractive materials, 268
   properties, 275
photorefractive sensitivity, 268, 276
photosensitivity, 267
photostability, 119
planarizing polymers, 155
plasmonic devices, 163
   plasmonic and metamaterial device architectures, 185
   plasmonic–organic hybrid (POH) devices, 163, 186
   stripline devices, 185
plasmonic loss, 163, 186
plasmonic waveguides, 121
   devices, 221
   losses, 121
plasmonics, 1, 285
plasticization, 121
p-nitroaniline (PNA), 53
Pockels effect, 3, 28, 53
   modulation, 163
   modulators, 169
point dipole approximation, 122, 126
polar materials, 11
polarizability, 40, 47, 52
polarization-optic effect, 29
poled polymers, 4
poling efficiency, 134
poling electrodes, 120, 217
poling field, 120
poling field/dipole moment interaction, 124
poling temperature, 121
poling-induced acentric order, 118, 126, 136
polycarbonates (PCs), 119
polyimide (PI), 119
polyimide polymers, 142
polymer host, 119
polymer waveguides, 216
polymeric or dendritic (macromolecular) material lattices, 118
polymers, 126–127
polymethylmethacrylate (PMMA), 119
post-Hartree–Fock (PHF) methods, 53
power consumption, 2, 284
power efficiencies, 163, 285

Index

prism coupling, 195
prism-based devices, 175
processing methodologies, 3
processing options, 154, 282
processing-associated loss, 146
prolate ellipsoid, 124
prolate ellipsoid approximation, 123
propagation constant, 191
propagation loss, 176
prototype devices, 118
pseudo-atomistic approach, 126
pseudo-atomistic Monte Carlo/molecular dynamical calculations, 126
pump-probe methods, 147
push–pull Mach–Zehnder interferometer, 176
push–pull modulator, 120

quadrupolar interactions, 138
quality factor, 201
quality, Q, 179
quantum and statistical mechanical modelling, 128
quantum mechanical calculations, 126
quantum mechanics, 126
quasi-phase matching, 22

radiofrequency (RF) photonic applications, 160
reactive ion etching (RIE), 147, 217
real-time holography, 266
real-time TD-DFT, 55
reconfigurable optical add/drop multiplexer (ROADM), 162
reflection grating geometry, 275
refractive index, 15
  change, 263
relative change of the effective index, 201
relative resonance shift, 201
reorientational effects, 262
resistivity, of metal electrodes, 176–177
resonance enhancement, 51
resonant devices, 158, 175, 179
resonator finesse, 201
response function, 51
reverse tapers, 162
rigid body Monte Carlo (RBMC) simulations, 124, 151
rigid π-electron segments, 126
ring microresonators, 154, 175
Runge–Gross theorem, 54

SBLD-1 (C1), 130
scaling of relaxation times, 144
scattering loss, 146
second hyperpolarizability, 47
second-harmonic generation, 13, 17, 32, 42, 53, 142, 149, 228
second-order hyperpolarizability, 40
second-order nonlinear optical effects, 15

self-assembled materials, 4, 75
  films, 70
  mono-and multi-layers, 75
self-focusing, 24
self-phase modulation, 24
Sellmeier equation, 34
sequential-synthesis/self-assembly, 119, 144
shadow mask, 218
shear modulation force microscopy (SM-FM), 131
side-chain relaxations or backbone relaxations, 131
signal routing, 178
silica fiber waveguide, 176
silicon modulators, 283
silicon photonic bandgap (slot wave) structures, 162
silicon photonics, 1, 160, 285
silicon slot waveguides, 121, 182
silicon–organic hybrid (SOH) devices, 2, 160, 285
silicon–organic hybrid waveguides, 221
single-beam reflection ellipsometry, 149
single-crystalline micro- and nano-wires, 113
single-crystalline organic microresonators, 215
single-crystalline thin films, 110
singlet biradical states, 57
singlet oxygen, 148
slab waveguides, 190
slotted microring resonator, 161
slowly varying envelope approximation, 17
soft/nano-imprint lithography, 2, 154
SOI waveguide structure, 222
sol-gel materials, 126
solution-deposited films, 75
solvatochromic shifts, 135, 146
solvent reaction field, 47
space-charge, 250
  field, 254
  field distribution, 260
spatial Kerr solitons, 24
spatial light modulation, 175, 182
spatially anisotropic intermolecular electrostatic interactions, 123, 136
spectral line broadening, 135
spectral width of the resonances at full-width half-maximum, 200
spherical chromophores, 124
spheroid approximation, 123
spin casting, 119, 147
  solvent, 119
spur free dynamic range (SFDR), 220
stability, 3
state-of-the-art theoretical methods, 3
statistical mechanical analysis, 123
steric interactions, 123
stilbazolium salts, 91
stripline devices, 175–176
stripline silicon–organ hybrid (SOH), 175
structuring methods, 208

Index

sublimation, 126
sum- and difference-frequency generation, 228
sum over states (SOS) method, 47, 51
sum-frequency generation, 15, 17, 230
supramolecular synthetic approach, 91
surface functionalization, 124
synthetic methods, 64
synthetic strategies for covalently-incorporated chromophore materials, 163

Taylor series convention, 47
TCNQ-doped COANP, 268
TE optical modes, 161, 192
Telcordia standards, 2, 120, 175
telecommunication wavelengths, 148
temperature tuning, 22
temperature-dependent change
 in density, ρ, 169
 in index of refraction, 169
Teng–Man technique, 149
terahertz conversion efficiency, 235
terahertz frequencies, 20
terahertz reflection images, 248
terahertz spectroscopy, 185, 244
terahertz wave generation, 15, 34, 228, 232
 by optical rectification, 238
terahertz-wave detection, 243, 286
thermal crosslinking, 165
thermal management, 1
thermal stability, 119, 141, 144
thermally activated transitions, 131
thermo-optic effect, 169
thermo-optic materials, 118, 169
thermo-optic polymers, 4
third-harmonic generation, 23
third-order nonlinear optical effects, 23
three-dimensional plot, of r_{33} vs. number density vs. dipole moment, 119
time-dependent DFT (TD-DFT), 54
time-dependent HF theory, 54
time-dependent quantum mechanical computer modeling, 149
time-independent perturbation theory, 46
titanium dioxide, 157
TM optical modes, 194
Tool–Narayanaswamy procedure, 144
transparency range, 228
traveling wave modulators, 38
trifluorovinyl ether moieties, 165
twisted bridge chromophores, 57
two-state model (TSM), 55
two-level model, 42
two-slit interference, 149
two-wave mixing, 264

type I phase matching, 21
type II phase matching, 21

ultra-stable oscillators, 178
undepleted pump approximation, 17
uniform distribution of chromophores, 124
united-atom approximation, 126
units and conversion factors, 7
urethane polymers, 126
UV-curable cladding materials, 147, 156

Van der Waals forces, 253
vapor crystal growth, 108
vapor-deposited films, 78
vapor deposition with chemical reactions, 83
variable angle polarization referenced absorption spectroscopy (VAPRAS), 130, 151
variable angle spectroscopic ellipsometry (VASE), 130, 152
variable optical attenuators (VOAs), 170
velocity matching, 34, 239
velocity mismatch, 284
 of propagating optical and electrical waves, 177
Vernier effect, 179
vertical-cavity surface emitting lasers (VCSELs), 283
Vilsmier–Haak, 68
viscoelasticity, 118
visible optical wavelengths, 148
VLSI electronic circuitry, 155
voltage-induced bandpass shifting, 179
voltage–length parameters, 163, 285
voltage–length performance, 169

walk-off angle, 22
wave equation, 14
wave vector, 15
waveguide loss, 121, 147
waveguide tapers, 196
wavelength dependence, 34
wavelength division multiplexing (WDM), 181
wavenumber, 15
Williams–Landel–Ferry (WLF) parameters, 144
Williamson ether synthesis, 164

X-shaped chromophores, 59

YLD124 chromophore, 135
YLD156 (F2), 130

ZINDO/PM6 methods, 58
ZnTe, 242
zwitterionic chromophores, 74
zwitterionic states, 57